Yuri Mnyukh

FUNDAMENTALS OF
SOLID-STATE PHASE TRANSITIONS,
FERROMAGNETISM AND
FERROELECTRICITY

SECOND EDITION
revised and complemented
by new sections and ten addenda

ISBN: 978-0-615-33972-6

"It was at this moment that Mr. Pickwick made that immortal discovery... The stone was uneven and broken,... but the following fragment of an inscription was clearly to be deciphered:

+

B I L S T

U M

P S H I

S. M.

A R K

...Mr. Pickwick lectured upon the discovery... and entered into a variety of ingenious and erudite speculations on the meaning of the inscription... [He] himself wrote a Pamphlet, containing ninety-six pages of very small print, and twenty-seven different readings of the inscription. ... [He] was elected an honorary member of seventeen native and foreign societies...

Mr. Blotton, indeed - and the name will be doomed to the undying contempt of those who cultivate the mysterious and the sublime - ...presumed to state a view of the case, as degrading as ridiculous. [He] ... observed... that he had seen the man from whom the stone was purchased; that the man presumed the stone to be ancient, but solemnly denied the antiquity of the inscription - inasmuch as he represented it to have been rudely carved by himself ...-

'BILL STUMPS, HIS MARK'...

But this base attempt to injure Mr. Pickwick recoiled upon the head of its calumnious author. The seventeen learned societies unanimously voted the presumptuous Blotton an ignorant meddler... And to this day the stone remains, an illegible monument of Mr. Pickwick's greatness..."

Charles Dickens

TABLE OF CONTENTS

INTRODUCTORY WORD FROM THE AUTHOR AND
ACKNOWLEDGMENTS

My experimental work on the problem of solid-state phase transitions began in mid 50's in Moscow, USSR, in the laboratory headed by late A. I. Kitaigorodskii, a noted expert in organic crystals. The work continued in the Crystallophysics Laboratory, which I headed, of the USSR Academy of Sciences Institute of Biophysics, and then, after my emigration in 1977 to U.S.A., in New York University. My initial idea of the study was in line with conventional views. Knowing that two crystal forms of simple organic substance p-dichlorobenzene were similar, with the unit cell parameters close to each other, it was decided to investigate the kind of "deformation" converting the initial structure into the resultant. Here I would like to tell everyone whose views on these phase transitions were incorrect as well: it was not my fault that the real nature of these phase transitions turned out profoundly different. The result was stunning: instead of "deformation," a growth of perfectly face-shaped single crystals was observed within the transparent solid medium. Moreover, they had arbitrary orientations notwithstanding the three-dimensional order of the medium. The existing theory was based on quite different premises and did not foresee these phenomena.

The state of the theory of solid-state phase transitions appears to me a rare one, if not unique. Rapid progress of modern science as a whole is beyond question and, at first glance, this may seem to hold true for its branch investigating solid-state phase transitions. In reality, however, this particular branch rolls back from an erroneous to an even more erroneous interpretation of the phenomenon it is to account for. Against the background of general scientific progress, this is a *regress*. Every time we are told that the phenomenon of phase transitions is even more complicated than it was thought to be. Many highly qualified and noted scientists contribute to this process by developing more and more sophisticated theories based on the same wrong assumptions to replace the previous simpler versions proved to be unsatisfactory. Yet, the regress is not total: there is also an accumulation of facts that have already come into fundamental conflict with the theory. The divergence between the theory and experiment is so great that the theory cannot be patched any more to absorb evidence. It ignores evidence, in particular that a solid-state phase transition is always a nucleation-and-growth, rather than a "critical" cooperative phenomenon.

Is this state of affairs generally unusual in science? In a sense, it is not. Scientific development proceeds by trial and error. An erroneous theory frequently tries to survive by modifying and complicating itself and diluting its original premises in the face of new evidence. What is, however, unusual in the case under consideration is that this resistance lingers already for many decades, while evidence is treated as if it were nonexistent. The consequences are ruinous for a large piece of solid-state physics owing to ramifications to many processes where solid-state rearrangements take place. One of the ramifications concerns ferromagnetism and ferroelectricity.

For years I was pursued by the questions: How and why did this situation come into existence? My reflections have led me to conclude that there was a unique combination of circumstances:

1) The fascinating topic of crystal phase transitions always attracted the attention of workers specializing in particular physical experimental techniques such as X-ray analysis, optical spectroscopy, neutron scattering, ESR, NMR, calorimetry, etc., as well as scientists in particular applied fields. Lacking an expertise specifically in the field of phase transitions, these workers, distinguished ones among them, not infrequently took initiative in interpreting the phenomenon in terms of "conventional wisdom."
2) In the early stages, physical metallurgists were especially active in developing the theory. The most difficult solids in which to investigate phase transitions are metals; the suggested theories were based on ideas not rooted in hard facts.
3) Even prior to any investigation, it is difficult not to succumb to the "self-evident" idea on cooperative orderly modification - a kind of distortion/deformation - of the initial structure into the resultant. Sometimes structural similarity found between the polymorphs bolsters such an approach.
4) Another idea, generally accepted as a matter of course, is that solid-state phase transitions are a subject of statistical mechanics. (In the catalogues of the U.S. scientific libraries the topic of phase transitions is presented as "phase transitions (statistical mechanics)"). This a priori mentality has resulted from two circumstances: (1) the success of statistical mechanics in accounting for other physical phenomena and (2) statistical mechanics is the only tool a theoretical physicist has for treating phase transitions. These circumstances turned out to be so powerful that phase transitions, at least most of them, were

assigned, no matter what, to occur under the action of fluctuations (phase fluctuations, fluctuations of order parameter, of density, of entropy, molecular fluctuations...). These fluctuations allegedly increase as the "critical temperature" is approached, at which the phase becomes unstable and suddenly turns into the alternative phase. But in reality a solid-state phase transition has nothing to do with statistical mechanics on a "cooperative" level, as will be proven in this book.

5) Due to mere coincidence, some experimental data has a misleading appearance and can be (and was) easily misinterpreted as supporting the statistical-dynamical approach. Thus, measurements of some physical properties *vs.* temperature through the transition sometimes exhibit "anomalies" which were regarded as a manifestation of "critical dynamics" of phase transitions. Similarly, comparison of the initial and final crystal structures in some cases led to the impression that the transition resulted from a cooperative distortion or deformation of the initial crystal, or from a displacement of the Bravais lattices. Sometimes an orientation relationship found between the initial and final crystal structures was an additional argument in favor of a cooperative reformation between the interdependent structures. The real origin of all these truly confusing effects will be elucidated in this book in all detail.

The reader may be surprised to find different problems of solid-state physics combined in the title of this book. The fact of the matter is that they are closely related. Crystal phase transitions turned out to be that neighboring scientific field which had to be clarified in order to lay down proper foundations of ferromagnetism and ferroelectricity. I could not miss the opportunity: this is the reason Chapter 4 is in this book. Every detail of the molecular mechanism of phase transitions described in the previous chapters becomes important. The result is a coherent unified picture that accounts for both ferromagnetism and ferroelectricity, instead of the failed classical theories by Weiss and Heisenberg. Ferromagnetism and ferroelectricity cannot serve as the last resort for "critical phenomena" in a solid state any more.

I expect the book to be relatively easy to read and understand. No sophisticated mathematics was required. The method employed usually involved collection and logical analysis of all facts, finding a self-consistent explanation and further its verification. The reader will discover that the phenomena of solid-state phase transitions, ferromagnetism and ferroelectricity are incomparably simpler than they were presented us previously. All processes can be imagined in the model form. Even a shadow of mystery is removed; no "continuous" transformations, no advance preparations of a phase to "flip-flop" into the different phase upon achieving the "critical point," no unexplained peaks of physical properties rocketing up into infinity...

Finally, a few notes about the literature I used. While the present book is anything but compilation, I would like to emphasize the great help provided to me by the following comprehensive review books: Rao & Rao on phase transitions [42], Vonsovskii [121] and Bozorth [65] on ferromagnetism, and Zheludev on ferroelectricity [157]. A review book, even if it is so voluminous as that by Vonsovskii (1260 pages), presents in a concise form the current (at the time it was written) state of affairs in the scientific field of interest, and makes it easier to find where the inconsistencies are.   I want to make it clear that when I find inconsistencies and quote these authors, I criticize not them, but the state of affairs they reflect. As to other literature, I was interested to learn when and how the science of solid-state phase transitions slipped onto the wrong track, as well as the history of disregarding "unwanted" data and right ideas. For this reason the most recent literature, with some exceptions, was of little interest, for it did not correct errors of the earlier years.

Listed below is a selection from the relevant publications. Not all of them are referred to in the text.

(1) Yu. Mnyukh. Laws of Phase Transformations in a Series of Normal Paraffins, *J.Phys.Chem.Solids* **24**, 631-40 (1963).
(2) A. Kitaigorodskii, Yu. Mnyukh and Yu. Asadov. Relationships for Single Crystal Growth During Polymorphic Transformation, *J.Phys.Chem.Solids* **26**, 463-72 (1965).
(3) Yu. Mnyukh, N. Petropavlov, and A. Kitaigorodskii. Stepwise Crystal Growth During Polymorphic Transformation, *Sov.Phys.-Doklady* **11**, 4-7 (1966).
(4) Yu. Mnyukh and N. Musaev. On the Mechanism of Polymorphic Transition from Crystalline to Rotational-Crystalline State, *Sov.Phys.-Doklady* **13**. 630-33 (1969).
(5) Yu. Mnyukh and N. Petropavlov. Polymorphic Transitions in Molecular Crystals -1. Orientations of Lattices and Interfaces, *J.Phys.Chem.Solids* **33,** 2079-87 (1972).
(6) Yu. Mnyukh and N. Panfilova. Polymorphic Transitions in Molecular Crystals – 2. Mechanism of Molecular Rearrangement at "Contact" Interface, *J.Phys.Chem.Solids* **34,** 159-70 (1973).

(7) Y. Mnyukh, N. Panfilova, N. Petropavlov, and N. Uchvatova. Polymorphic Transitions in Molecular Crystals - 3. Transitions Exhibiting Unusual Behavior, *J.Phys.Chem.Solids* **36,** 127-44 (1975).

(8) Yu. Mnyukh. Polymorphic Transitions in Crystals: Nucleation, *J.Crystal Growth* **32,** 371-77 (1976).

(9) Yu. Mnyukh. Polymorphic Transitions in Crystals: Kinetics, *Molec.Cryst.Liq.Cryst.* **62,** 201-18 (1979).

(10) Yu. Mnyukh. Molecular Mechanism of Polymorphic Transitions, *Molec.Cryst.Liq. Cryst.* **62,** 163-200 (1979).

(11) Yu. Mnyukh. On the Origin and Role of the Domain Structure in Ferroelectrics and Ferromagnetics, *Sp.Sci.Tech.* **4,** 485-500 (1981).

(12) Yu. Mnyukh. The Cause of Lambda Anomalies in NH_4Cl, *Sp.Sci.Tech.* **6,** 275-85 (1983).

This book might not have been written if the New York University physicist Professor M. Pope had not extended a helping hand to me at the critical moment of my life and scientific career. It was 1978, soon after my emigration from the former Soviet Union over my "disagreement" with its communist regime. I found United States in a period of serious recession, when there were no openings for employment of an "overqualified" 50-year-old Russian scientist, speaking, what is more, very poor English. At that time Professor M. Pope provided me with the opportunity and conditions to renew my research in his NYU Solid-State Laboratory, even though his own scientific interest was different from mine. This was followed by my employment at NYU. I always will be grateful to Professor M. Pope for that disinterested assistance.

I owe much to my dear friend, the Auburn University Professor V. Vodyanoy who, in spite of his own busy schedule, found time to read the manuscript for the first edition from A to Z and suggested a number of improvements and corrections. His word was approving and encouraging.

I would also like to thank my daughter Anna for valuable help in arranging publication of the second edition.

Yuri Mnyukh
Connecticut, U.S.A.

SYMBOLS, NOTATIONS, ABBREVIATIONS

Abbreviations

C	- Crystal
SC	- Single crystal
PC	- Polycrystal
PDC	- Polydomain ("single") crystal
ODC	- Orientation-disordered crystal
OM	- Optimum microcavity
OR	- Orientation relationship between crystal structures of two polymorphs
TH	- Temperature hysteresis of phase transition
2-D	- Two-dimensional
3-D	- Three-dimensional
VA	- Vacancy aggregation
VAs	- Vacancy aggregations
DSC	- Differential scanning calorimetry
ESR	- Electron spin resonance
NMR	- Nuclear magnetic resonance
FE	- Ferroelectric
AE	- Antiferroelectric
PE	- Paraelectric
DHL	- Double hysteresis loop

Solid - state phases

i	- Initial phase
r	- Resultant phase
L	- Low-temperature phase
H	- High-temperature phase
$\alpha, \beta, \gamma, \delta$	- Polymorphs
I, II, III,	- Polymorphs

Temperatures

T	- Temperature
T_0	- "Equilibrium temperature" (at which the free energies are equal)
T_{tr}	- Actual temperature of phase transition
T_n	- Temperature of 3-D nucleation
T_{3-D}	- Temperature of 3-D nucleation
T_{2-D}	- Temperature of 2-D nucleation

T_c	- Critical / Curie point
T_m	- Melting point
T_{th}	- Temperature in thermostat
T_r	- Room temperature (20 °C)
ΔT	- Overheating or overcooling (= $T_{tr} - T_o$)
$\Delta T (+)$	- Overheating
$\Delta T (-)$	- Overcooling

Thermodynamic

F	- Free energy
E_a	- Activation energy
E_s	- Sublimation energy
Q	- Latent heat of phase transition
U	- Potential energy of crystal lattice
S	- Entropy

Crystallographic

$a, b, c, \alpha, \beta, \gamma$	- Unit cell parameters
d_{hkl}	- Spacing between crystal planes (h,k,l)
Z	- The number of molecules per unit cell

Physical parameters and properties

t	- Time
p	- Pressure
v	- Specific volume
m	- Mass fraction of a phase in two-phase system
P	- Physical property (in general terms)
C_p	- Specific heat / heat capacity
y	- Calorimetric signal
K	- Reaction equilibrium constant
K	- The number of powder particles or polycrystal grains
N	- The number of free radicals
R	- Universal gas constant
R	- Electric resistance
I	- Intensity of light
ℓ	- Length or distance
C	- Concentration of VAs
α	- Volume coefficient of thermal expansion
δ	- Density

ε	- Dielectric constant
ρ	- Electric conductivity
α_i, α_r	- Sublimation coefficients
k	- The number of molecular layers
n	- Grade of crystal quality
V	- Velocity of interface motion

Ferroic-related

E	- Electric field
H	- Magnetic field
P	- Electric polarization
M	- Magnetization
P_s	- Spontaneous polarization
M_s	- Spontaneous / saturation magnetization
θ	- Angle between directions of M measurement and H
φ	- Angle between directions of M and H

Substances

PDB	- p-Dichlorobenzene
PDI	- p-Diiodobenzene
HCE	- Hexachloroethane
MLA	- Malonic acid
GLA	- Glutaric acid
OHA	- Octahydroanthracene
HMB	- Hexamethyl benzene
DL-N	- dl-Norleucine
TCB	- 1,2,4,5-Tetrachlorobenzene
AHB	- Aniline hydrobromide
KDP	- K_2PO_4

CHAPTER 1. CRITICAL SURVEY

1.1 Critical approach to "critical phenomena"

Solid-state phase transitions are treated in scientific literature predominantly as a *critical phenomenon*. Such an approach, it may seem, rests on a solid experimental basis: anomalies of some physical properties (especially in the form of peaks) were observed in the temperature region of "critical points" (or "Curie points") T_c where the phase transitions occur (Fig. 1.1a). A conspicuous example of such an anomaly is "λ"-shaped peaks of specific heat C_p (T). Similar peaks have also been found in the measurements of some other properties *vs.* temperature: coefficient of thermal expansion, light scattering ("critical opalescence"), neutron scattering, dielectric constant and some others. Anomalies of this type were frequently referred to as "lambda-anomalies" or "λ-peaks." For decades these peaks fascinated the researchers who investigated or in any way touched on phase transitions in their work. The λ-problem was listed in the "Physics and Astronomy Classification Scheme."

λ-Peaks were regarded as a key to the internal mechanism of solid-state phase transitions based on a cooperative participation of all molecules (or atoms) in the process. Countless efforts have been made to experimentally delineate these anomalies and develop a theory accounting for such singular behavior [1-30]. Even though no adequate theory has been developed (and cannot be, as will be shown), there was a consensus that λ-peaks result from increasing "phase fluctuations" as the "critical temperature" is approached. The theories of solid-state phase transitions as "critical phenomena," a branch of statistical mechanics, are among the most intricate (and fruitless) physical theories.

Our experimental study and analysis of the "λ-anomalies" led us to conclude unequivocally that these "anomalies" are not genuine. Not only the observed peaks do not represent "critical phenomena," they do not even represent the corresponding physical properties. Thus, "heat capacity λ-peaks" do not represent true heat capacity, "dielectric constant peaks" do not represent true dielectric constant, "λ-peaks of thermal expansion coefficient" do not represent this coefficient, and so on. It is demonstrated in full detail in this book that these "anomalies" have been misinterpreted. In fact, they are merely "side effects" of the process of *nucleation and growth* according to the particular molecular mechanisms described in the following chapters; this process is an antithesis of a "critical" behavior.

It will be shown that "critical phenomena" are simply nonexistent in solid-state phase transitions. The source of the phenomena mistaken for "critical phenomena" will be demonstrated later in detail. Briefly, it is a *heterophase* state of the specimen in the temperature range where the "anomaly" is observed. A fundamental fact, previously overlooked or disregarded, is that in this range the matter *always* consists of two distinct crystal phases separated by interfaces. Taking this into account, the data on different physical properties in the two-phase transitional range were revised. It became clear not only why certain properties of crystal solids exhibit peaks upon phase transitions, but also why others do not, showing instead a different type of apparent anomalous change referred below to as "sigmoid" (or "S"-) curves (Fig. 1.1b). Several reexamined "classical" phase transitions, previously exemplified the above-described anomalies, will be presented to prove our point.

Representation of solid-state phase transitions as "critical phenomenon" is a great misconception in the modern physics. In almost every particular case the already available data point at the process by nucleation and growth. The actual mechanisms of the nucleation and growth had to be the main subject of the research in order to account for solid-state phase transitions. In the meantime, their careful investigation was largely neglected, since one's attention was distracted by the "critical phenomena" concept. The purpose of the investigation presented in Chapter 2 was to correct this state of affairs. The real features of the nucleation and growth turned out to differ from the common views on the subject. These features easily account for all the facts, including the "anomalies."

The clarification of the nature of solid-state phase transitions had more far-reaching consequences than was initially expected. In particular, it has given rise to the fundamentals of ferromagnetism and ferroelectricity, free of the contradictions and imperfections of the presently existing theories. A major point of these new foundations is the fact that ferromagnetic and ferroelectric phase transitions are not a critical phenomenon either, and rather occur by the same nucleation and growth as all other solid-state phase transitions. These results are presented in Chapter 4.

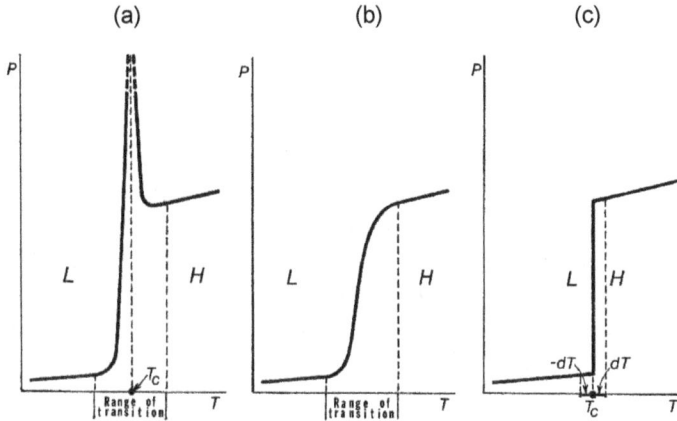

Fig. 1.1 A physical property P *vs.* temperature T in the region of phase transition; L stands for low-temperature phase, H for high-temperature phase.

1. A "λ"-peak that appears upon recording some physical properties. One assumes such a peak to be indicative of *critical* behavior, the location of the maximum pointing out the *critical temperature* T_c .

2. A sigmoid- ("S"-) like gradual change characteristic of the properties not exhibiting the λ-peaks. But one may notice that the first derivative of this curve will produce λ-peak.

3. First-order phase transitions by their common definition as those experiencing an abrupt jump in physical properties at a critical point T_c.

1.2   First- and second-order phase transitions

When contemplating possible mechanisms of phase transitions, it should be first realized that they have, as minimum, to meet the following condition in order to comply with thermodynamics.   An infinitesimal change of the thermodynamic parameter (dT in case of temperature) may produce only two results: either (A) an infinitesimal *quantity* of the new phase emerges, with the structure and properties changed by finite values, or (B) a physically infinitesimal "*qualitative*" change occurs uniformly throughout the whole macroscopic volume [31]. This condition, however, is only a necessary one: it does not guarantee both versions to be found in nature.

The version 'A' is, evidently, an abstract description of the actually observed phase transitions by nucleation and growth. Every input of a minuscule quantity of heat δQ either creates a nucleus or, if it exists, shifts the interface position by a minuscule length $\delta \ell$. The issue is thus reduced to whether version 'B' can materialize. As far back as 1933, Ehrenfest formally classified phase transitions by *first-order* and *second-order* in terms of "continuity" or "discontinuity" in their certain thermodynamic functions [32]. The validity of the classification was disputed by Justi and Laue by asserting that there is no thermodynamic or experimental justification for second-order phase transitions [33]. Judging from the absence of references in subsequent literature, their objections were ignored. It became widely accepted that there are "discontinuous" first-order phase transitions, exhibiting jumps in their physical properties, as well as "continuous" second-order phase transitions, showing no such jumps.* The latter are to be identified with the version 'B', for they fit that particular version and, besides, no other option exists.

Landau [34,35] developed a theory of second-order phase transitions. Landau and Lifshitz in their book "Statistical Physics" [36] devoted a special chapter to them. There are a number of items in that chapter that need to be highlighted:

(1) Two points need to be emphasized in the statement "Transition between different crystal modifications occurs usually by phase transition at which jump-like rearrangement of the crystal lattice takes place and state of the matter changes abruptly. Along with such jump-like transitions, however, another type of transitions may also exist related to change in symmetry." (a) First-order phase transitions are recognized real and usual, while second-order transitions only *may* exist. No real examples of second-order transitions were given, except $BaTiO_3$, which turned out to be of the first order. (b) The description of first-order transitions, although basically not incorrect, created a strong impression that the process occurs instantly over the bulk. The latter would be inconsistent with the thermodynamic condition 'A' given in the beginning of this section. As seen from the pertinent literature, a theoretical treatment of first-order phase transitions as an instant structural rearrangement at the transition point (or critical point, or Curie point) is rather typical.

---

* It is frequently forgotten that, according to the theory, certain functions (C_p, for example) are "discontinuous" in second-order phase transitions too.

(2) A description of the structural changes upon second-order transitions was given: atomic positions in the crystal lattice change continuously over the transition point (or critical point, or Curie point, or λ-point), while only the crystal symmetry experiences a jump.

(3) It was stated unequivocally that neither overcooling nor overheating are possible upon the second-order phase transitions, nor liberation or absorption of heat can take place.

(4) Several characteristics of first-order phase transitions were given in order to distinguish them from those of the second order:

    a. the symmetries of two phases are not related in any way and can be quite different;

    b. phase transition is accompanied by absorption or release of latent heat;

    c. overheating or overcooling is possible.

After correcting the last point (not only *possible*, but *inevitable*), these characteristics are a good tool to identify the phase transition "order." Detecting one of them suffices to identify a first-order transition and means that all must be present, whether they are observed or not.

Since then, second-order phase transitions were customarily defined as a cooperative, *continuous*, homogeneous process, involving simultaneously all constituent particles in the bulk, and became almost exclusively the subject of interest. However, first-order phase transitions, in spite of representing (at least) the majority of cases, attracted much less interest. In effect, they were left to be associated with the possibility 'A' above, to occur by *nucleation and growth*, but this important fact was rarely given due attention, or even noticed. As a rule, the interpretation of first order phase transitions is reduced to their abstract definition as those exhibiting *discontinuity* in the first derivative of thermodynamic potential, or even to the primitive view as exhibiting jumps in their physical properties upon transition (Fig. 1.1c).

It is the latter overly formal definition and treatment, which lacks a reference to the factual physical process, that caused difficulties and confusion in the application of the classification in practice. It is rather common to treat a phase transition as of the second-order in spite of its first-order characteristics. It is deemed acceptable especially in the cases when the "jump" in one or another physical property is considered "small." Also, in some cases the jumps were overlooked, not being detected with the available techniques. Moreover, the "jumps" themselves in the experimental plots *vs.* temperature are frequently found to be "continuous," smeared over a temperature "range of transition" (Fig. 1.1b). One regards these "smeared jumps" as another singularity, and the transition to be "continuous" or "almost continuous." As a result of all these uncertainties, many first order phase transitions have long been classified as second-order. A slow, but steady, process of second-to-first order reclassification is going on. Sometimes a new "second-order" phase transition is reported, but only until it is investigated in more detail. No justifiable reclassification is known in the opposite direction.

Two trends are to be noted regarding the contemporary state of the classification in question. (1) Persistent efforts to find good examples of second-order phase transitions failed to the point that "100% pure" second-order transitions hardly exist. (2) One circumvents the restriction clearly expressed by Landau that a theory of second-order transitions is not applicable to first-order transitions. Obviously, this trend is the consequence of the first one.

Here are the "justifications" given by different authors to a treatment of first-order transitions as if they were of second order:

(1) All crystal phase transitions, including well-recognized first-order transitions, are merely treated in terms of the notions and theory of second-order phase transitions (as critical phenomena) with no justification suggested. For instance, this "method" was used by Bruce and Cowley in their monograph on phase transitions [19], who avoided the "order" problem by simple replacement of the original Landau's heading [36] "Phase Transitions of the Second Kind" (*i.e.*, second order) by the "Landau Theory."

(2) The theory is applied to many first-order transitions on the grounds that they are "almost," or "nearly," or "close to," second-order. Or, as Buerger specified, "90% second-order and 10% first-order" [37]. Such statement as "Although the Landau theory assumes continuous second-order phase transitions, it can be applied to weakly first-order transitions" [38] is typical. Even the above-cited book by Landau and Lifshitz has not escaped this misconception. The following footnote was placed there about $BaTiO_3$ which they used to exemplify the structural mechanism of a second-order transition: "To avoid misunderstanding it should be noted that in the particular case of $BaTiO_3$ atomic shifts experience a finite jump, although a small one, so that the transition is still that of first order." A size of the jump is irrelevant: the phase transition proceeds by nucleation and growth, rather than by cooperative atomic shifts.

The real ferroelectric phase transition in $BaTiO_3$ does not have any resemblance to the description given to it by Landau and Lifshitz (see Sec. 4.7).

(3) A phase transition is considered "continuous" (second-order), while the observed first-order "jump" is ignored as unimportant temporary interruption of the continuous process.

(4) First-order transitions are interpreted as a cooperative process of displacement, distortion, rotation, etc.

In all these "justifications" such inseparable attributes of first-order phase transitions as nucleation, moving interfaces and a temperature range of two-phase coexistence are missing. The first-second-order classification, still being recognized *de jure*, is almost abandoned *de facto*. The original intent to *distinguish* two antipodal types of phase transitions has been nullified by blurring all boundaries between them. *First-order phase transitions occur by rearrangement at interfaces, and this simple fact invalidates all theoretical attempts to regard them as resulted from fluctuations over the bulk.*

Not a single well-proven second-order crystal phase transition has been found. In practice, all the transitions occur by nucleation and growth. But they were not treated as such. They were artificially substituted with a homogeneous, continuous, cooperative process by which second-order phase transitions are assumed to occur. No theory developed for one mechanism is suitable for the other, even as a loose approximation. In spite of that, a desire to treat all phase transitions as of the second order has proven to be irresistible. The source of this desire is easy to see. The theoretical physicists wanted to apply their powerful tool: statistical mechanics. The fact is, however, that it is inapplicable to solid-state phase transitions - in spite of the great power of this theoretical tool proven in other scientific applications.

Coexistence of the two phases in the process of first-order phase transitions makes the classification clear and easily verifiable. Indeed, a mere observation of the interfaces or just a heterophase state at any temperature is sufficient to identify a first-order transition. On the contrary, if the matter always remains homogeneous, the transition is of a higher-order, whatever name it is given: λ-type, second-order, higher-order, continuous, cooperative, displacive, soft-mode, or the like. By using this criterion, only one problem with the classification by first and second order will remain, namely, to find at least one well-proven crystal phase transition that would be not of the first order.

1.3   First-order phase transitions cannot have a "critical point"

The prevalent interpretation of first-order phase transitions as those exhibiting an abrupt jump in their properties at the "critical point" T_c runs counter to thermodynamics. A critical point according to its very meaning is one of absolute *instability* of the macroscopic system. It is a *fixed* point, whether it is approached upon heating or cooling. The concept of a critical point assumes simultaneous change in the whole volume. But the abrupt jumps in physical properties, being combined with a critical point, would make *perpetual motion* possible. The finite abrupt changes (Fig. 1.1c) can be reversibly triggered with infinitesimal temperature variations $T_c \pm dT$. Infinitesimal portions of energy will do that and be capable of producing a finite work. Thus, if the changing physical value is a crystal dimension, the macroscopic jumps produced in this manner can be utilized to move an engine piston and generate more energy than was spent to trigger the cyclic phase transitions.

1.4   General considerations on lattice energy, polymorphism and phase transitions

1.4.1   Taking into account definitions of a solid phase and a phase transition

We agree with the definition by Megaw [39]: "A *phase*, in the solid state, is characterized by its structure. A solid-state *phase transition* is therefore a transition involving a change of structure, which can be specified in geometrical terms."

The direct consequence of this definition is that *any* solid-state phase transition is *structural*. In the meantime, the term *structural phase transitions* has been chosen as a title for several books [*e.g.*, 18, 19, 40, 41] for the exactly opposite reason, namely to limit the consideration to only those solid-state phase transitions that are "structural." Here we deal with one more (implicit) classification: "structural" and "non-structural" phase transitions. How does it relate to *first* and *second* order? Which transitions are *structural*, and which are not? The answers remained in obscurity. Previously one distinguished between *structural*, *ferromagnetic*, *ferroelectric*, *superconducting* and *order-disorder* phase transitions. At present, however, all of the above are classified into the *first* and *second* order. Both the consideration in Sec. 1.2 and the Megaw's definition make it clear that every first-order phase transition is structural. Therefore, the *first-order* ferroelectric, ferromagnetic, superconductive and order-disorder transitions are *structural*. But what about those designated as of the *second* order? According to the Landau's theory, the crystal phases before and after second-order phase transition have different crystal symmetry, and the Megaw's definition regards it, evidently, sufficient to qualify as a structural change. Therefore, all above-listed "non-structural" phase transitions are *structural*. What's more, after second-order phase transitions are recognized as nonexistent (which, we predict, will eventually be the case, see also Secs. 4.2.1 and 4.2.2), this conclusion will be simplified to the following: *all* solid-state phase transitions are structural rearrangements by nucleation and growth.

1.4.2 Controlling factor: similarity or lowest energy?

We shall examine in a general schematic way the relationship between lattice energy, polymorphism and phase transition. This should dispel some common misconceptions regarding solid-solid phase transitions and lay basis for a better understanding of the experimental results described in the subsequent chapters. For clarity, a molecular crystal is chosen for the illustration.

The same substance can be represented by an unlimited variety of hypothetic crystal structures, each one being characterized by its free energy F. The latter is given by:

(a) $F = U + E - TS$

where U is lattice potential energy, E is molecular vibration energy, and S is entropy. In turn, U can be represented as a sum of the contributions Ul, U2, and U3 from the Van der Waals' interactions, hydrogen bonding and electrostatic interactions:

(b) $F = U1 + U2 + U3 + E - TS$

Given temperature T and pressure p, every term on the right is a function of the unit cell parameters a, b, c, α, β, γ and the coordinates X_i, Y_i, Z_i of all i atoms in it:

$$U1 = U1\ (a,\ b,\ c,\ \alpha,\ \beta,\ \gamma,\ X_i,\ Y_i,\ Z_i\ )$$
$$U2 = U2\ (a,\ b,\ c,\ \alpha,\ \beta,\ \gamma,\ X_i,\ Y_i,\ Z_i\ )$$
$$\text{(c)}\qquad U3 = U3\ (a,\ b,\ c,\ \alpha,\ \beta,\ \gamma,\ X_i,\ Y_i,\ Z_i\ )$$
$$E = E\ (a,\ b,\ c,\ \alpha,\ \beta,\ \gamma,\ X_i,\ Y_i,\ Z_i\ )$$
$$S = S\ (a,\ b,\ c,\ \alpha,\ \beta,\ \gamma,\ X_i,\ Y_i,\ Z_i\ )$$

To generate a new hypothetical structure the independent variables in the parentheses have to be changed. Any such change will create in eq. (b) a new set of the free energy components U1, U2, U3, E and S, increasing some of them and decreasing others. We can think of them as variables, but not independent of one another. Choosing a particular mode of molecular packing will affect all the F components. In one hypothetical structure, for instance, the arrangement offering the strongest hydrogen bonding can be achieved only at the expense of a somewhat looser molecular packing, that is, of a somewhat higher energy due to weaker Van der Waals' molecular interactions. Another arrangement may provide the closest molecular packing, but requires a parallel dipole alignment, which gives rise to a higher potential energy due to electrostatic dipole repulsion. And so on.

Thus, optimizing one component will in general preclude others from being minimal. The number of the hypothetic phases with different F is unlimited. Among them there are a few with especially low F. These hypothetic optimum structures result from intricate competition between different components of F, so that, in general, neither is at its exact minimum after all. In this competition no restrictions exist regarding crystal formation. Two crystal structures of the lowest F can have quite different symmetry, cell parameters, and modes of molecular packing. If structures 1 and 2 with the lowest free energies F happened to have $F1 \cong F2$, they will be neighbors on the phase diagram p -- T, thus providing the condition for phase transition.

Only free energy F will determine which polymorph will be stable under one or another particular conditions (p, T). In nature's game, with many variables in play to determine the structures with the lowest F, there is no room for any factor limiting the law of minimum free energy. The habitual assumption of a structural "similarity" of polymorphs is such a factor. Which polymorphs will come into being: those with the lowest F but very different structures, or with similar structures, but higher F? Obviously, the former. In general, any concept involving the idea of structural interdependence implicitly challenges the priority of the law that a system with minimal free energy is that to materialize.

The foregoing consideration is, in fact, a thermodynamic justification of first-order phase transitions: their structural parameters and symmetries of the phases are not only different, but not even related. It also should help to figure out why the idea of second-order phase transitions was unrealistic. The most rigid condition preventing the minimum free energy law to determine a polymorphism is the requirement that the two structural polymorphs were completely identical (and not just "similar") at a certain "critical" point (p, T) of the phase diagram, where the phase transition has to occur. This would reduce the probability of the phase transition to zero.

1.4.3 Why polymorphs sometimes look similar

One of the most persistent and widespread misconceptions is a belief that polymorphs are structurally interrelated and transform into each other by some sort of modification. Therefore, one always looks for similarity between the two structures. Reflecting this approach, such terms as *displacement, shift, distortion, deformation,* and *displacive transformations* are in common use. These views are not based on any scientific evidence: they were taken for granted as "self-evident." Second-order phase transitions are the ultimate expression of this idea: the two polymorphs must be not only "similar," but become *identical* at the transition point, while the phase transition resulted from continuous atomic shifts. This imaginary way of phase transition has nothing to do with the real process, which always is nucleation and growth (see also Sec. 3.7.1).

Then, why is it not unusual for polymorphs to be similar in one or another way or just exhibit a resemblance to each other? Any similarity is neither required, nor prohibited by the principle of minimum free energy. The source of the similarity, when found, is the same molecules that the two structures are built from. Sometimes certain structural elements, *e.g.*, molecular arrays or layers, are so much advantageous as to their packing energy that they reappear (with only slightly different parameters) in every polymorph. It will be demonstrated that even such structures change into each other by nucleation and growth. Another source of the "similarity" is human imagination.

1.5 Other classifications

If the *first/second-order* classification had been satisfactory, no further classification attempts would be needed. However, the history of research of solid-state phase transitions is a chain of such attempts - each time after disappointment with the previous one. Their failure is rooted in the belief that the phase transitions are a kind of a cooperative shift, displacement, distortion, deformation, or disordering of the initial crystal structure. This section contains a brief characterization and critical remarks with respect to some of them in the context of our solution of the problem. Otherwise, a more detailed description of the major classifications and suggested mechanisms of phase transitions can be found elsewhere (*e.g.*, [42]). All of them are in a more or less advanced stage of fading. On the other hand, the earlier classification attempts have later exploded into so many types of phase transitions (refer to our collection, Appendix 1) that even their brief consideration would be impractical. In addition to the classifications discussed in this section, two more are considered in Sections 1.6 and 1.7.

1.5.1 Martensitic – and something not martensitic

One of the oldest was the classification by *martensitic* and ... an alternative. It came from physical metallurgists who studied formation of a phase called *martensitic* in iron alloys from the higher-temperature phases. This classification was later claimed to cover many other solid-solid phase transitions*. According to the initial premises, a martensitic transformation is a strictly orderly process localized at a straight interface called "habit plane." There the two crystal structures exactly match with one another, the adjacent lattices on both sides of the habit plane being under local elastic distortions to provide this matching. A martensitic transformation occurs at a specific temperature T_M which is neither T_O, nor T_C. The velocity of the interface propagation is that of a sound wave, rather than a function of temperature. There must be a certain rigorous orientation relationship between the crystal lattices prior to and after the transformation. The theory of martensitic transformation approximated this process by a uniform transformation in the bulk. Since observation of phase transitions in iron and its alloys is an extremely difficult task, the suggested "martensitic" mechanism was based more on imagination than on solid facts.

---

* The term "transformation," rather than "transition," is always used in this classification, probably to emphasize a structural interrelation of the initial and resultant phases.

The alternative to martensitic transformations was sometimes called *diffusional*, but diffusion is too slow a process to account for the rates of "non-martensitic" transitions. Then the terms "usual" or "nucleation and growth" were used. As to the term "usual," it was as informative as a question mark. The term "nucleation and growth" was also inappropriate, since it has been agreed that martensitic transformations also occur by nucleation and (very fast) growth. Since both alternatives in the *martensitic / usual* classification were nucleation and growth and recognized to be first order, there was no room for second-order phase transitions.

The more "martensitic transformations" were investigated, the more it became clear that they do not have a single specific experimental characteristic separating them from what was claimed to be their alternative. They start from nucleation; their actual speed was lower than that of sound propagation and depended on temperature; temperature hysteresis was not their specific feature either;* crystal orientation relationship was not always as expected, was not strict, or was absent. All attempts to find a specific characteristic of a martensitic mechanism have failed. The basic assumptions of a martensitic mechanism were not substantiated by experiment.

Once dominant over a significant part of literature on solid-state phase transitions, the *martensitic transformation*, as a phase transition mechanism, was fading for a period of time until it was recently resurrected in relation to the shape memory effect**. Now it was taken for granted; the problems with its introduction and definition that were discussed decades ago were forgotten.

1.5.2 Displacive – and reconstructive

The classification *displacive-reconstructive* was put forward by Buerger [37, 43]. It was exclusively based on a comparison of the crystal structures before and after a phase transition. This author shared a rather common belief that one can make judgments about the *process* of phase changes from a comparative analysis of the initial and the resultant crystal structures. In [37] Buerger postulated that the

---

* On the origin of the hysteresis see Sec. 2.6.6.

** There is no reason to bound shape memory effect to martensitic mechanism (see Addendum F).

number of different mechanisms of phase transitions is more than one and outlined the goals of his classification. They are to explain the "fundamental experimental results," namely:

(1) energy change upon transition can be continuous or discontinuous,
(2) it can be large or small,
(3) some transitions are fast, others are sluggish,
(4) the symmetry can change or be retained.

Neither quantitative, nor even semi-quantitative estimates to the notions "large," "small," "fast," "sluggish" were given. The energy change is neither continuous, nor discontinuous: it follows the change of the phase ratio in the heterophase sample during the phase transition. The other three differences do not represent different mechanisms. They are merely a result of the structural independence of the polymorphs. The previous part of this chapter is devoted to the cause of showing that, whatever the difference or similarity between the polymorphs, the way of structural rearrangement is nucleation and growth.

Buerger suggested that structures can change into one another in two ways. If they are similar, the transition does not involve breakdown of the original bonding and is *displacive*. But, if there is no way to reform the initial crystal without breaking the existing bonding net, the transition must be *reconstructive*. The descriptions given to these two mechanisms were ambiguous. The *reconstructive* transitions are of the first-order. "Their structures are so different that the only way a transformation can be effected is by disintegrating one structure into small units and constructing a new edifice from the units"; such a transition is "sluggish," because the substance must pass through the intermediate state of a higher energy. It suffices to note that at the time of that presentation there was plenty of experimental data showing phase transitions proceeding by propagation of interfaces. Nowhere in the classification was this fact taken into account.

The description of *displacive* phase transitions was not less problematic. They are fast, barrierless, involving only a small displacement of one or more kinds of the atoms. We were also informed that many *displacive* transitions exhibit a small energy jump, certainly indicating first-order phase transition, but the physical rearrangement could still proceed as in second-order phase transitions. Not repeating the arguments against phase transitions by "displacements," it should be stated that such hybrid "firstsecond"-order phase transitions are not permitted by thermodynamics.

There were more drawbacks. The introduction of the two distinct types - *displacive* and *reconstructive* - turned out to be only a headline for a rather cumbersome classification. It was found impossible to relate these two types with the changes in the first and second coordination. Several mechanisms, such as "dilatational" and "rotational," were added. They were neither quite *displacive*, nor quite *reconstructive*. Finally, the predicted velocities of phase transitions ("rapid" or "sluggish") did not correlate with experiment. As McCrone [44] pointed out, "one should always be ready to meet unforeseen velocities."

The above "structural" classification was a failure, as was its "refined" version by Megaw [39]. The method itself of assuming the *process* of the structural change simply by comparing the two structures is flawed. Available experimental observations of nucleation and growth were disregarded. None of the phase transitions considered in Chapter 2 belongs to any of the types described by Buerger. There are no displacive transitions. All phase transitions are reconstructive in a sense, but of a different type.

1.5.3 Displacive – and order-disorder

Some authors adopted the term "displacive" to divide phase transitions into two broad types: *order-disorder* and *displacive*, assuming the former to proceed by "disordering," and the latter by "displacement." The Buerger's *reconstructive* transitions were missing there. In particular, this classification was used in the theory of ferroelectrics. It will be shown (Sec. 2.7) that "order-disorder" phase transitions proceed by nucleation and growth.

1.5.4 Phenomenological approach: I, 2I, ... etc. types

The problems with classification of phase transitions in crystals by their molecular mechanism led to attempts to empirically sort them out into phenomenological groups. McCullough [45,46] did it on the basis of the shape of singularities (λ-peaks) observed in the experimental plots of heat capacities C_p vs. temperature. He suggested seven types denoted as I, 2I, 3I, 2N, 3N, H and G. Some justification to this approach would be to show that other properties fit the same graduation as well. This has not been done, and could not be done. The reason that the classification turned out worthless is that the so-called "heat capacity λ-peaks" are neither a heat capacity nor a singularity. Their origin is disclosed in Sec. 3.3 and Appendices 2 and 3.

1.5.5 Another phenomenological classification (Pippard)

The classification suggested by Pippard [47] was from the same category. Again, the $C_P(T)$ curves were divided into a number of distinct types, this time in terms of the idea that C_P and dC_P/dT, approaching the transition point, can appear either finite or infinite from one or both sides of the temperature scale. This led to subdivision of the original Ehrenfest's second-order and third-order transitions into four types of each [the Ehrenfest's classification did not deal with the λ-peaks of $C_P(T)$]. This classification, like the previous one, is useless due to the misinterpretation of λ-peaks.

1.5.6 Quantum – and classical

It is assumed in this classification that all solid-state phase transitions, except those at or near absolute zero temperature, are "classical" and occur at their critical temperature points T_c. The fact that at least the majority of them are first order and do not occur at critical points is ignored. It is claimed instead that the crystal order in the "classical" phase transition is destroyed by thermal fluctuations.

According to this classification, phase transitions near 0 °K occur under the action of only quantum, rather than thermal, fluctuations, which makes them somehow specific. They are called "quantum phase transitions." See Addendum B.

1.6 The soft-mode concept

The soft-mode concept was put forward in about 1960 to explain the mechanism of *displacive* ferroelectric transitions, but then it was also applied to *order-disorder* ferroelectric transitions and, finally, to all "structural" phase transitions. According to the developed theory, a structural phase transition is a cooperative *distortion* of the initial crystal structure as a result of atomic shifts (displacements). This distortion is produced by one of the "soft" (*i.e.*, low-frequency) optical modes, which "softens" toward the transition temperature. When the soft-mode wavelength becomes comparable with the crystal dimensions, the cooperative displacement of certain atoms makes the crystal unstable, the displacement suddenly becomes "frozen" and the crystal switches into the alternative phase.

The soft-mode concept was developed, tested and demonstrated by using ferroelectric $BaTiO_3$ as an example; even *jumps* in the physical properties at the Curie point were calculated [48]. The same $BaTiO_3$ was used by Landau to illustrate a *continuous* second-order phase transition. Evidently, at least one of the two conflicting approaches must be incorrect. Nevertheless, in this section we deal with soft modes. A first-order ferroelectric phase transition in $BaTiO_3$ is well established, including large hysteresis of the transition temperature. Consequently, (a) the only way it can proceed is nucleation and growth (Sec. 1.2), (b) no cooperative jumps at the Curie point are possible (Sec. 1.3), and (c) the Curie point does not exist there.

In 1970's, the *soft-mode* theory became a popular interpretation of solid-state phase transitions.* Optical and neutron spectroscopic experiments were aimed at finding a soft mode for every phase transition. In 1973 Shirane [53] distinguished two groups of phase transitions in solids: (1) magnetic and superconducting, which he regarded not to be "structural," and maintained (unfoundedly) that they "were already reasonably understood," and (2) "a large variety of other phase transitions," such as in SiO_2, Nb_2Sn and those in ferroelectrics and antiferroelectrics. He contended that "the generalized soft mode concept covers the essential mechanism of phase transitions in solids" and that "the soft mode concept brings a unified picture" of how phase transitions take place in the whole second group known as *structural* phase transitions.

Such generalization conflicted with well-established facts. Even if the *soft-mode* concept suggested the physical picture, correct or not, for some phase transitions, its application to all "structural" phase transition was not justified. The available experimental literature would reveal that (a) there were quite a few examples of phase transitions with the polymorphs so different in their symmetry, lattice parameters and atomic arrangements that their mutual reformation by *distortion* (displacement, shift) was unimaginable (Buerger classified them as

---

* For more on the *soft-mode* concept see [41, 48-53].

reconstructive, Sec. 1.5.2), and (b) there was a host of observations of phase transitions by nucleation and growth - certainly not the subject for the soft-mode treatment.

The *soft-mode* mechanism had certainly nothing to do at least with the majority of solid-state phase transitions. Moreover, it is not even applicable to ferroelectric phase transitions, since "only very few ferroelectrics... have critical or near critical transitions... the majority having first-order transitions" [54]; "most ferroelectric phase transitions are not of second order but first [55]." To that it remains to add that *all* ferroelectric phase transitions are of the first order and occur by nucleation and growth (Sec. 4.2.1).

If so, how can the evidence presented in support of the soft-mode mechanism be explained? It was not definitive at all. In some cases rather "soft" modes were indeed found in the corresponding spectra of a phase, but in many other cases, including almost all molecular crystals [56], no soft modes were detected. Selection of a soft mode that "softens" toward the transition temperature was arbitrary and regarded sufficient to declare the phase transition of the soft-mode type. "Soft" modes, as any vibrational modes, can be found in many crystals with or without phase transitions. Like all crystal properties, a soft mode is temperature-dependent and occasionally can show "softening" in the "desirable" direction. This in no way proves that it has any part in the phase transition, if there is one.

The soft-mode concept has not justified the hopes of its inventors. It still exists as one of the possible approaches to some solid-state phase transitions. A truly unified picture of how *all* solid-state phase transitions occur will emerge from the fact that they all belong to the nucleation-growth type.

1.7 The imaginary "incommensurate" solid state

It had been well established that condensed matter can be in a liquid, crystalline, mesomorphic (liquid-crystalline or orientation-disordered-crystalline) state, or be amorphous. Then the new solid state, called *incommensurate*, was introduced and for a decade or so became very popular in certain circles of research scientists [57-61].

There was something unusual in this discovery. Dozens of volumes of the reference book "Structure Reports" had already been filled with descriptions of thousands solved crystal structures and never a structure of the "incommensurate" type was reported. Therefore, finding a new type of solid state had to be, it seemed, a matter of great scientific and technological significance. One would expect an avalanche of intensive studies of the physical properties and practical usefulness of the new solid matter. It has not happened. This new solid state was not the subject of interest *per se*. The *incommensurate solid state* was invented as a remedy to cure the ailing *soft-mode* model of solid-state phase transitions. Pynn [59] asserted in the 1979 review that "the discovery and study of incommensurably distorted structures is a milestone in the investigation of structural phase transitions." In spite of the word "discovery," no evidence of the *incommensurate* state was presented in that review. As a matter of fact, no hard evidence has ever been found. Yet, the *incommensurate* solid state was accepted as a reality.

According to the initial *soft-mode* model, a phase transition occurs under the action of a soft mode whose frequency "softens" toward the transition temperature where it turns into zero. There was a problem, however: in most real cases such an optical mode was not found. This increased doubts in the validity of the soft-mode mechanism or, at least, limited its applicability. The new idea was to "soften" requirements to the soft-mode lattice modulation. Now it did not have to "soften" further or even be a rational multiple of a dimension of the crystal unit cell. Now "the new phase does not at all possess any periodicity along the coordinate axis ... and is referred to as incommensurate. Incommensurability may, naturally, occur along two or three coordinate axes... The fundamental feature of the crystalline state is lost" [59]. The incommensurate phase transition occurs by a "distortion" of the *underlying* ("prototype," "basic," "mother," "undistorted," "symmetrical") higher-temperature phase.

All attention in the literature was directed at the proposed new mechanism of phase transitions. No attention was paid to the resultant peculiar solid state where the displacement of every particular atom had to be unique, so that the resultant structure lacked translation symmetry. Such a solid state defies logic, our knowledge about solid state, and thermodynamics. It cannot exist for each of the following reasons:

(1) The fundamental assumption that structural phase transitions occur by a displacement (distortion, shift) is in error, for they occur by nucleation and growth. The relation of the soft-mode and incommensurate transitions to the first/second order classification

deserved more attention than a common statement to which class one or another transition belongs. Being a *cooperative* phenomenon, they are usually regarded second-order phase transitions, but applied to first-order and "partly first-order" as well. A first-order incommensurate phase transition is an oxymoron and will not be discussed further. It cannot be of second order either: like the soft-mode transitions (see Sec. 1.6), it should occur by a *finite* structural jump between the polymorphs and would comply neither with the second-order transitions, which are continuous, nor with thermodynamics.

(2) The theory of a commensurate → incommensurate transition assumes that the modulating wave becomes "frozen-in" in the resultant phase. The reverse transition should "unfreeze" it with exactly the same mode. However, the vibrational spectrum of the resultant phase is different and does not have that particular mode any more. Thus, the conclusion has to be made that this type of transition is intrinsically irreversible. The theory, however, is not limited by irreversible transitions.

(3) The polymorphs are structurally independent (Sec. 1.4). But the incommensurate phase transitions assume all the lower-temperature phases of a substance to be derivatives of a "prototype" phase. Suppose there is a prototype high-temperature phase H which changes by a distortion into the lower-temperature *incommensurate* phase L. The same phase L can also be obtained by growing it from a solution or the vapor at the same lower temperature where it is stable. We come then to the absurd results: (a) the grown L crystal will have "incommensurate" rather than normal crystal structure, and (b) the grown crystal L will be a *modulated* phase H. Why does the L structure have to be "incommensurate" if the way it came into being had nothing to do with distortion of the "prototype" phase by a vibration mode? What is the source of the "intellect" that enables the crystal grown from solution to know that it must be a distorted version of another phase that can exist at a higher temperature?

(4) The alleged "incommensurate" structure cannot materialize due to a violation of the *close packing principle* valid towards metallic, ionic and molecular crystals. Violation of this principle is equivalent to rejection of the universal principle of minimum free energy in the formation of a structure. Molecular crystals are especially pictorial to illustrate the principle of close molecular packing [62]. The cause

behind the principle is minimization of energy of the Van der Waals' interactions in a crystal. By encircling the molecular "skeleton" with the standard Van der Waals' radii, an organic molecule can be assigned a particular shape, as shown in Fig. 1.2a for biphenyl. Any real organic crystal belongs to one of the most closely packed structures of the molecules defined in this way. For an illustration, the molecular packing of the high-temperature phase of thiourea is shown in Fig. 1.2b. Crystals that disobey the principle of close packing in the "incommensurate" manner are unknown. Incommensurate modulation of a prototype structure by a soft mode will cause individual molecular displacements without regard for the resultant intermolecular distances. Molecules in this structure would penetrate into one another, leaving the adjacent areas vacant. All accumulated experience to date shows that such a structure cannot exist; the polymorphs always represent two different versions of the most closely packed molecules.

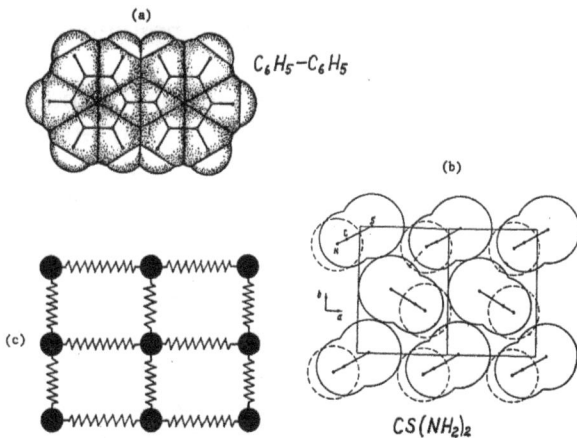

Fig. 1.2 (a) The model of a biphenyl molecule constructed by encircling the molecular "skeleton" with the intermolecular radii (by Kitaigorodskii, [62]). (b) The close molecular packing in the high-temperature phase of thiourea. Nitrogen atoms (broken lines) are off the plane ab shown by solid lines. The two shown inner molecules have eight "contacts" with the surrounding neighbors (i.e. positioned at the optimum Van der Waals' distances). (c) Any irregular displacements of the balls in this model (equivalent to disturbing the network of standard interatomic distances by an "incommensurate" soft mode in an atomic crystal) will result in returning it into the shown original state. Only rearrangement leading to a new network of standard distances is plausible.

To illustrate the point farther, let us turn to the mechanical model of an atomic crystal where balls represent atoms, and springs their bonds (Fig. 1.2c). To assume that it is possible to produce an "incommensurate" structure from this undistorted structure is equivalent to the assumption that one can displace the balls in different directions (that is, arbitrarily change the lengths of interatomic couplings in the crystal lattice) and the balls will not return into their initial equilibrium positions (*i.e.*, the distortions will be "frozen-in," as a proponent of the incommensurate phase transition would say).

Any particular "incommensurate distortion" depends on the wavelength of the mode that caused this transition ("frozen-in wave"). However, no specific mechanism of phase transition can *impose* the resultant state. The latter is determined by the minimal free energy. Any equilibrium phase state in the p -T phase diagram is the exclusive function of these parameters. If the diagram shows the existence of two different crystal phases, the only function of the phase transition, whatever its mechanism is, is to change the above phases into each other.

1.8   300 mechanisms of one phenomenon

Appendix 1 is a collection of 300 different types of solid-state phase transitions found in literature. Even if they are sorted out into groups, the number of mechanisms does not lend credibility to all of them; rather it indicates the failure to identify one general mechanism. Such a state of affairs is in keen contrast with what is known about nature's laws. Nature is extremely thrifty. There is a single equilibrium state of any solid matter, be it a metal, ionic, or organic substance: it is a *crystal state.* Crystals can come into being from vapors, melts, solutions, or other crystals. There is only one general mechanism by which crystals of any nature can be developed from any solution, vapor, or melt: it is a *nucleation and growth.* This is hardly consistent with the idea that the same process in a solid medium requires scores of diverse mechanisms. A major goal of the present book is to show that the nucleation and growth principle is the only one by which a crystal state develops from a solid medium. Ferromagnetic, ferroelectric and superconducting phase transitions (Secs. 4.2.1 and 4.2.2 and 4.15) are also covered by this universal process.

"Transition" means *process*: passage from one state or condition to another. Giving a name to a phase transition means identification of the specific *mechanism* of passage from one phase to another. This reminder is needed when looking at the collection of 300 different "mechanisms" in question. Some of that chaos of names can be conditionally sorted out into groups:

- Names tending to somehow reflect on the process (mechanism) of the phase transition: nucleation-and-growth, displacive, order-disorder, cooperative, diffusional, distortive, catastrophic, spin-flop, cation ordering, continuous... It is usually assumed that the phase transition is basically or exclusively reduced to atomic/molecular displacements, structural distortion, spin-flopping, *etc.*
- Names not reflecting the mechanism directly, but have a more or less established description of the mechanism (however erroneous) in literature: martensitic, soft mode, incommensurate, second order, quantum.
- Names carrying no characteristic at all, except being *not* something: "usual" are not martensitic, "classical" are not quantum, "diffusionless" are not diffusional, "structural" are not ferromagnetic or ferroelectric or superconductive. So are "ordinary," "normal" and "simple."
- Simply names of particular authors: Kastelein, Jahn-Teller, Mott, Anderson-Mott, Kosterlitz-Thouless, Berezinskii-Kosterlitz-Thouless, Ising, Lifshitz, Oguchi, Wilson, Stenley-Kaplan, Gardner, Neel, Peierls, Potts, Salam, Verway. This is a convenient way to name phase transitions: it is prestigious to those authors, shifts the responsibility to define them ...and impedes scrutiny.
- Names identifying the assumed driving forces, evidently in the belief that this is the same as identifying the specific phase transition mechanism: density-driven, density-driven quantum, electronically driven, driven by soft-shear acoustic mode, driven by soft mode, current-induced, pressure-induced, shock-induced, stress-induced, field-induced. The idea advanced in the present book is the opposite: the mechanism is always the same, and the function of a driving force, whichever it is, is to move the phase across the dividing line on the phase diagram.
- A loose group of names that are too formal to reflect meaningfully on the mechanism: first order (showing "jumps" in physical properties), lambda (showing singularity of a physical property reminiscent to letter 'lambda'), infinite order, weak-order, non-weak, isothermal, thermodynamic, non-thermodynamic, volume-change, symmetry-breaking, symmetric-antisymmetric.
- Names indicating the resultant phase by its prominent property: ferromagnetic, ferroelectric, superconductive. At that, it is automatically assumed that there are, accordingly, specific phase transition mechanisms leading to them. The idea that in every case they are simply *the result* of a structural rearrangement was missing.

All these names reflect the difference in the physical properties of the phases that resulted from the change of crystal structure rather than the process of that change, which is always crystal rearrangement by nucleation and growth.

1.9   Ferromagnetism without Weiss-Heisenberg molecular / exchange field

The content of this section is basically applied to ferroelectrics as well. Let us turn again to Eq. (b) from Sec. 1.4

$$F = U1 + U2 + U3 + ... + E - TS$$

Contributions to the free energy F from U1, U2, U3, etc., are far from being equal: potential energy of atomic or intermolecular bonding will always tower. This is the case even in molecular crystals where, as known, the (Van der Waals') intermolecular bonding is relatively weak. Due to the domination of the intermolecular bonding energy, any molecular crystal is a structure of closely packed molecules [62], even though there can be other contributors to the crystal energy. As estimated by Kitaigorodskii [63], the energy of electrostatic interactions in the case of strong dipole interactions in organic crystals does not exceed about 5% of the sublimation energy. The part played by such a "small contributor" is not negligible, but limited. It can determine which of the possible close packings will prevail, so that the real molecular close packing may be not the closest one. An important point is, however, that by itself it is unable to impose any crystal structure; the structure must be one of those allowed by Van der Waal's bonding. Thus, the contribution of electric dipoles to the free energy of ferroelectric crystals is small. Calculations of other authors were in accord with this conclusion. The same holds true for magnetic dipoles in ferromagnetics: it has been estimated that magnetic interaction is too weak to have a measurable effect on the lattice parameters.

This conclusion is important. The conventional theory incorrectly reduced the source of internal energy of a ferromagnetic crystal to the "exchange" interaction of its magnetic dipoles. The energy of Fe or Ni crystals is primarily due to the same *chemical bonds* as in nonmagnetic metals Cu or Al. The energy of true magnetic interaction constitutes just a small part of the total crystal energy. The magnetic "exchange" energy is either insignificant or zero. These assumptions will be carried throughout Chapter 4 and will produce the consistent and simple fundamentals of ferromagnetism.

In the beginning of the 20th century, the nature of ferromagnetism was a great enigma. It was recognized that a ferromagnetic state was a stable state and that it resulted from a *parallel mutual alignment* of elementary magnetic moments. The puzzling question was: why are they parallel? It seemed that the repulsive magnetic forces of the parallel magnetic dipoles had to destabilize the system.

A solution was suggested by Weiss: since a stable existence of ferromagnets was a reality, there had to be an additional factor making parallel alignment preferable. This additional factor was postulated under the name "molecular field." Another reason to introduce this additional field was the real temperatures of ferromagnetic phase transitions (called "Curie points") that the theory had to account for - otherwise these "points" were placed by the theory in the region close to 0 °K. The Weiss theory had been accepted as the solution having no apparent alternative and became the "classic theory of ferromagnetism," even though the nature of the molecular field has never been explained within the framework of classical physics. Later on, Heisenberg suggested ("discovered") that the molecular field has a quantum-mechanical "electronic exchange" origin.

The Weiss' "classic" theory is described in many detailed college courses of physics as valid and irreproachable. What these textbooks do not reveal is that neither the classic theory, nor its quantum-mechanical version, could adequately account for the observed ferromagnetic phenomena, thus misleading new generations of specialists and building a psychological barrier to discovering the true nature of ferromagnetism. Still, a number of authoritative statements were made, pointing at serious failings of the theory (Appendix 4). What is the real status of the theory if "even for such a 'simple' ferromagnetic substance as nickel it is not possible to 'squeeze' the experimental results into the Weiss-Heisenberg theory" [64]? The answer to this question is that "the Weiss theory in either its original or modified form is quite inadequate" [65].*

---

* This criticism is not intended to diminish Weiss' contribution to understanding ferromagnetism. Weiss was the first to suggest the existence of domain structure and spontaneous magnetization of the domains. Rather, it is his successors who had to reject the "molecular field" instead of embracing it.

The failure of the Weiss-Heisenberg theory, including the subsequent efforts to improve it, is rooted in the fact that a system of magnetic moments was basically analyzed as an independent system rather than a property of the crystal. It is a *crystal* field*, and not a *molecular* field, that determines one or another specific order of the elementary magnetic moments, be it parallel, antiparallel, or any other. This holds true for the dipoles in ferroelectrics as well, which eliminates a scientific mystery stemming from the fact that no "molecular field" was concluded to exist there. It is the *crystal field* that makes them a dielectric analog of ferromagnetics. To reflect their common nature the term "ferroics" will sometimes be used.

A parallel system of magnetic or electric dipoles by itself will *increase* the free energy F of the crystal. But the dipole interaction energy alone does not determine whether a ferroic structure is stable or not. The crystal free energy F as a whole does. The dipole contribution is small. A ferroic phase is stable *in spite* of the disadvantage of their parallel alignment; it so happens that F for any hypothetic crystal structure with antiparallel magnetic moments is higher notwithstanding the advantage of their antiparallel alignment.

The foregoing will be illustrated by a hypothetic numerical example. Suppose there is a molecular crystal of dipole molecules. A parallel alignment of the dipoles destabilizes the crystal by 1 kcal/mole, while antiparallel increases its stability by the same value. The resultant internal energy is a total of the energy U of the molecular close packing and the contribution from the dipole interactions. Let us assume that the bonding energy of the closest packing U' = 20 kcal/mole, that this structure (No.1) imposes the alignment $\uparrow\uparrow$, and that the alignment in the alternative structure (No.2) with the packing energy U" is $\uparrow\downarrow$. If U" = 17 kcal/mole, the total bonding energy in the competing structures will be

[No.1 ($\uparrow\uparrow$)]   U' -1 kcal/mole = 19 kcal/mole
[No.2 ($\uparrow\downarrow$)]   U"+1 kcal/mole = 18 kcal/mole

Molecules in No.1 are bounded stronger. Therefore, No.1 will come into being *in spite* of the dipole repulsive energy. This crystal will be *ferroelectric*. But if U" = 19 kcal/mole then

[No.2 ($\uparrow\downarrow$)]   U"+1 kcal/mole = 20 kcal/mole

---

* The term "crystal field" in this book denotes the field of coupling forces that bound the constituent atoms and molecules into the three-dimensional crystal order.

In such a case, No.2 is preferable and will come into being. This crystal will be *antiferroelectric*. The same idea is illustrated in Fig. 1.3 in a diagram form. An important element of the above concept is that a parallel, antiparallel, or any other dipole order is *imposed* by the crystal structure, in this example by the mode of molecular close packing.

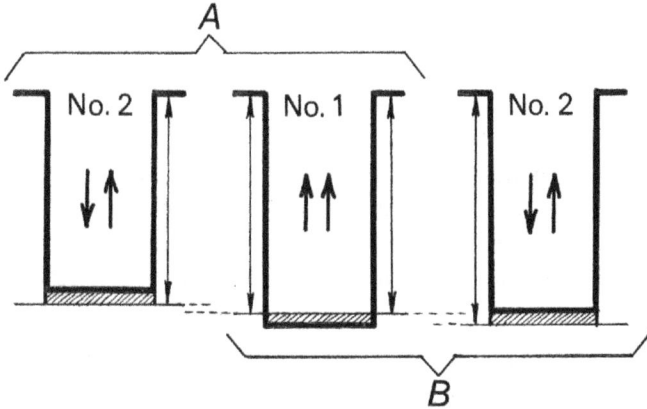

Fig. 1.3   A diagram explaining thermodynamic stability of ferromagnetics and ferroelectrics without any "molecular / exchange field" involved. The bold lines represent potential wells owing to crystal (chemical) bonding only. Contributions from the dipole interactions are shown by the shadowed areas.
  A. While the dipole interaction in No.1 enhances the energy of the system, and in No.2 it lowers it, the total energy level of the system (bilateral arrows) in No.1 remains lower, so that a ferroic (ferromagnetic / ferroelectric) order is realized.
  B. In this case the situation is opposite: antiferroic dipole order is preferable.

Our account for ferromagnetism and ferroelectricity involves four major elements:

(1) The first has been considered above. A simple answer was given to the fundamental question why all kinds of ferroic structures (ferro-, antiferro-, ferri-, and any other magnetics and "electrics") can be quite stable. The conventional theory was unable to consistently explain this phenomenon.
(2) The second is the origin of ferromagnetic and ferroelectric domain structures. The generally accepted view is that the domain structure is a specific property of ferromagnetic and ferroelectric interactions. Our solution is that the opposite is the case: the domain structure is essentially of crystal-structural origin (Sec. 2.8.6) and a condition for a remagnetization / repolarization in the applied fields.
(3) The third is origin and formation of ferroic hysteresis loops. The primary role of the heterogeneous nucleation lags - as described in Sec. 2.5 - is demonstrated.

(4) The fourth element is concerned with remagnetization and repolarization process. The common view is that this can occur by a dipole rotation either in the crystal lattice or over the thick domain boundaries having a "Bloch wall" structure. Our solution is as follows. There is a single universal mechanism both of magnetization / polarization upon ferroic phase transitions and remagnetization / repolarization in the applied fields: it is a structural rearrangement at the contact interfaces (Secs. 2.4.3 and 2.8), the dipole orientations being a *result* of this restructuring.

This section may be viewed as the introduction to Chapter 4, where both the origin of ferromagnetism and ferroelectricity and their major manifestations will be consistently accounted for. To do so, all details of the mechanism of structural rearrangements in solids presented in Chapters 2 and 3 will be necessary. As known, the solution of a difficult scientific problem sometimes comes from an adjacent field. This time it has come from the investigation of solid-state phase transitions.

CHAPTER 2. THE MOLECULAR MECHANISM OF SOLID-STATE PHASE TRANSITIONS

2.1 Summary of the problems

As discussed in Chapter 1, no attempt to account for the molecular mechanism of solid-state phase transitions has been adequate. Therefore, it is reasonable to start all over again. First the problems to be solved must be formulated. We want to discover the *molecular mechanism* of the phenomenon, namely, what actually happens in time and space to the constituent particles (atoms, molecules, ions) of an initial phase in the process of changing into the resultant phase. The knowledge of the real mechanism should allow us to consistently account for all relevant experimental facts.

The direction of our experimental investigation of the molecular mechanism of solid-state phase transitions is summarized in the flowchart 1. Naturally, it was not so definite from the inception of this work as it is now presented in the chart. Rather, the problems and their sequence were clarified as the work advanced. In other words, while the flowchart could serve as a guide for the investigation, it already partially reflects its results. The flowchart separates the real problems (denoted by arrows) from the imaginary (dotted lines) and explains the purpose and structure of the chapter.

Only two ways of crystal rearrangement are conceivable. One is a molecular rearrangement at the boundaries between coexisting phases (first-order phase transitions). The other is a concerted, cooperative, homogeneous, continuous change in the original crystal structure (second / higher-order phase transitions). Only one of them - the former - materializes. The whole problem boils down to finding out the mechanism of first-order phase transitions. It splits into two major branches represented in the flowchart as "nucleation" and "interface."

Speaking of first-order transitions, it is impossible for a macroscopic block of a crystal to change instantaneously, as a whole, into the alternative phase: the transition must start from the formation of a microscopic nucleus of the resultant phase in the body of the initial phase. Nucleation in solid state was a subject of special, for the most part theoretical, investigations, predominantly by metallurgical physicists [66]. One assumes two major ways, called *homogeneous* and *heterogeneous*, for the nuclei to form. The former is supposed to occur at any arbitrary point of the bulk, the latter only at crystal defects. In the case of heterogeneous nucleation, a number of specific questions arise regarding the type of crystal defects, the particular mode of the nuclei formation at the defects, etc.

FLOWCHART 1

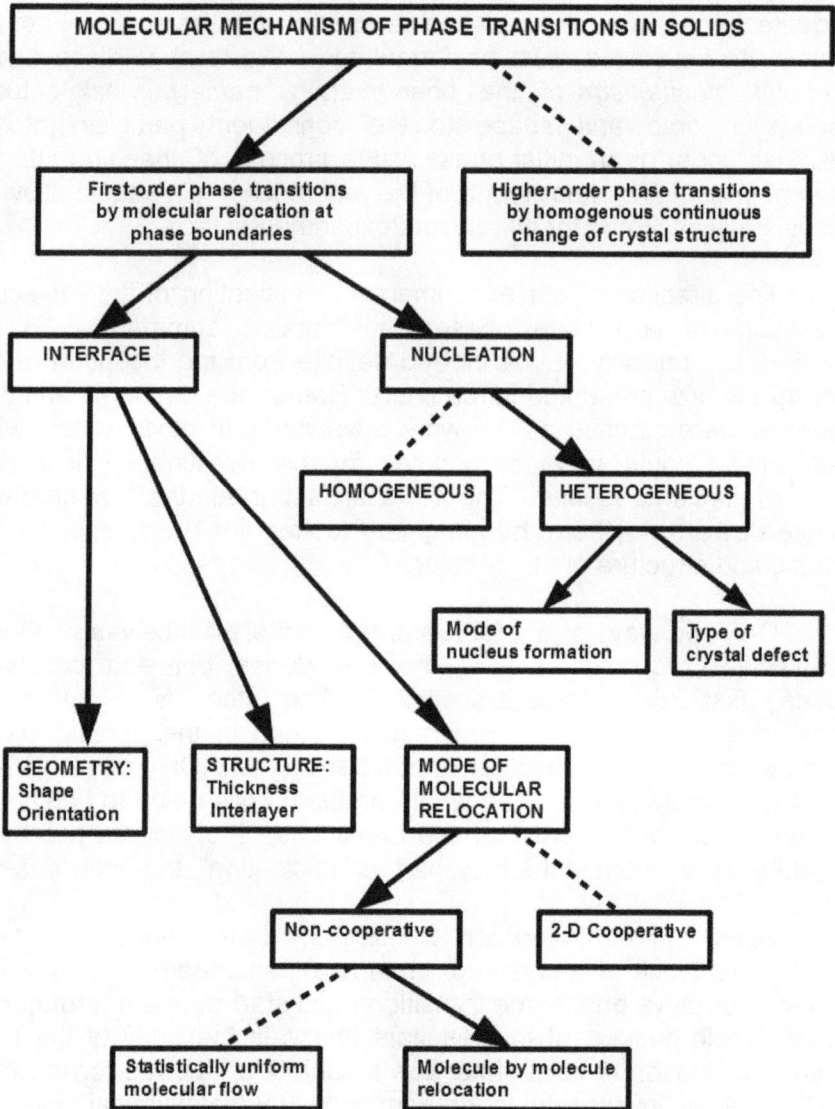

Once the nuclei of the resultant phase have been developed, they grow by consuming the substance of the initial crystal. This means the two phases coexist while the interfaces are propagating. Therefore, the second part of the molecular mechanism of phase transitions amounts to the shape of the interface, its orientation relative to both crystal lattices, the structure of the interface (thickness, any transitional layer), and the mode of molecular relocation over the interface.

As for the molecular relocation, two possibilities exist: cooperative throughout the interface, *i.e.* comprising the whole molecular layer at a time (2-dimensional cooperative, layer by layer), and non-cooperative, with the interface molecules changing their phase adherence not simultaneously. In the latter case the process might either be a statistically uniform molecular flow, or "molecule by molecule" relocation, with the molecules building up the new phase one by one in succession.

2.2 Crystal growth in a crystal medium: the crystallographic meaning of interfaces

2.2.1 The experimental approach

The experimental study described below was initiated in 1960 and soon after led us to conclude that the available theory had nothing to do with real phase transitions. A decision was made to begin with verification of the very basics. This decision determined the general approach: choosing the most convenient and simple objects, laying emphasis on visual observations, carrying out phase transitions under carefully controlled conditions. To this end, any non-transparent, polycrystalline, and multicomponent systems were rejected, as well as too imperfect crystals. Thus, neither individual metals, nor their alloys were appropriate. Large crystals were abandoned as containing more crystal defects than tiny ones. A convenient temperature of phase transition - not too high or too low - was an important factor in the selection.

Selected were *p*-dichlorobenzene and several other organic crystals. They provided certain diversity within this type of solids (Table 2.1): the list included aromatic and non-aromatic substances, structures with Van der Waals' bonding only and those with hydrogen bonding, phase transitions between two crystal states and between crystal and orientation-disordered crystal state. Small (~1 mm), pure, clean, transparent, well-bounded single crystals were grown from a solution or vapor phase and studied by optical microscopy and X-ray diffraction.

TABLE 2.1

Substance and symbol	Structural formula	T_0 (°C)	T_m (°C)	Structural data
Paradichlorobenzene (PDB)	Cl—⬡—Cl	30.8	53.2	[67, 68]
Hexachloroethane (HCE)	Cl—C(Cl)(Cl)—C(Cl)(Cl)—Cl	43.6 71.7*	186	
Malonic Acid (MLA)	HO(O)C—CH₂—C(O)OH	78.0	135	[69]
Glutaric Acid (GLA)	HO(O)C—CH₂—CH₂—C(O)OH	64.0	97.5	[70]
Octahydroanthracene (HMB)	(octahydroanthracene structure)	58.4	73	
Hexamethyl benzene (HMB)	C₆(CH₃)₆	110.8	166	[71,72]
Tetrabromomethane (CBr₄)	CBr₄	46.9	93.5	
DL-Norleucine (DL-N)	CH₃—CH₂—CH₂—CH₂—CH(NH₂)—C(O)OH	117.2		[73]

* There are two phase transitions

Single crystals of *p*-dichlorobenzene (PDB) were the first to be studied under a standard optical microscope equipped with a hot stage and polarizers. In many respects PDB turned out to be most convenient in the investigation of basic phenomena of solid-state phase transitions. A small experimental predicament, its sublimation, had been overcome by placing the crystals in a tiny transparent cell filled with glycerol serving as an inert medium. The literature data on the phase transition temperature varied from 30 °C up to the melting point $T_m = 53$ °C. The cause of such a divergence will be explained later. For now, it suffices to state that the two PDB crystal phases (low-temperature L and high-temperature H) are in their mutual equilibrium at $T_0 = 30.8\ °C$, even though the transition does not occur at T_0. As the only temperature constant of a particular phase transition, T_0 should have a specific name; it will be "equilibrium temperature."*

In PDB the unit cell parameters for the monoclinic phase L (a=14.80; b=5.78; c=3.99 Å; β=113°; Z=2) and triclinic phase H (a=7.32; b=5.95; c=3.98 Å; α= 93°10'; β=113°35'; γ=93°30'; Z=1) are close to each other. The molecular packings in some respects are similar too [67,68]. For these reasons the phase transition was earlier interpreted [69] as orderly modification of one crystal form into the other.

2.2.2 Growth of face-bounded single crystals in crystal medium

From the outset, the microscopic observations revealed striking parallelism between solid-state phase transitions and common crystallization from liquids and gases, particularly from a melt.

MELT CRYSTALLIZATION	SOLID-STATE PHASE TRANSITION
Requires overcooling as a necessary condition	Requires overcooling (H → L) or overeating (L→ H) as a necessary condition
Originates from a foreign object or seed	Originates from a crystal defect – being present or intentionally inflicted
Requires greater overcooling for melt of higher purity	Requires greater overcooling (H → L) or overheating (L → H) for crystals of higher perfection
The greater overcooling, the faster the crystallization.	The greater overcooling (H → L) or overheating (L → H), the faster the phase transition

These initial observations led to the basic idea of thoroughly investigating the analogy between the molecular mechanisms of melt crystallization and solid-state phase transitions. Also, the cause of the discord in the reported temperatures of many phase transitions was immediately clarified. The experiments with the PDB crystals revealed that the divergence in the starting temperatures of phase transitions is rooted in their very nature: what is usually believed to be *temperature of*

---

* Temperature of phase transition will be the subject of Sec. 2.6.

transition or *critical temperature* was, in fact, the *temperature of nucleation.* In PDB, nucleation in L → H transitions occurred somewhere between 34 °C and the melting point 53.2 °C: the higher the crystal perfection, the higher the temperature of nucleation. In the reverse direction, nucleation usually occurred below 27 °C and, again, the crystals of a greater perfection required stronger overcooling. The observations made it obvious that nucleation was responsible for both the hysteresis (higher transition temperature upon heating than upon cooling) and the observed scatter in the transition temperature.

The equilibrium temperature T_0 was, of course, somewhere between 27 °C and 34 °C. It is possible to narrow this interval with the following procedure. Suppose nucleation in the L → H transition occurred at 40 °C and interfaces started propagating from the nucleus. It is possible to stop their motion by lowering the temperature quickly enough and stabilizing the coexistence of the two phases at 30 °C to 32 °C. Then, changing the temperature up and down slightly, a smaller temperature range (about ±0.5°) could be found in which the interface would not move in either direction. The midpoint was T_0 = 30.8 °C. Since the temperature of phase transitions was commonly measured only in one direction, and not by interpolation, the result contained a systematic error: all the temperatures found in the heating experiments are higher than T_0, and those found in the cooling experiments are lower than T_0.

When subjected to a slow heating on the hot stage of an optical microscope, even as slow as 1 °C to 6 °C per hour, the crystals would usually exhibit a rather disorderly picture of the phase transition (Fig. 2.1). While such observations were not new, their features deserve to be noted. *First*, the transition starts at a temperature higher than T_0, but otherwise unpredictable. *Second*, it proceeds by the propagation of a single unified interface. *Third*, the interface is unshapely. *Fourth*, this interface moves quickly, completing the process within a minute or few minutes. *Fifth*, the initially transparent single crystal loses its perfection behind the interface, becoming cloudy.

First of all, it had to be understood why a phase transition in a 3-D ordered system proceeded in a disorderly manner. The analogy with melt crystallization suggests the direction in which to look for the answer. A small amount of a very pure melt can show a similar result upon cooling down. Significant overcooling may be required for its solidification. The latter proceeds by a fast-moving unshapely interface, leaving behind an opaque polycrystalline material. On the other hand, if the melt contains impurities and foreign mechanical particles, the overcooling will be smaller, but the abundance of the crystallization sites will again prevent a slow and orderly crystallization. Therefore, in order to grow a good single crystal, both purity of the matter and small overcooling are required.

Quite the same situation takes place in a polymorphic transition. The two conditions have to be observed simultaneously to produce a more definite shape of the interfaces and thus obtain more information about their geometry: high quality single crystals and a very slow phase transition under very small overheating. These conditions are contradictory, since high quality small single crystals require higher overcoolings / overheatings to initiate the transitions. They would herefore proceed too quickly. The way toward meeting both these requirements is having a "seed" in such a crystal. Sometimes the "seed" is already present as a casual local lattice defect. It can also be introduced by a very slight prick with a needle or glass thread. But even the "seed" would not reduce the speed of interface motion to the desirable level, such as 0.1 mm per hour. In order to establish a sufficiently low speed, an additional action was usually needed: quick cooling down of the crystal toward T_0 as soon as the first sign of nucleation was noticed. To achieve this, however, an experimental predicament had to be overcome, considering that the tiny crystal (0.2 to 2 mm size) was mounted on a massive microscopic heating stage.

Once the above conditions were met, the phenomenon presented in Figs. 2.2-2.8 was discovered. *A solid-state phase transition turned out to be a growth of face-bounded single crystals of the new phase in the initial crystal medium.* As in any crystal growth, the slower the process is, the higher the perfection of the resultant crystals. Experimenting with PDB revealed the following fundamental features:

(1) Resultant crystals can grow in any orientation relative to the crystal lattice of the initial phase, *i.e.*, there is no specific orientation relationship between the phases.
(2) The facets bounding the single crystals growing in the initial crystal medium are their own natural faces. In other words, they are low-index crystallographic planes of the resultant crystal, having no rational orientation in the initial crystal lattice.

These features are in a remarkable contradiction with any earlier suggested mechanism of a solid-state phase transition, in particular with the routine assumptions of a "displacement" and "distortion" of the

Fig. 2.1 Phase transition L → H in a single crystal by motion (from left to right) of a thick shapeless phase boundary (arrow) leaving behind a non-uniform and less transparent material. The viewing field is 1.4 mm. (Malonic acid).

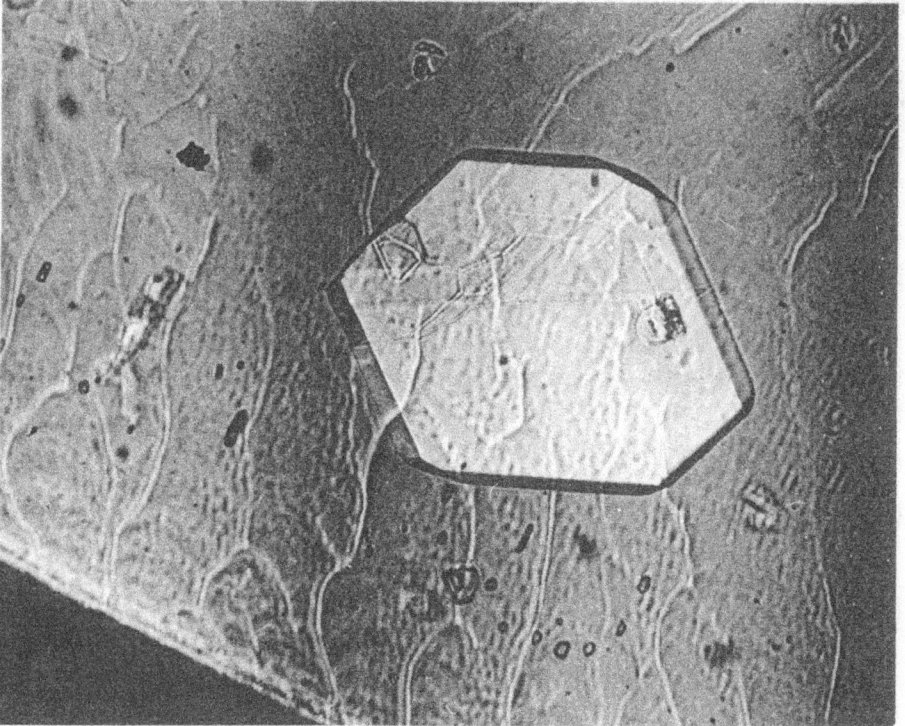

Fig. 2.2 Growth of well-bounded H single crystal in the single-crystalline PDB plate. A part of the natural edge of the L crystal is visible at the left lower corner. The real diameter of the H crystal is 0.4 mm. Partially polarized light chosen to optimize the optical contrast between the phases.

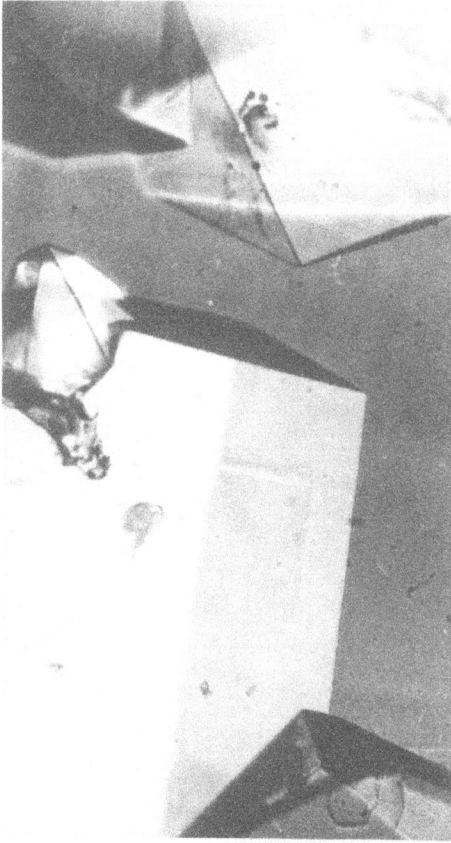

Fig. 2.3 (left)   Phase transition L →H in PDB by growth of several independent well-bounded single crystals of H phase in the same original L single crystal. No orientation correlation is visible between the crystals. The faces of the L crystal are outside of the picture. Partially polarized light. The optical magnification is the same as in Fig. 2.2.

Fig. 2.4 (below) Phase transition L → H in PDB. Four separate H single crystals of different orientations are seen within the L single crystal. Two largest ones have grown into one another as a result of competing for the building material that the surrounding L crystal is. The picture is a demonstration of the phase boundaries (interfaces) to be natural faces of the H crystals. In such a case, the interfaces do not have rational crystallographic orientations in the L lattice. Partially polarized light.

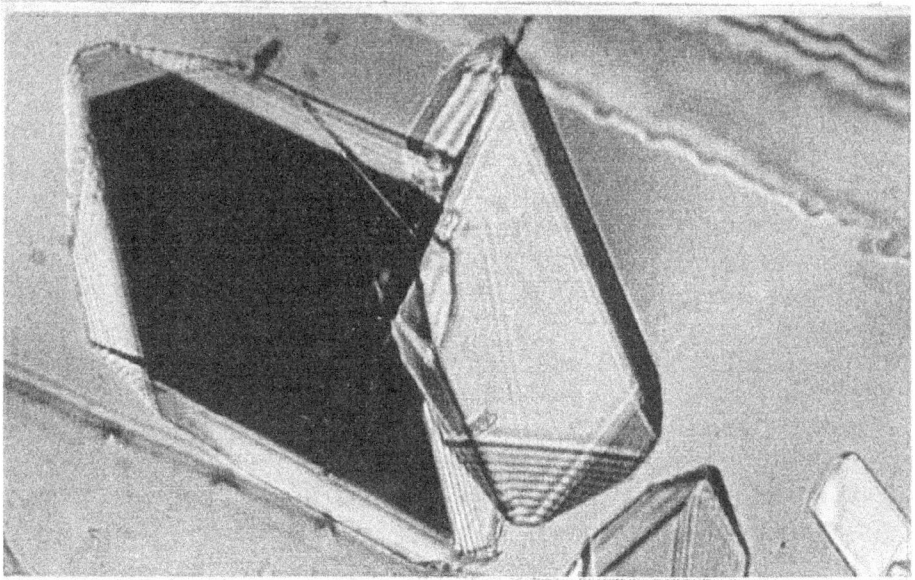

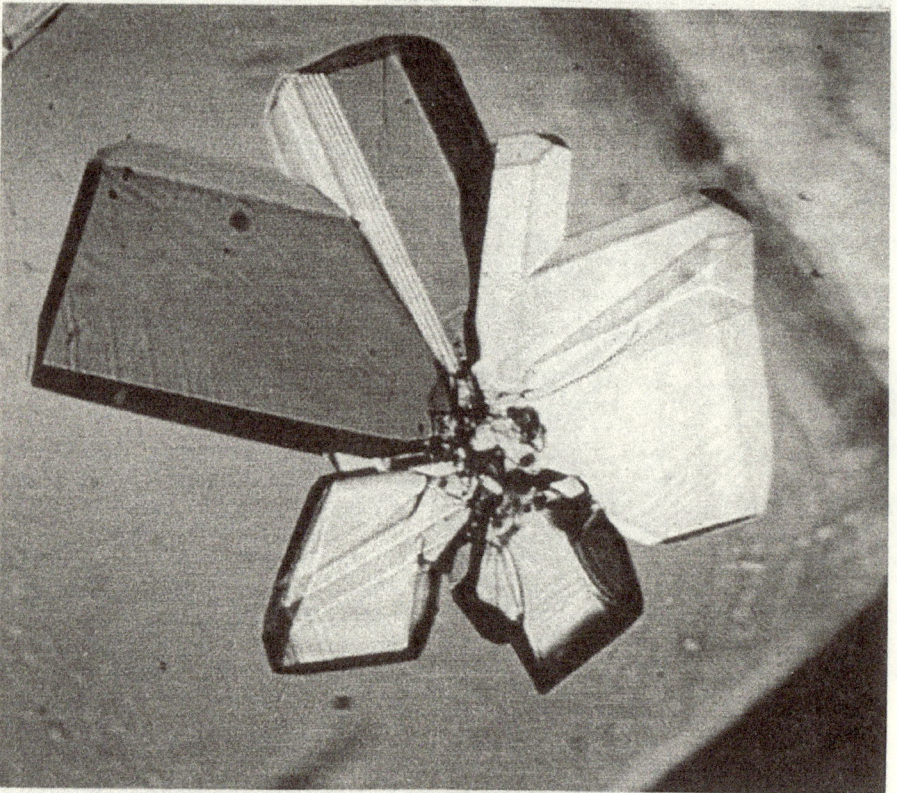

Fig. 2.5   Phase transition L → H in PDB. It was initiated by a slight prick of the L crystal. The prick created a local damaged area with the nucleation sites for a cluster of the H crystals growing in different directions. Each crystal of the cluster has its own set of natural faces not appearing in others. The natural faces of the L crystal are seen at the both corners on the right. Partially polarized light.

original structure. In Fig. 2.2 the notable features of the growing higher-temperature (H) single crystal are the perfectly straight faces, the convex shape, the duplication of the faces and angles on its opposite sides. All the features are characteristic of a single crystal exhibiting its own natural faces. Yet, the picture does not offer solid proof that the facets are the crystal *natural* faces. Although it seems unlikely, they still could be crystallographically rational in the initial phase. The significance of their exact identification resides in the fact that they are *phase interfaces*, the location where all the structural rearrangements occur. The standard technique for a crystallographic identification of crystal faces - optical goniometry - can hardly be applied in this case, for we deal with the "crystal within crystal" which, in addition, would lose its current shape at the slightest change of its temperature. The only remaining practical way was observation with a microscope and

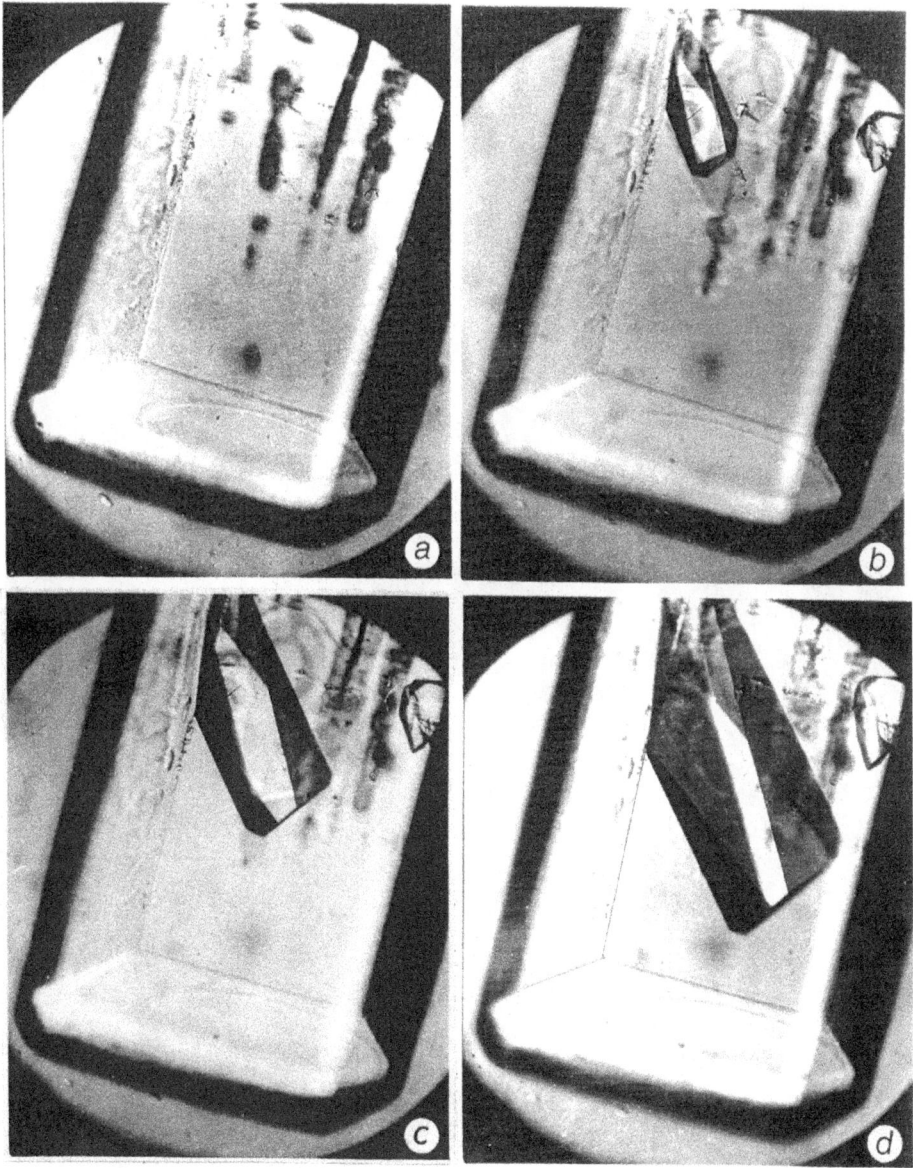

Fig. 2.6   Phase transition L → H in PDB. (a) Prior to the transition. (b-g) Successive stages of growth of a well-bounded H crystal. At stage 'g', only two small darkened pieces at the corners represent the initial phase, while the resultant structure fills out the rest of the initial frame. (h) Growth of the L single crystal after the transition has been reversed. The view field is 1.8 mm. Partially polarized light.

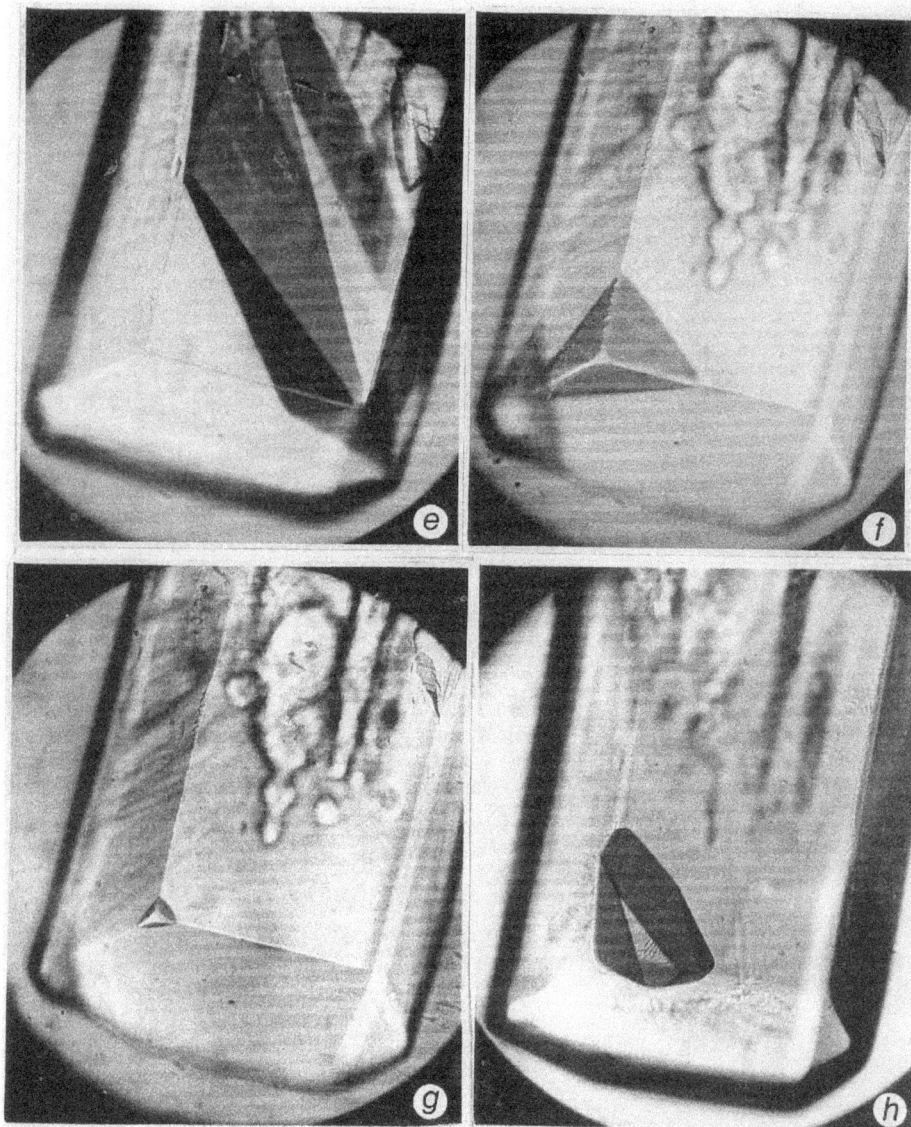

photography. Photography, however, provides only one geometric projection of the 3-D oriented forms. The initial study of many similar single crystals led to the conclusion that there was a great variety of interface orientations relative to the original crystal lattice. It was suggested that any interface orientation is possible.

When several single crystals grew simultaneously in the same single crystal (Fig. 2.3), their orientations could be compared by their extinction in polarized light. Again, no correlation of orientations was found between them, or with the crystal medium. Everything suggested that the phase transition was a growth of single crystals in random orientations in the initial crystal. The two phases coexist independently, while the phase transition takes place at their interfaces. Thus, the main problem of the phase transition mechanism was reduced to the molecular structure of the interface and the mode of molecular rearrangement there. In order to disclose the molecular structure of interface, its crystallographic meaning had to be determined first. Identification of the interfaces as the natural faces of the resultant single crystals, as well as the arbitrary orientation of these crystals in the original crystal medium, had to be confirmed.

The pictures in Figs. 2.4 and 2.5 show a simultaneous growth of several well-bounded resultant crystals, each having its own individual set of parallel faces with no orientation correlation with the other growing crystals. The only reasonable conclusions that could be drawn are that (a) the phase boundaries are the natural faces of the resultant crystals, and (b) the spatial orientations of the growing crystals have no rational relationship with the initial crystal lattice. Yet, even more straightforward proof to these crucial points will be produced in Sec. 2.2.3.

Figs. 2.6,a-h complete the presentation of the phenomenon in PDB. To better visualize the process of phase transition, several successive stages of crystal growth representing a phase transition *single crystal* → *single crystal* are shown. The only rationale for this type of transition to occur was the growth from a single site. In the same PDB the transition could well be of a *single crystal* → *polycrystal* type if the crystal growth proceeded from multiple sites.

The phenomenon described above, initially observed in PDB, would then be consistently found in all phase transitions listed in Table 1. The resultant crystals are not always as well-bounded as was demonstrated with PDB. But the nature and main characteristics of the phenomenon are the same. Two additional examples of phase transitions - in hexachloroethane (HCE) and octahydroanthracene (OHA) - are shown in Figs. 2.7 and 2.8. More examples of crystal growth in crystal medium will be given later.

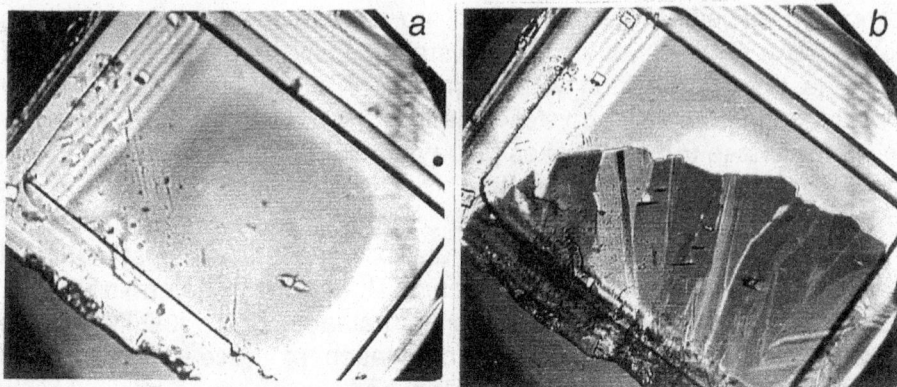

Fig. 2.7 Phase transition L → H in hexachloroethane (HCE). (a) Prior to the transition (b) During the transition; a growing cluster of the H single-crystal blocks bounded from outside by flat faces. The 1.4 mm view field. Partially polarized light.

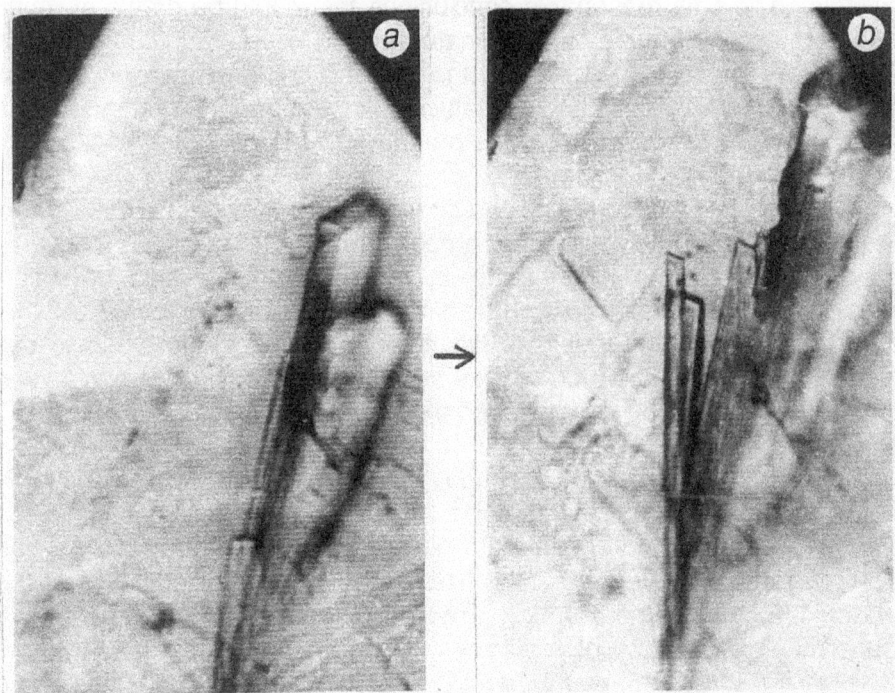

Fig. 2.8 Phase transition L → H in octahydroanthracene (OGA); 'a' and 'b' are two successive stages. A cluster of similar rod-like H crystals growing in divergent directions in the rhombus-shaped L single crystal. Partially polarized light.

2.2.3 Using thin crystal plates to identify interfaces

In the following the initial phase will be denoted by 'i', and that after the phase transition (resultant) by 'r'. In view of the facts demonstrated above, all the 'r' crystals in a phase transition are expected to be bounded with the same set of low-index crystallographic planes $(h,k,\ell)_r$. If this is indeed the case, the set of solid angles in all 'r' crystals must also be the same, *i.e.* comply with the crystallographic *law of constant angles*. In the case of PDB phase transition, however, we were unable to prove that: solid angles could not be measured from a photographic projection owing to a 3-D random orientation of the 'r' crystals. What's more, the real picture seen in the photos is a combination of the true crystal edges of the 'r' crystals and those formed by 'r' intersections with 'i'. As a result, even a simple rectangular form of 'r' crystals would be unrecognizable. For example, neither of the solid angles of the 'r' crystal seen in Fig. 2.2 is characteristic of its structure: all resulted from its intersection with the plate-like 'i' crystal. Measuring them would be useless, even if it were possible.

The problem of comparing different 'r' crystals to one another was circumvented by turning to very thin 'i' plates. The latter have two advantages. One advantage is relaxation of the internal strains accumulating during interface motion. These strains are caused by a difference in the densities of the phases. With thin plates most of the strains can "escape" on the surface, thus greatly improving the conditions for growing single crystals in the solid medium. As known, the task of growing well-shaped single crystals even from liquids and gases, where such adverse strains are absent, is often a delicate and capricious matter involving a good deal of effort and art. Another advantage of very thin plates is their ability to "select" certain orientations of the 'r' crystals. The latter usually tend to acquire the form of thin plates as well. If they start to grow in random orientations within a thin 'I' plate, only those that happened to be parallel, or almost so, to the 'i' plate can develop into real plates. All other 'r' crystals quickly cease growing, becoming residuals from the intersection of the two plates. Therefore, any 'r' crystals developed into a platy form are parallel to the 'i' plate and to one another. This geometrical selection allows the comparison of their shape in the photograph; any other forms that do not duplicate themselves in the picture must be ignored.

The photographs of a phase transition in malonic acid (MLA) in Fig. 2.9 show the growth of two 'r' crystals in a thin (few microns) 'i' plate. The two lie in the 'i' plane, but otherwise have different unspecified turns about its normal. Only three of a long set of the successive photographs are shown. A comparison of the r_1 and r_2 crystals in the accompanying drawing reveals compliance with the law of constant angles, confirming that the observed faces are their *natural* faces. The emergence of the same set of natural faces in differently oriented 'r' crystals makes it clear that the phase transition is a *crystal growth* - in spite of the fact that it proceeds in a rigid anisotropic crystal medium, rather than in a mobile isotropic medium of liquids and gases.

Fig. 2.10a-c shows a phase transition in a very thin crystal plate of glutaric acid (GLA). A number of rather well-bounded growing H crystals scattered and oriented at random are seen. In the whole original L crystal plate of 2 X 2 mm (Fig. 2.10 represents only its part) there were 30 well-bounded H crystals, of which 23 were lamellar hexagons with the same set of angles (measured in the plane of picture) 127° and 116-117°. The experiment was extended by one step further: a single crystal of the H phase was grown on the microscopic hot stage from a drop of the melt. The melt-grown single crystal was found to be also a thin-plate hexagon (Fig. 2.10d) with the same set of angles 127°, 117° and 116°. This experiment revealed the trend by crystals to assume their own structure and natural faces even when they grow from a solid state - additional evidence that the role of an 'i' crystal is to be just a supplier of the building material.

2.2.4   More on orientations of lattices and interfaces

As demonstrated in Chapter 1, almost every interpretation of solid-state phase transitions involves the idea of an orderly modification of the initial crystal structure into the resultant structure by "displacement," "deformation," etc. In particular, this is taken for granted if (a) the two structures appear similar, (b) the transition is from single crystal to single crystal (SC→SC), and (c) the transition is from crystal to orientation-disordered crystal state (C→ODC). If they are such orderly transformations indeed, a specific structural *orientation relationship* (OR) between 'i' and 'r' crystals, or at least a set of discrete orientations, would be imminent. But if there is no reproducible OR, the above misinterpretation will be exposed. Thus, OR can provide important information. For this reason, OR was studied in more detail and complemented with the data on the interface orientations. The data were collected in three ways: X-ray Laue patterns, extinction in polarized light, measurements from photographs. The results on the latter two are presented in the form of "orientation diagrams."

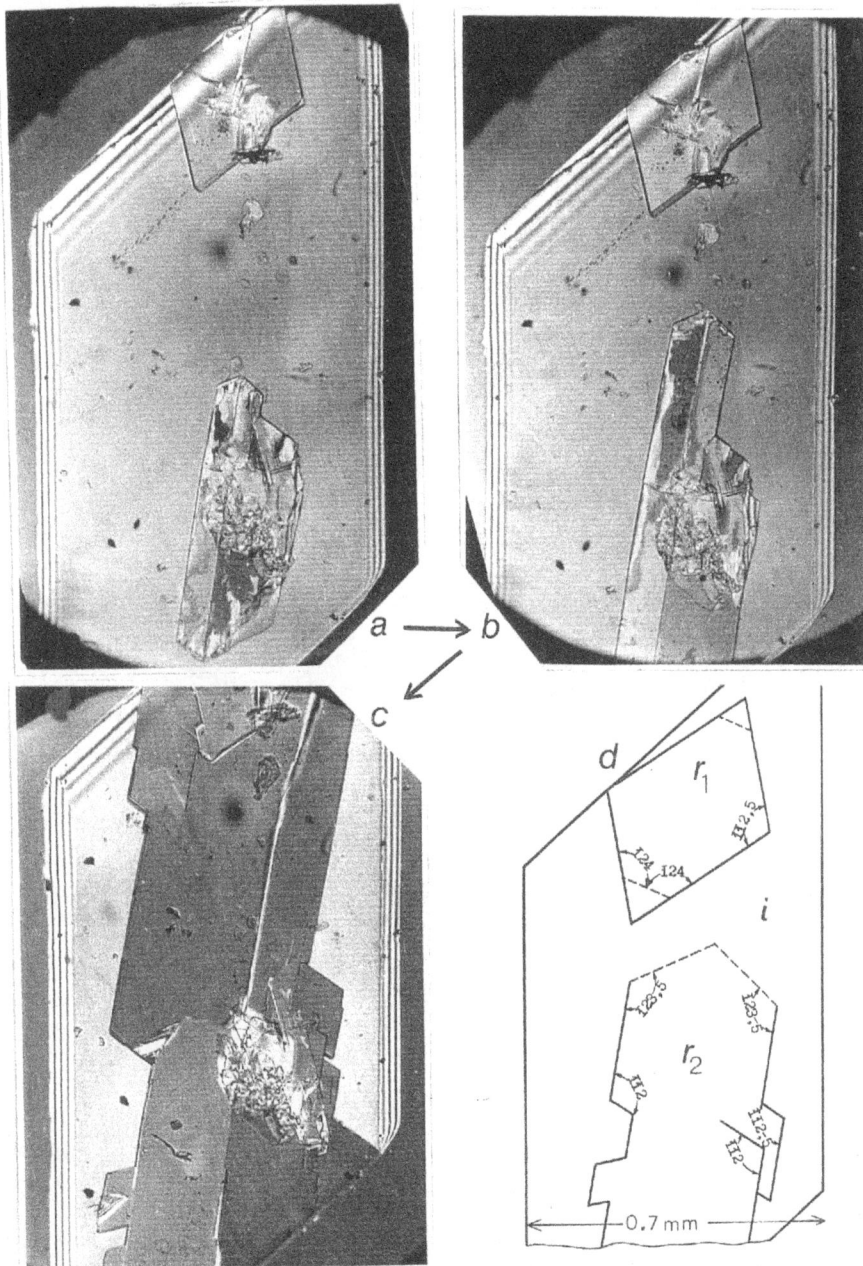

Fig. 2.9 Three successive stages of phase transition in a very thin (~ 2 µm) single crystal plate of malonic acid (MLA). The drawing (d) is to visualize the angles (measured in the plane of picture) of the resultant crystals r_1 and r_2. As seen, the crystallographic law of constant angles is observed. The angles between a fast-growing (broken line) and a slow-growing (solid line) faces are 124°, while those between two slow-growing faces are 112.5°. The fast-growing faces of r_1 were observed at earlier stages of its growth (not shown).

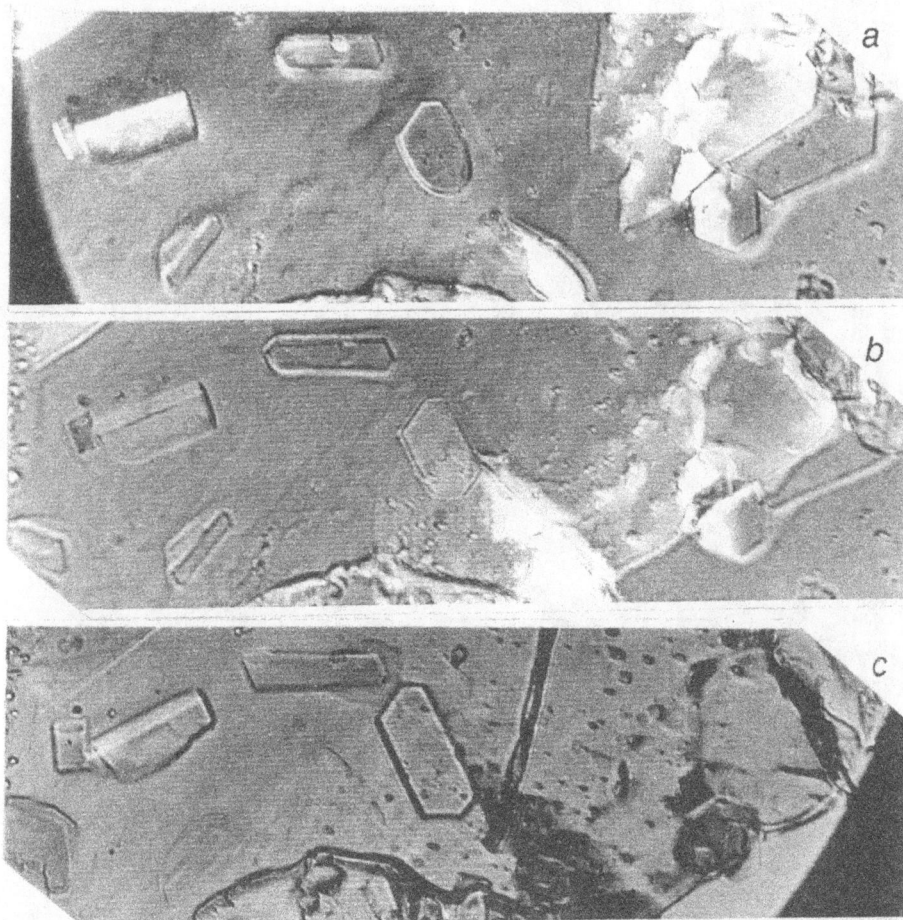

Fig. 2.10 (a-c) A fragment of a thin single crystal plate of glutaric acid (GLA) in which phase transition proceeded very slowly (the shown stages are separated in time by hours). Several growing flat-bounded H crystals are seen scattered over the field of view. Those of a hexagon shape were oriented at random about a normal to the plate, but had the same angles 127° and 116-117°. The angles were measured in all the r crystals scattered over the initial crystal at those stages of their growth when the hexagons acquired their best shape. (d) A melt-grown GLA crystal that proved to have the angles 127°, 116° and 117°.

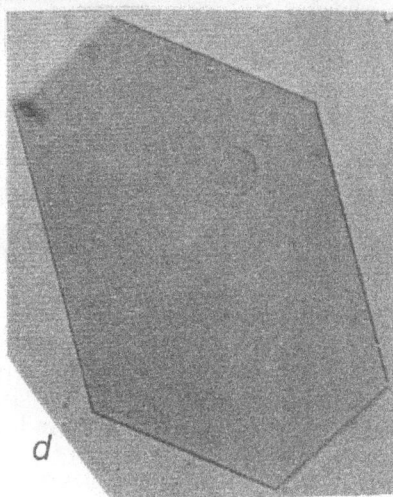

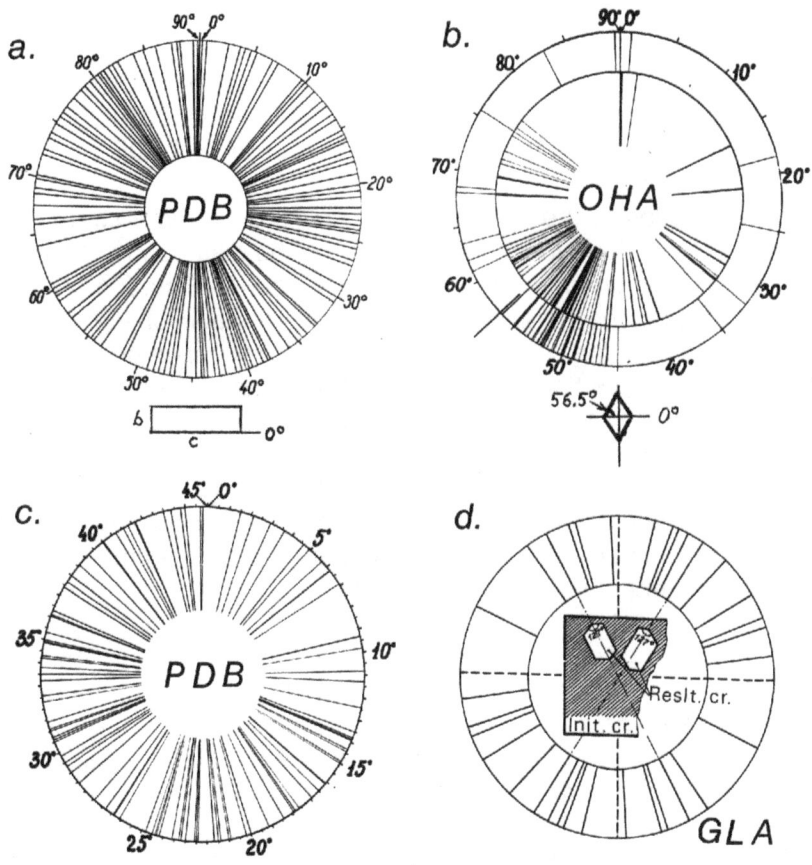

Fig. 2.11 Orientation diagrams for *p*-dichlorobenzene (PDB), octahydroanthracene (OGA) and glutaric acid (GLA).

(a, b) The diagrams representing the observed directions of rectilinear interfaces (shown as the radii) relative to a certain direction in the initial crystal. Many initial crystals were used to collect the data for each diagram. (a) For PDB; the angles were read off from the *c* edge. (b) For OHA; the angles were read off from the bisector of the rhomb's obtuse angle. The external ring is to show an additional quantity of the measurements.

(c, d) The diagrams of lattice orientations of the growing resultant crystals. (c) The diagram composed of the directions of maximum extinction of the 'r' crystals (of the kind shown in Figs. 2.2 to 2.6), when viewed in crossed polarizers, relative to a chosen "zero" direction in the initial crystals. Many crystals were used to collect the data. (d) The relative orientations (in the plane of picture) of the 16 best-bounded resultant hexagonal crystals observed in the initial thin-plate single crystal shown in part in Fig. 2.10. The bisector of the 127° angle was taken as the zero direction. The directions of the side faces of the 'i' crystal are shown by dashed lines.

The orientation diagrams in Fig. 2.11a,b represent interface directions. The diagram for PDB reveals almost equal probability for the interfaces to appear under any given angle. On the other hand, the diagram for OHA exhibits an anisotropic distribution. The analogous diagrams for MLA and HCE (not shown) may be placed somewhere in between regarding the angular distribution of the observed interface directions. Neither one showed a definite reproducible OR.

Extinction of the 'r' crystals in polarized light is a straightforward and simple technique for collecting data on the distribution of their lattice orientations. When the crystals, like those seen in Figs. 2.4 and 2.5, are observed under the polarizing microscope, rotation of the microscope stage will cause their periodic extinction. The crystals of the same orientation will be extinguished simultaneously, but their orientations are not identical if they are extinguished at different angles. The diagram for PDB, compiled from the extinction directions, is shown in Fig. 2.11c. It exhibits a rather random angular distribution of the 'r' crystals in accord with the interface diagram. It is essential to note that any particular 'r' crystal kept its orientation unchanged during the whole time of its growth from the moment of its inception.

The diagram for GLA in Fig. 2.11d represents the angular distribution of the 16 plane hexagons of the H phase around their mutual normal. This is the same example that was discussed in Sec. 2.2.3 and shown in part in Fig. 2.10a-c. All the orientations are different and appear distributed at random.

The X-ray study consisted of mounting the initial L crystals on the X-ray goniometric head in a certain oriented position and taking Laue pictures before and after the phase transitions caused by heating or cooling without disturbing the crystal. In some cases the experiments involved rather long sets of the Laue pictures taken in the consecutive cyclic phase transitions L $\longleftrightarrow$ H. The character of the Laue patterns made it possible to estimate whether they were of a single crystal type, polycrystal type, or intermediate, as well as whether the crystal orientations are repeated in the successive transitions. One such set is partially reproduced in Fig. 2.12.

The X-ray study of PDB revealed that:

(a) The phase transition can be SC $\rightarrow$ SC in some cases and SC $\rightarrow$ PC (PC stands for polycrystal) in others, or it can be of an intermediate SC $\rightarrow$ BLOCKS character.

(b) Phase transitions PC → SC also occur. Growth of a single crystal from the medium of differently oriented surrounded crystal grains deserves special attention.

(c) In general, transitions SC → SC from the same initial orientation do not give rise to the same (reproducible) 'r' orientations. One exception was the effect of *orientation memory,* when some orientation in the cyclic transitions returns a few times in succession, only to be eventually "forgotten."

One frequently observed picture of phase transition is illustrated in Fig. 2.13. Several similar GLA crystals were mounted in succession in strictly the same orientation shown in the complementary sketch. Two Laue patterns were taken every time, one from the original single crystal (only one of the identical 'i' patterns is shown, Fig. 2-13a), and the other after the phase transition L → H (two characteristic ones are shown). All the Laue patterns of the H phase revealed change of the single crystals to polycrystals, SC → PC, and exhibited textures, always different. Such a result corresponds to a "disorderly" picture of phase transition -- like the one in Fig. 2.1. The similar X-ray technique was applied to HCE and OHA as well, and it also exhibited SC → PC patterns, thus indicating growth of numerous 'r' crystals in random orientations.

2.2.5   The "odd" requirements to the molecular mechanism of crystal phase transitions

All described phenomena, facts, and observations will be explained in terms of the universal molecular mechanism of solid-state phase transitions to be put forward. In turn, it will rest on those findings. Below their significance and implications are considered in some detail.

Geometry of the interfaces. Unlike other physical theorists who assert solid-state phase transitions to be a *critical phenomenon* (Sec. 1.1), metallurgical physicists treat the phase transitions (calling them "transformations") as nucleation and growth by propagation of a specific interface [74]. Direct observation of an interface in a bulk metal is impossible, for it is not transparent. The theories are based on the idea of a cooperative transformation localized at the interface. The latter must be a *coherent interface*, a plane mutual to both crystal lattices. This plane moves forward as a whole. The perfect matching of two different crystal lattices at the coherent interface is achieved at the expense of elastic deformations on its sides. If the crystal parameters are too different for the coherent interface to be imagined, the theory

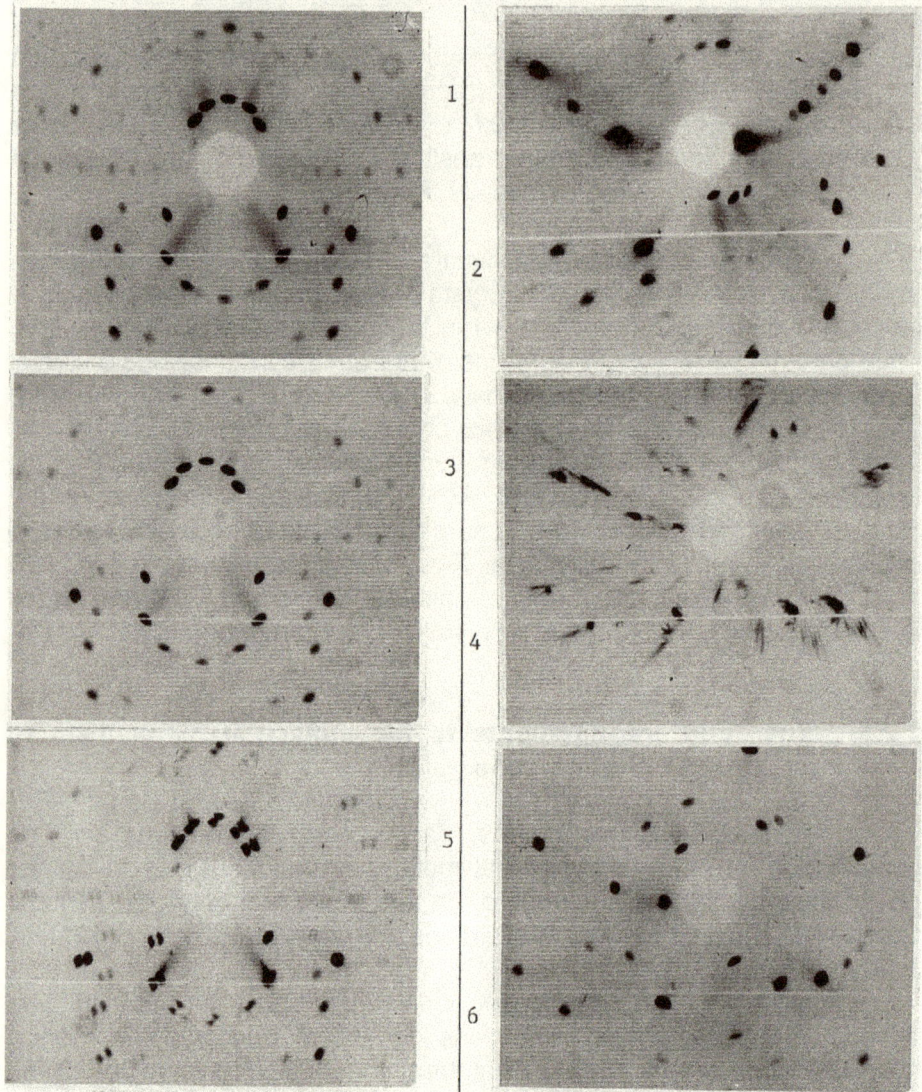

Fig. 2.12   Sequence of the Laue patterns from a stationary single crystal of PDB in which the following cyclic phase transitions were induced by heating and cooling:
L → H → L → H → L → H → (L → H →)L → H → L → H → L → H.
1   2   3   4   5   6   7   8   9   10   11   12   13   14

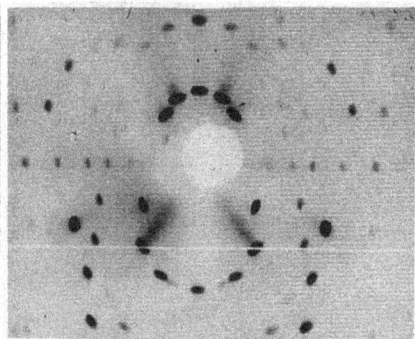

9

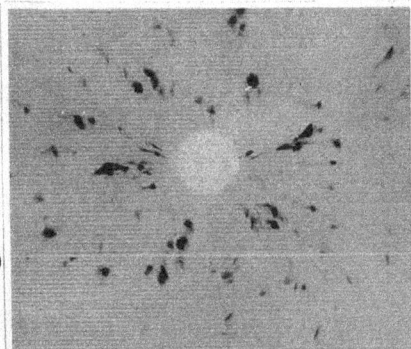

10

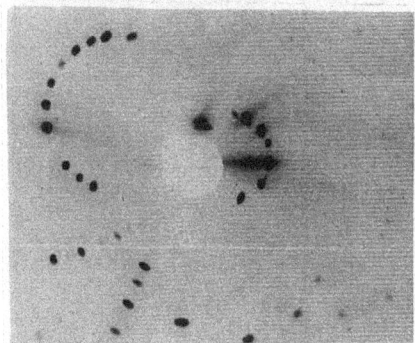

11

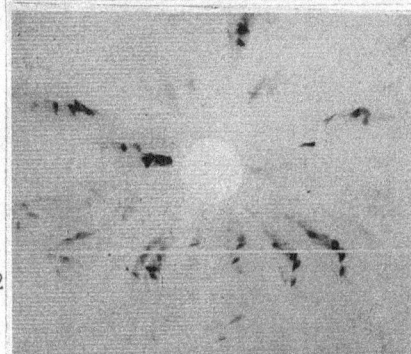

12

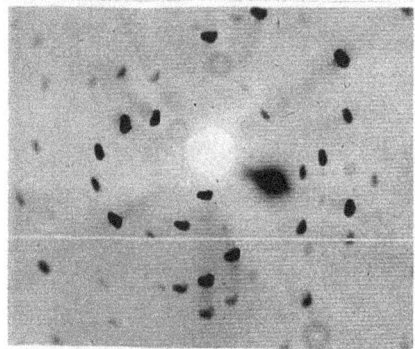

13

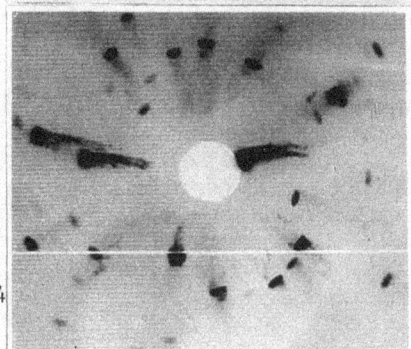

14

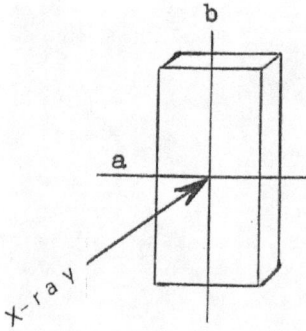

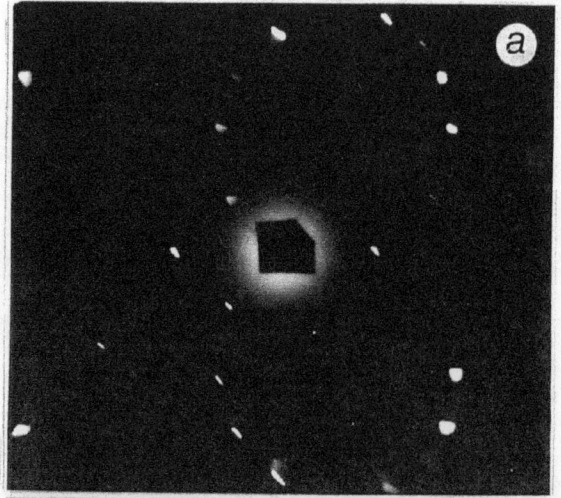

Fig. 2.13 X-Ray Laue patterns of GLA crystals prior to and after phase transition.

(a) The pattern of all initial single crystals before L → H phase transition; the corresponding crystal orientation is shown in the supplemental drawing.

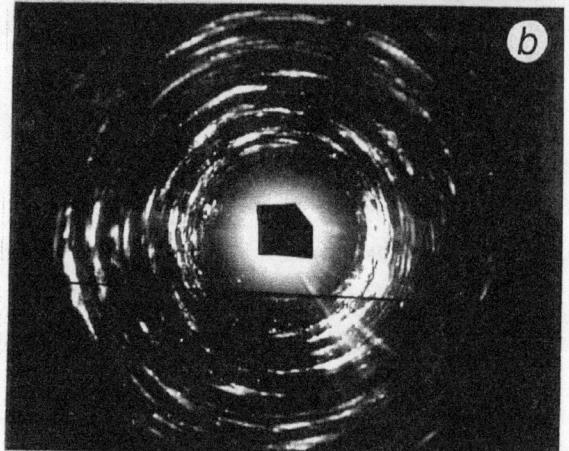

(b, c) The Laue patters for two (of many investigated) crystals after phase transitions activated by heating the specimens while in the X-ray chamber. Every time the phase transitions resulted in different polycrystalline textures.

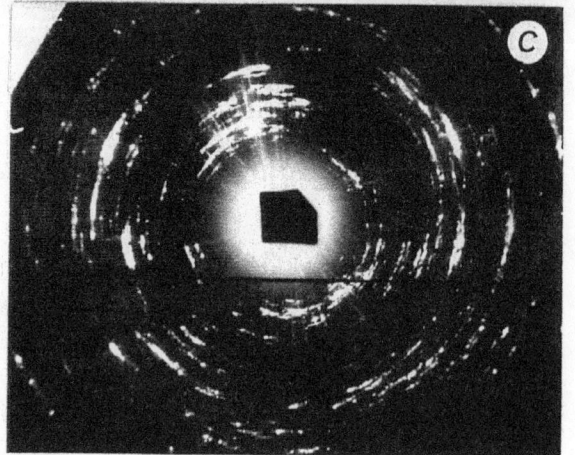

suggests a *semi-coherent* and *non-coherent* types of interface. The former is thought of as the coherent interface, breaking up periodically to eliminate the mismatch, and the latter requires a transient amorphous layer between the phases. But the theory assumed all these interfaces to be coherent anyway. To make this approach more acceptable, the idea of a "best fit" of the two crystals at their interface was proposed [75]: the interface was supposed to choose the direction where the parameters of the two different structures fit most closely to one another. In such a case, the interface is generally irrational in both crystal lattices. No data exist confirming that interface takes the "best fit" direction. The following ideas on the interface geometry can be found in the literature:

1. A rational plane in the two structures:
   $(h, k, \ell)_i$ & $(h, k, \ell)_r$.
2. A rational plane in 'i', while irrational in 'r':
   $(h, k, \ell)_i$ & $(h, k, \ell)_r$.
3. An irrational plane in both 'i' and 'r':
   $(h, k, \ell)_i$ & $(h, k, \ell)_r$.
4. Not a plane (curved or irregular shape).

The only version not mentioned in the literature was the one found in our study: a rational plane in 'r', while irrational in 'i':

$$(h, k, \ell)_i \ \& \ (h, k, \ell)_r$$

This is the first "odd" feature the mechanism of phase transitions will have to incorporate.

Random lattice orientations. The general pattern of the orientation relationship (OR) between two crystal phases appears as follows. No definite OR was found in the instances considered to this point. The quantity of different mutual orientations was unlimited, representing, evidently, a *continuous* angular distribution rather than a discrete set. A *probability* to find one or another orientation is another matter, and it bears little importance in the present context. The actual OR distribution can vary from quite isotropic, to slightly anisotropic, to strongly anisotropic, to strictly definite OR. (The latter case will be considered in Sec. 2.8 and shown to be in line with the general pattern). The point is, however, that the structure of the proposed interface should not require any particular OR.

Rearrangement at the interfaces. Regarding the molecular rearrangement at the interfaces, the latter conclusion means that it has to proceed at whatever OR happened to be. This "odd" requirement will not seem so odd if one recalls a well-known phenomenon of solid-state recrystallization in a polycrystalline matter. There the crystal grains are oriented at random, yet some grains grow at the expense of others. In both cases - recrystallization and phase transition - one deals with the same process of molecular rearrangement at interfaces without any specific OR. Both cases represent crystal growth where the only function of the surrounding crystalline medium is to supply the growing crystal with the molecular material, no matter how this material is packed. The two processes have undoubtedly the same molecular mechanism, differing only in their driving forces.

Nucleation of the resultant phase. Now it is appropriate to raise the question as to why the 'r' crystals grow in unspecific and unpredictable orientations. The answer to this question, as well as to why the orientation anisotropy occurs, resides in the peculiarities of nucleation (Sec. 2.5). It is the original orientation of the nucleus that determines the orientation of the 'r' crystal. Nucleation in solids is commonly assumed to be *coherent*, producing the nuclei in a certain orientation determined by the 'i' structure [66]. The random OR found in PDB and other phase transitions runs counter to this view. Formation of an 'r' crystal in the 'i' lattice without any specific OR is the additional "odd" property that needs to be integrated into the molecular mechanism of phase transitions that will be put forward later.

2.3 Edgewise mode of interface motion

2.3.1 Microscopic observation under high optical resolution*

Thus far a deep-seated similarity between solid-state phase transitions and crystal growth from liquids and gases was demonstrated in a number of ways. The next issue is to find out *how deep* it is. Growth of naturally face-bounded crystals in a crystal medium in random orientations was already a striking manifestation that the two processes are related. But is it extended to the intimate molecular process of this growth, considering the differences between a crystal medium, with its 3-D order of firmly bonded molecules, and gas or liquid with their

---

* In 1966, when the study was performed [76], optical microscopes of the resolution exceeding 0.2 µm were still not available.

disordered structure and molecular mobility? With this in mind, observation was undertaken of the interface and the mode of its motion under maximum resolution attainable with a regular optical microscope.

Although the experimental arrangement was rather simple (Fig. 2.14a), the obstacles to be overcome may explain why the result of this study proved to be the first of its kind [76]. The observation required an immersion objective lens (X 95) and application of immersion liquid between the lens and the cell containing a crystal sample submerged into an inert liquid. Since the lens-to-sample distance was only 0.46 mm, a heat flow from the tiny sample to the massive objective lens developed. This hindered setting the required temperature just when it had to be controlled extremely carefully. A typical procedure involved "catching" the moving interface in the very small viewing field before the phase transition is over, regulating and stabilizing the temperature to stop its motion, waiting until the interface becomes strictly linear, then creating a slight heating pulse and filming what happens.

With this technique, an *edgewise* mode of interface motion was detected. The interface motion was found to proceed by edgewise shuttle-like strokes of small steps, each one adding a thin layer to the growing 'r' crystal (Fig. 2.14b). Fig. 2.14c is a sequence of the cinematographic frames (mounted, except the first frame, into a compact column) showing an area of the PDB crystal surface during an L$\rightarrow$ H phase transition. The *layer-by-layer* growth of the H phase is obvious. The height of the steps is 2.5 μm. The interface line seen is that of intersection of the oblique interface and the observed face of the 'i' crystal. In other words, the interface is seen "in profile." A number of different pictures, both "in profile" and "in full face," showing the layer-by-layer mode of the phase transition have been obtained, leaving no doubt of the general significance of the phenomenon. The height of the steps varied in a wide range, the lowest being down to the limit of the optical resolving power.

Fig. 2.14d is a photomicrograph taken with the maximum resolution. The smallest discernible step in it (viewing along the interface under small angle to the plane of picture) is 0.2 to 0.3 μm. The diffuse bands on the side of the 'i' (L) phase are interference bands, indicating that the interface forms a very narrow wedge with the surface under observation. It can be concluded from this that the real height of the smallest steps is about one order lower, *i.e.* 200 or 300 Å. The higher steps in the picture are rather rounded, but their shape suggests that they represent a good deal of smaller ones. Estimation of the steps

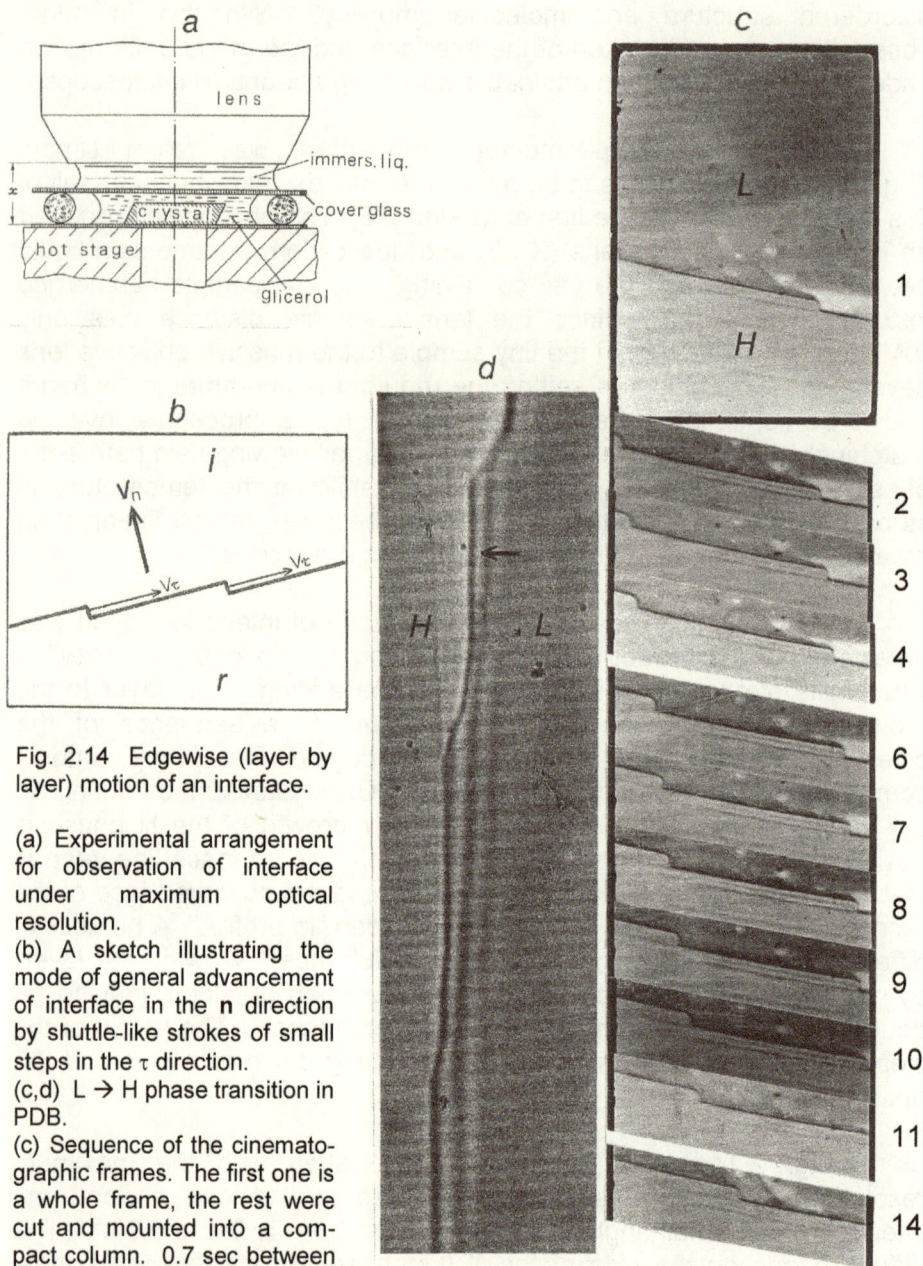

Fig. 2.14 Edgewise (layer by layer) motion of an interface.

(a) Experimental arrangement for observation of interface under maximum optical resolution.

(b) A sketch illustrating the mode of general advancement of interface in the n direction by shuttle-like strokes of small steps in the τ direction.

(c,d) L → H phase transition in PDB.

(c) Sequence of the cinemato-graphic frames. The first one is a whole frame, the rest were cut and mounted into a com-pact column. 0.7 sec between two frames of successive num-bers. The steps are 2.5 µm high.

(d) A photomicrograph taken with the ultimate optical resolution (X 1600). The smallest discernible step (against the arrow) is as low as 0.2 to 0.3 µm. The interface should not be confused with the equal-thickness interference diffuse bands to the right of it.

from the pictures taken "in full face" [76] yielded an average height as small as 700 Å. A reasonable guess is that much smaller steps had to be among them.

The microscopic technique in question could be applied only to the phase transitions with T_0 that doesn't much exceed room temperature. The edgewise mode of phase transition in HCE was detected too, even though its $T_0 = 43.6$ °C was already far beyond the optimum experimental conditions.

2.3.2 The implication of the edgewise mechanism

Crystal growth from liquids and gases by the lateral movement of tiny steps on a flat crystal surface towards the crystal edges, discovered in the first quarter of the 20th century, was a decisive breakthrough in the understanding of crystal growth. Relatively high steps (microns) were initially observed, but later proved to reflect the edgewise growth at the molecular level. A simple model interpretation of the phenomenon was suggested by Kossel [77] and Stranski [78] (Fig. 2.15).

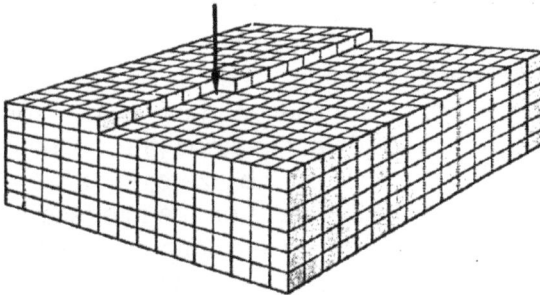

Fig. 2.15 The classic model by Kossel [77] and Stranski [78] of a growing low-index crystal face. The molecules are shown as cubes. There is a molecular step on the growing face and a kink (arrow) in the step. Molecules approaching the surface will prefer sites with a maximum number of nearest neighbors. Therefore, the kink is the most preferable place for the next molecule to be attached. The crystal growth proceeds "row by row" and "layer by layer." Every completed layer requires formation of a two-dimensional nucleus on its closely packed surface if the growth is to continue.

Extension of the edgewise mechanism of crystal growth to phase transitions in solids should have a comparable significance for understanding their nature. A simple generalization comes to light: any structural rearrangement resulting in a crystal state, whichever the initial phase is - gas, liquid, or solid - is crystal growth. It proceeds by development and layer-by-layer edgewise growth of the natural crystal faces. There is, of course, a certain feature of crystal growth from a solid medium, which is the topic of our subsequent consideration. Otherwise, the difference is semantic: crystal growth from gaseous and liquid phases is called *crystal growth*, while crystal growth from a solid phase is called *phase transformation* or *phase transition*.

It is to be noted that a layer-by-layer interface advancement upon $\beta \rightarrow \alpha$ phase transition in thin polycrystalline sulfur films confined between two microscopic slides was noticed by Hartshorne and Roberts [79]. Mentioned among many details, this observation has not attracted any attention. Now it can be taken as substantiation of a general significance of the edgewise mode of phase transitions. Finding naturally-face-bounded "crystals in crystals" in sulfur phase transitions is just a matter of time.

The idea of "coherent" interfaces advancing as a whole, never having been proved experimentally, seems unrealistic in the light of the above evidence. With a great degree of confidence, we can state that a solid-state structural rearrangement is a *molecule-by-molecule* process, rather than a 3-D or 2-D *cooperative* phenomenon.

Other consequences of importance can be inferred from the edgewise mode of interface motion. There is no equal probability for the molecules at the interface to change their phase adherence. The probability is high for molecules at the steps (the activation energy E_a of phase transition is low), but low, possibly zero, for those at the flat areas (E_a is high). The quantity of the steps on the interface is a major factor in controlling the overall speed of its motion. The kinetics of phase transition, therefore, depends on the conditions responsible for formation of those steps.

One more inference is that any unstable interlayer of the excited molecules between the two phases cannot exist. This follows from two facts: (1) a phase transition proceeds only under overheating or overcooling relative to the T_o, and (2) the phase transition is localized at the steps, while nothing occurs at the flat interface areas during finite periods of times. The interlayer in question cannot exist as there would be plenty of time for the excited molecules, if any, of the flat areas to return to their normal state by joining the stable phase.

Now we turn to the physical model of a solid-solid interface and the manner of molecular rearrangement at this interface. The edgewise mode of its motion adds a significant feature which the model must incorporate.

2.4   Phase transition at the contact interface

2.4.1   Recalling the Hartshorne's effort

A properly aimed attempt to solve the problem of solid-state phase transitions was the work by Hartshorne and a number of associated authors, initiated in the1930's and terminated in the late 1960's [79-91]. These researchers directed their investigations at the core of the problem which is the molecular structure of the interface and the mode of molecular passage across it. Meaningful experimental material had been collected. The central Hartshorne's idea of a *vapor gap* between the phases had ultimately failed, but these researchers were closer to the correct interpretation of solid-state phase transitions than their orthodox contemporaries. Their contribution was ignored. For instance, the results of that essential research have not even been mentioned in the review book *Phase Transitions in Solids* [42] by Rao and Rao who declared as their goal "to bring about a sufficiently broad-based book [which] covers the many facets of a fascinating subject." The work by Hartshorne and associates was not in line with the mainstream which was dominated by "self-evident" assumptions on "continuous," "displacive," "soft-mode" and other such phase transitions.

The work by Hartshorne *et al.* is instructive both by its achievements and failures described by Bradley (a participant in the studies) in his sincere and well-reasoned review article [84]. Both the positive and negative conclusions of that research turned out to confirm our *contact* molecular mechanism of phase transitions [92].

Hartshorne and co-authors measured the velocity of interface motion V as a function of temperature T. Representing V with the Arrhenius equation $V = A \exp(-E_a/RT)$, the activation energy E_a can be estimated from the plot $\log V = f(1/T)$ which is expected to be a straight line. Since Hartshorne *et al.* [80] found E_a in *o*-nitroaniline to be approximately equal to sublimation energy E_s, the idea of a vapor interlayer emerged. According to it, a phase transition involves two steps: sublimation of the 'i' material into the vapor interlayer and condensation on its opposite side occupied by the 'r' phase. In terms of the standard formula for evaporation from a solid surface, Hartshorne has derived an expression for V as function of T and sublimation coefficients α_i and α_r, as well as of sublimation heats E_i and E_r. The formula did not make any allowance for closeness of the phases to each other at the interface. Rather, it was a formula for recrystallization through the vapor phase when two separate crystal phases are contained within the same sealed volume.

The dependence $V = f(T)$ in phase transitions of o-nitroaniline, azoxybenzene, and between α, β and γ-forms of sulfur has been studied, the latter especially in detail. A grain of a solid substance was melted on a microscopic slide, covered with another slide and let harden. Such a sample was a thin (< 0.1 mm) polycrystalline film enclosed between two glass plates. Once a phase transition has been initiated in it, the velocity of interface motion was measured either over a short length of interface within a grain, or on the average over the whole interface across the slide. The measured velocities differed as much as up to 50 times. The authors themselves noted that not all the factors were under control (the factors they were not aware of are identified in Sec. 2.9). But the temperature dependence in a logarithmic scale showed a rather reproducible slope, so E_a could be estimated, as they believed, with accuracy of 10%.

Aside from the details, the outcome of the whole work can be reduced to two major points. (1) There is an approximate equality between the activation energy E_a and the heat of sublimation E_s in a number of molecular solids (Table 2.2); this seemed to confirm the Hartshorne's vapor gap theory. (2) The actual velocity of the phase transitions is 10^3 to 10^5 times higher than that predicted by the Hartshorne's sublimation-condensation formula; this was inconsistent with the vapor gap idea. Bradley [84] summarized the state of affairs as follows: "We have therefore the impasse - the activation energy for a reaction such as $\beta \rightarrow \alpha$ sulfur suggests a vapor transition layer, but the experimental rates are too fast for such a mechanism."

Table 2.2. E_a and E_s (after Ref. [84])

Substance	Phase transition	Activation energy	Heat of sublim. of unstable form
Sulfur	$\beta \rightarrow \alpha$	22-23 kcal	23.24 kcal
o-Nitroaniline	$\alpha \rightarrow \beta$	17-18 kcal	19 kcal
Azoxybenzene	II $\rightarrow$ I	22.5 kcal	23 kcal
Mercuric Iodide	Yellow $\rightarrow$ Red	10-11 kcal	19 kcal
Tin	White $\rightarrow$ Grey	6 kcal	78 kcal

No satisfactory solution to the problem (it will be called *Hartshorne's paradox*) had been found. The *contact* mechanism, which will now be introduced, solves this apparent paradox.

2.4.2 Designing the contact molecular model of interface

Bradley correctly predicted that "most progress could be made by studies on single crystals." The findings presented in Secs. 2.2 and 2.3 were made with single crystals and have constituted the experimental basis for designing the *contact* model of interface. The requirements to the model were as follows:
  (1) The interface is a low-index natural face of 'r'.
  (2) The interface, in general, is an irrational plane in 'i'. (A rational orientation is not excluded as a particular case.)
  (3) No restrictions regarding mutual lattice orientation.
  (4) No transitional layer between two phases.
  (5) Edgewise, layer-by-layer, mode of molecular relocation at the interface.*

One can notice that the first four requirements are of a static nature, and the last one involves dynamics of the molecular relocation.

The *contact* interface, which meets the four static requirements, is shown in Fig. 2.16 with a 2-D model. Two crystal phases are simply in contact with each other, coupled by the forces of molecular attraction. The arrangement bears some resemblance to two highly polished glass plates firmly attached to each other by means of the *optical contact*.** There is no transitional layer of any kind between the two phases: every molecule belongs to either one phase or the other. The *contact* structure of the interface requires neither any correlation between the structural parameters of the two phases, nor any particular orientation relationship (OR). From the 'r' side, the interface consists of molecules closely packed into a layer of low crystallographic indices (h, k, ℓ). In the 'i' lattice this direction is irrational. As a result, there is a net of microcavities at the interface that cannot accommodate additional

---

* This assumption was based on the author's experiments of limited optical resolution [Soviet Physics - Doklady, 11, 4 (1966)]. The first direct recording of the edgewise (or "stepwise") propagation of interface at real molecular level was reported 37 years later [K.S. Novoselov et al., Nature 426, 812 (2003)].

** A technique to join glass parts without using a cement. The two joining surfaces must be perfectly flat and polished to the highest degree. The bonding is provided by the molecular attraction.

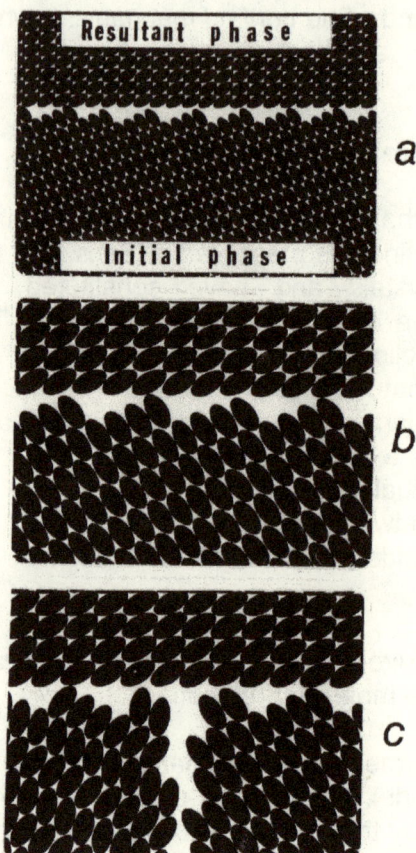

Fig. 2.16 A molecular model exhibiting the main features of the *contact* interface (frames of the animated film). No phase transition is shown yet. Frame *b* is an enlarged part of *a*. Frame *c* illustrates that no restrictions exist regarding structural differences and mutual orientations of the phases. It represents frame *b* after the left half of the initial phase has been substituted by an arbitrary structure and the new *contact* interface turned out to be as good as the previous one.

molecules. There is a certain degree of order in the distribution of the molecular contacts, but the microscopic environment for every particular contact is unique. The intermolecular distances at the points of contact are close, although not necessarily exactly equal, to the standard bonding values. No essential lattice distortions exist.

2.4.3 Phase transition: molecular relocation at contact interface

The physical picture of a phase transition at the *contact* interface that meets the last of the above-listed requirements is shown in Fig. 2.17. The figure is a set of successive frames from an animated film showing how the interface advances. The frames 'a' to 'e' illustrate the edgewise movement of a molecular step. Molecule by molecule detaches from one side to build up a closely packed layer on the opposite side. Once one layer is completed, the interface has advanced

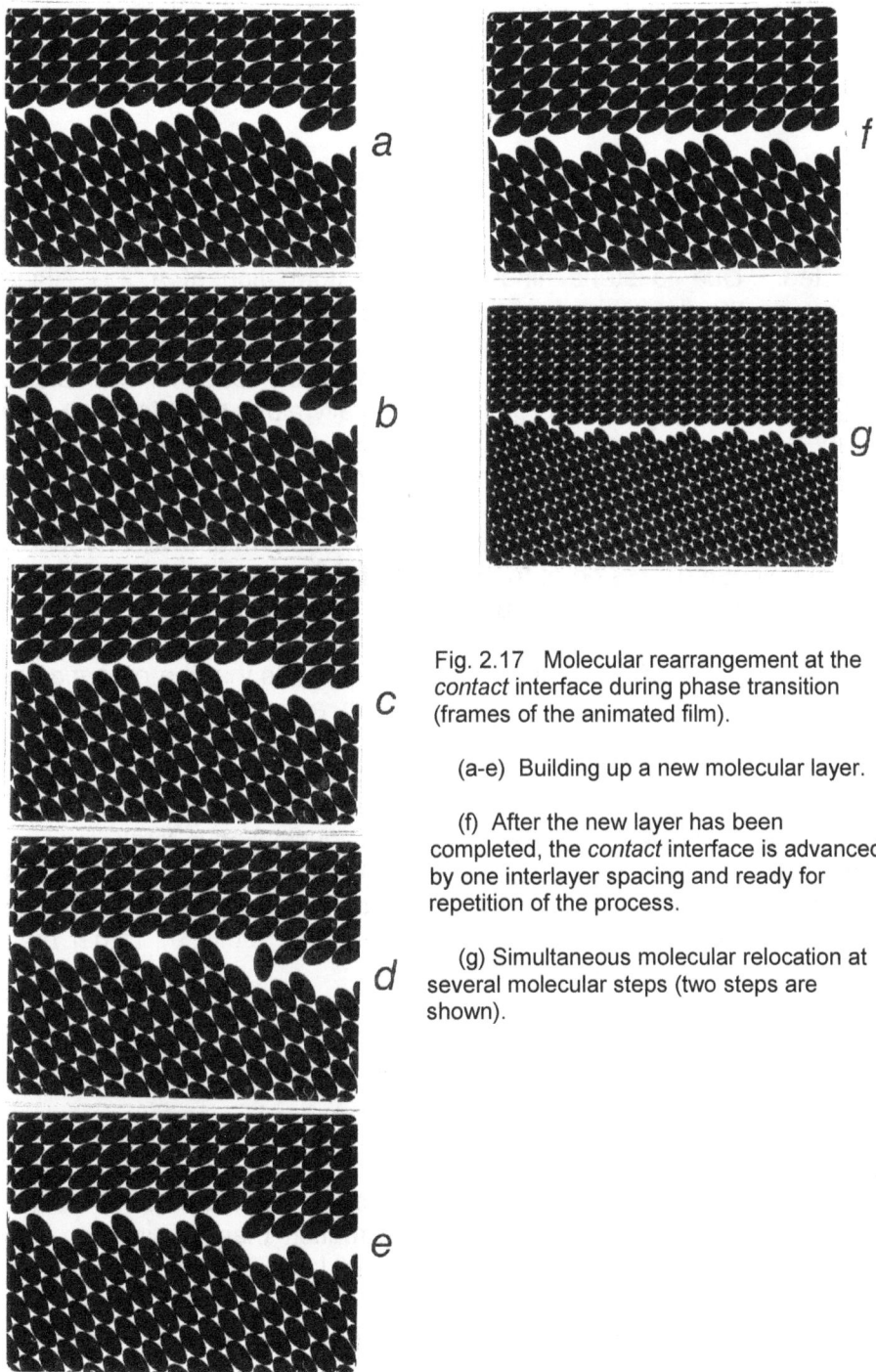

Fig. 2.17 Molecular rearrangement at the *contact* interface during phase transition (frames of the animated film).

(a-e) Building up a new molecular layer.

(f) After the new layer has been completed, the *contact* interface is advanced by one interlayer spacing and ready for repetition of the process.

(g) Simultaneous molecular relocation at several molecular steps (two steps are shown).

by one interlayer spacing, while the "contact" structure of the interface at its new position is preserved (frame 'f'). The process is thus localized at the molecular steps. Of course, there can be more than one moving step on the same interface (frame 'g').

For the phase transition to continue, a new 2-D nucleus has to form on the surface every time it becomes completely flat. Here, as compared to crystal growth from liquids and gases, a certain feature is expected. Due to steric hindrances, the formation of a 2-D island nucleus on the flat regions of the contact interface is hardly possible. The model suggested that an extra free space, such as a vacancy or, possibly, cluster of them, is required. If the extra space is available, a new molecular step can emerge to run laterally along the interface. The vacant space will accompany the running step, providing for steric freedom just where the molecular relocation occurs at the moment. The space will ultimately come out on the crystal surface. Formation of a new step will require another lattice defect in the form of a vacant space. Such a defect must reside at the new interface position or migrate to it if the phase transition is to continue.

The model of the *contact* interface suggests that the *presence of crystal defects of a certain quality and in a sufficient quantity is a necessary condition for a phase transition to proceed.* Evidence will later be presented that the availability of crystal defects of a particular kind is a major controlling factor. This appears to be a specific feature of crystal growth in a crystal medium, as 2-D nucleation is rarely realized upon crystal growth from gases and liquids where screw dislocations provide self-perpetuating steps. Evidence in question supports the kinetics based on a distribution of point defects (Sec. 2.9), rather than on the presence of the self-perpetuating steps.

2.4.4   The solution of the Hartshorne's paradox

To solve the Hartshorne's paradox is to account for a combination of two facts: (1) the proximity of the activation energy E_a of a phase transition to the heat of sublimation E_s, and (2) the 10^3 to 10^5 times higher speed of the interface propagation than sublimation can offer. Any hypothesis assuming a transitional layer between two phases is bound to regard molecular rearrangement at the interface as two separate transitions: from 'i' to the interlayer, and from the latter to 'r'. In the *vapor gap* theory they are sublimation and condensation respectively. In contrast, molecular rearrangement at the *contact* interface is a one-step molecular relocation 'i' → 'r' under the attractive action of 'r'. This process can be termed "stimulated sublimation." Accordingly, E_a is lower than E_s by the attraction energy E_{attr}

$$E_a = E_s - E_{attr}.$$

To evaluate E_a quantitatively, computer calculations have been carried out, using benzene structure [93] as the model. The Van der Waals' molecule--crystal interaction energy E has been estimated as a function of the distance between the molecule and the crystal face (001). This was done by a direct atom-atom summation, using the potential function [94] inferred mainly from the benzene structure. The summation radius was 30 Å.

The result is shown in Fig. 2.18. The X-axis is a distance (expressed as the number k of missing molecular layers) from the molecule to the face (001). The distance $k = 0$ corresponds to the energy required to remove a molecule from the (001) surface to infinity. This energy was found to be 7.36 Kcal/mole, which is somewhat less than the standard value 10.7 Kcal/mole for the heat of sublimation. The difference is easily explainable, as real sublimation involves molecular departure from two-side and three-side corners rather than from a flat surface. This deviation from the real conditions was of little importance since the calculations were aimed at estimating a *relative* decrease in E_a due to the attraction by the 'r' phase.

Now, one can see from Figs. 2.16 and 2.17 that the gap between two phases at the *contact* interface is equal, on the average, to $k = 0.5$. It follows from the calculated curve in Fig. 2.18 that

$$E_{k=0.5} \cong 0.3 \, E_{k=0},$$

that is, an activation energy of a phase transition, not departing too far from the sublimation energy, is still lower by $\sim 30\%$

$$E_a \cong 0.7 \, E_s$$

in a reasonable accord with the data by Hartshorne and his colleagues for sulfur and two molecular substances (Table 2.2). Considering the uncertainties in deriving E_a from the experimental data, those authors may have underestimated the experimental error as being within 10%. Besides undetermined objective factors that may give rise to a slightly exaggerated E_a, a subjective systematic error was possible too.

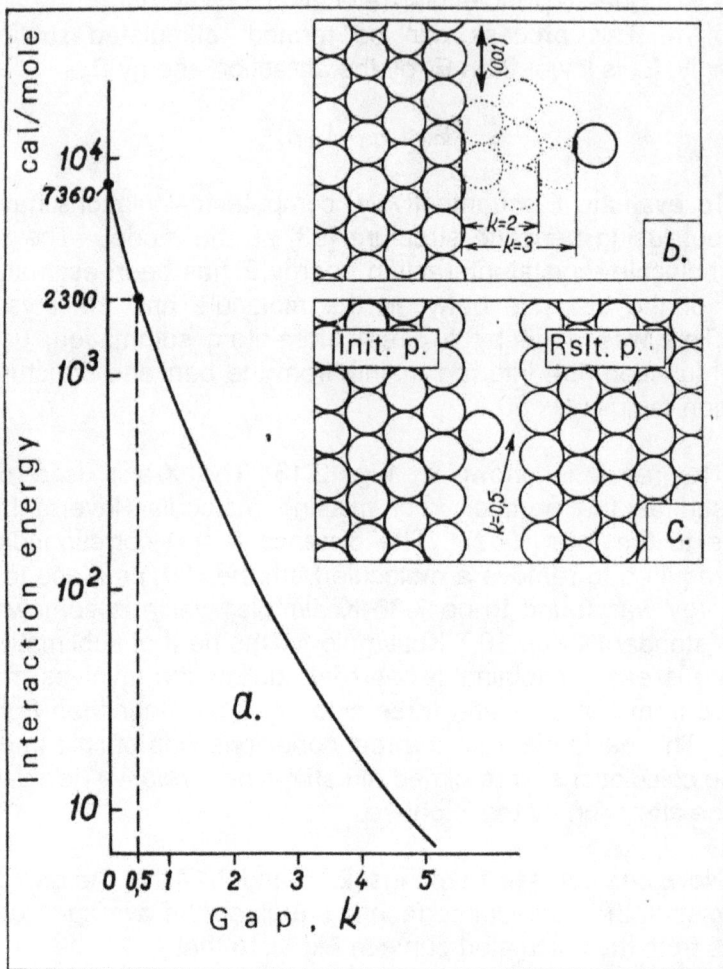

Fig. 2.18 The interaction energy 'molecule-crystal' for benzene (serving as a model) *vs.* distance between the molecule and (*001*) crystal surface. The distance is counted in the number 'k' of missing molecular layers (*001*), as shown in *b*. The curve *a* was calculated in order to estimate "sublimation" from the 'i'-phase under the attraction from the 'r'-phase at an interface, as shown schematically in *c*.

Believing that E_a has to be equal to E_s, some measurements leading to the "too low" E_a might be rejected as erroneous. Yet, the E_a of the three substances in Table 2.2 could not possibly be "pulled up" to the E_s level exactly: they remain somewhat lower (by ~ 5% on average) than E_s.

The activation energies found by Hartshorne and colleagues are almost as consistent with the *contact* interface as with the *vapor gap* interface. However, only the *contact* interface accounts for the 10^3 to

10^5 times faster movement of an interface than sublimation can offer. It is the above-estimated ~ 30% reduction in E_a as compared to E_s that gives rise to the much higher speed of phase transitions. If one takes values close to the real (see Table 2.2), such as $E_s = 20$ Kcal/mole and $T = 300$ °K, then $E_a = 0.7$ $E_s = 14$ Kcal/mole, and the ratio

$$\frac{V_{tr}}{V_s} = \frac{\exp(-E_a/RT)}{\exp(-E_s/RT)} \cong 10^4$$

The *contact* interface has solved the Hartshorne's paradox.

2.4.5 The coupling at the contact interface

As another test, a coupling between two polymorphs at their interface was measured and compared with that predicted by the *contact* structure of interface. Even prior to the quantitative measurements, the very fact of dealing with single crystals, rather than polycrystals or films between slides, provided the opportunity to make valuable qualitative observations:

(1) There is an essential coupling at the interface. Handling a "hybrid" crystal crossed by a flat interface does not result in its separation into two detached parts, as it would be in the case of a vapor gap between the phases.
(2) Yet, the interface is the weakest section. Application of an external force breaks the "hybrid" crystal exactly along the interface into two separate individual phases.
(3) Breaking down the crystal into two detached phases instantly halts the phase transition, indicating that not even a minute amount of the 'r' phase is left attached to the 'i' phase to serve as a seed for continuation of the phase transition.

These observations are quite consistent with the *contact* structure of the interface. Being a kind of thin gap between two phases, it is the weakest section of the specimen. But the gap is only about 0.5 k, so there still is enough coupling to keep the two phases together. The purpose of the study described below was to quantitatively estimate whether the real coupling corresponds to the 0.5 k effective gap.

Computer calculations have been carried out to estimate the Van der Waals' attractive forces between two parts of a benzene crystal (taken as a model) divided along a plane (*001*) as they move apart from $k = 0.5$ to $k = 5$. It was done by an atom to atom summation of the normal components of the attractive forces, using the first derivative of the potential function already mentioned in Sec. 2.4.4. The resultant curve is given in Fig. 2.19. The curve was needed to appreciate the effective "interphase" gap corresponding to the experimental value of the coupling force.

The measurements involved breaking miniature PDB single crystals with a miniature spring dynamometer when the interface across the whole crystal was moving along it. The breaking force was then recalculated into the ultimate normal stress, that is, into the coupling force. The experimental arrangement and the formula to calculate the normal stress are presented in Fig. 2.20. The final values are given in Table 2.3. As can be seen, the coupling strongly depends on the thickness of the crystal. This dependence was caused by the contribution of internal strains that always appear in the process of a solid-state phase transition. (In fact, these strains sometimes break specimens with no application of any external force, but this was not the case with PDB). The thinner a crystal, the better are the conditions for relaxation of the internal strains. Therefore, the number in Table 2.3 to use is that for the thinnest crystal, *i.e.* 1500 g/mm^2. In the calculated curve this value corresponds to the gap less than 1 *k*, thus supporting the *contact* structure of the interface. Should the coupling force be measured in the crystals thinner than 0.25 mm (which was beyond the experimental capabilities), it would certainly correspond to the gap closer to 0.5 *k*.

Table 2.3   Coupling force at interface as measured in PDB

Crystal thickness, mm	0.9 - 0.7	0.55	0.30	0.25
Coupling force, g/mm^2	130 -160	200	700	1500

2.4.6   General significance and predicting power of the contact mechanism

Scientific advancements are rarely quite straightforward. When current theories fail to account for a phenomenon, it is commonly declared more complicated than previously believed; the failed theories are modified, still not becoming consistent; the phenomenon is declared even more intricate... the theory becomes even more cumbersome...

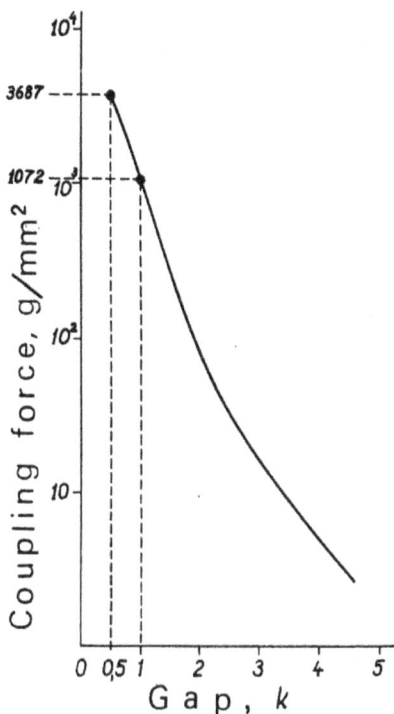

Fig. 2.19 The coupling force between two parts of a benzene crystal (serving as a model) vs. the number k of missing molecular layers (001) between them. The curve was calculated to verify whether the actual coupling at a phase interface corresponds to the effective gap $k = 0.5$ as the *contact* model of the interface suggested.

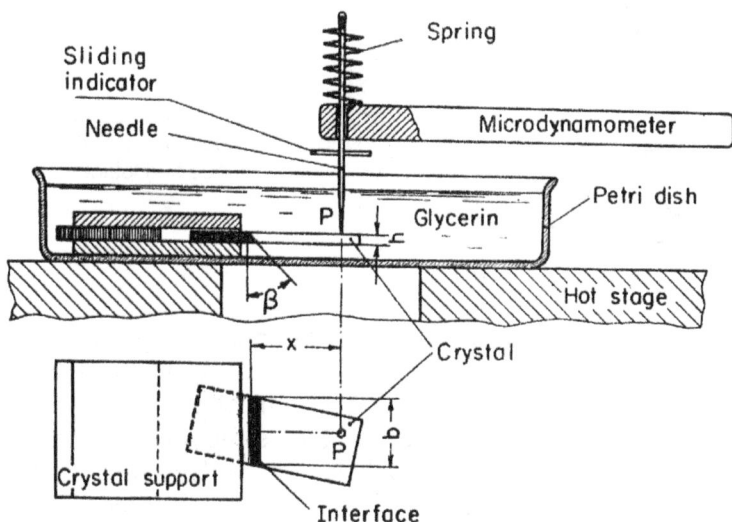

Fig. 2.20 The experimental arrangement in the breaking tests during phase transition in PDB to measure the coupling strength at the interface. The parameters in the calculation of the ultimate normal stress $\sigma_n = (6\,P \cdot x \cdot \cos^2\beta)\,/\,b \cdot h^2$ are shown.

This "vicious helix" is carried on and on until the simple general solution is found, and sometimes long after that. Solid-state phase transitions are an example of such a development. The *contact mechanism* is such a solution. In the broad sense, the *crystal growth concept* is the general solution, while the *contact mechanism* is the mode of its implementation in the specific case of crystal growth in a solid medium.

It will be our major objective to demonstrate that the *contact mechanism* is universal, that it is consistent with all reliable evidence, that it has predicting power, that it invalidates the whole variety of previously suggested mechanisms, theories and interpretations, and that it accounts for all apparent "anomalies." This removes the atmosphere of mystery around the problem of phase transitions. The presentation of corresponding evidence is already under way. In the case of phase transitions in sulfur, for instance, the contact mechanism accounts for the:

- E_a value
- Velocity of interface propagation
- Edgewise mode of interface propagation
- Indifference of moving interface to orientations of "consumed" grains (Fig. 2.21).

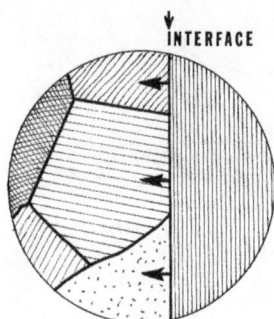

INTERFACE

Fig. 2.21 Interface in sulfur phase transitions has been observed [79] moving through grains of the original phase regardless of their boundaries and lattice orientations.

Besides solid-state phase transitions, there are other major phenomena where structural rearrangements at the interfaces play a decisive part. Recrystallization and chemical reactions in solids are among them. In the latter case the structural rearrangement is accompanied by a chemical reaction, so the "phases" differ not only structurally, but also chemically. From this point of view, a solid-state phase transition is "the most elementary chemical reaction," when the initial and final substances are identical. Whatever the type of structural rearrangement, there is good reason to believe that the *contact mechanism* is general to all of them. It is energetically advantageous to the interface molecules to relocate under the attractive action of the opposite interface side. It is also energetically advantageous to the interface to be rational in the 'r' lattice and irrational in the 'i' lattice, for it provides lower activation energy in the relocation of the interface molecules. These molecules are in the potential field of the two adherent phases: the relocation occurs from the irrational surface where the molecular bounding is weaker to the steps on the closely packed crystal face where it is stronger.

The contact interface offers a new insight into recrystallization (migration of grain boundaries) of polycrystalline solids. The recrystallization process was actually considered separately from that of a solid-state phase transition. The latter deemed to be an orderly transformation of the original phase, requiring a certain orientation relationship (rarely verified, however). Random grain orientation in a polycrystal could not be denied; therefore two neighboring grains did not seem akin to different phases. But the grain boundaries do migrate. Since the random orientation of two contacting phases is now proven, it is hard to find any reason to assume one molecular mechanism for interface motion and another for migration of grain boundaries. The difference is in their driving forces rather than in the molecular mechanisms. In recrystallization, the force driving grain boundaries is to minimize their surface energy and/or substitute a more perfect lattice for a less perfect one. In all other respects it is the same crystal growth that takes place in phase transitions.

Recrystallization in itself is a large topic, a branch of physical metallurgy and some other applied sciences. It is touched upon here to illustrate the general applicability of the *contact* mechanism. It tells us in which direction a grain boundary moves: from the grain where it has rational (h,k,ℓ) to the grain where it is irrational. It follows that a major component of the recrystallization driving force is the elimination of irrationally oriented boundaries. But there is more to it. The boundary will migrate only if the two neighboring grains have different orientations. Since all grains have the same crystal structure, the boundary between two grains of the same orientation will be either equally rational with the same (h,k,ℓ), or equally irrational. In such a case, no driving force to instigate the molecular relocation in one or the other direction exists: E_a in either direction is the same. The boundary remains still. In such a straightforward manner the *contact* mechanism accounts for one of the unexplained "recrystallization laws" which states that the boundary between two grains of the same orientation does not migrate [95].

Here is one more example in the field of recrystallization. The following fact perplexed observers: sometimes one part of the boundary between grains A and B migrates from A to B, while another part migrates from B to A (Fig. 2.22). In terms of the contact mechanism, the cause of the phenomenon is simply in the directions of the boundaries: in the former case, the boundary is a rational crystal plane in A and irrational in B, but the other way around in the latter case.

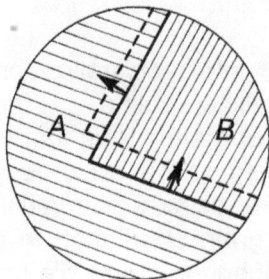

Fig. 2.22 In solid-state recrystallization, the A → B and B → A migration of the boundary between grains A and B can proceed simultaneously [44]. Dashed line indicates a subsequent position of the grain boundary.

So far the term *contact mechanism* was related to the structure of interface and the mode of molecular relocation at it. This, however, is only a part of the molecular mechanism of phase transitions, more exactly, its second part. Its first part is *nucleation* of the new phase.

2.5 Nucleation in crystals

2.5.1 Problems with the standard interpretation

Do solid-state phase transitions involve nucleation of the new phase? We answer: *all* do. Some authors would probably answer that *many* do. Domb and Lebowitz - editors of *Phase Transitions and Critical Phenomena* in many volumes - associated nucleation with first-order phase transitions [96], as did Landau and Lifshitz [36] before them. But the assertions by Rao and Rao are more rigorous [42]: "A large majority of polymorphic transformations... are described by a process known as nucleation and growth... Nucleation is common to all types of solid-state transformations."

It would seem it is hardly possible not to recognize nucleation as one of the main elements of solid-state phase transitions. Nevertheless, the literature commonly treated the transitions as if the phenomenon of nucleation (and subsequent growth) was nonexistent. As only a few examples, nucleation is missing in three (unrelated) books *Structural Phase Transitions* [18,19,40], in books [4,39,54,55,57,97,98,99], in many volumes of *Phase Transitions and Critical Phenomena* [1], in review articles [59,100], and in numerous other articles ([101] is a typical example).

The statement by Rao and Rao in their review book [42] that "nucleation is common to all types of solid-state transformations" deserves clarification, since its authors failed to disclose its real significance. In fact, it implies (see Sec. 1.2) that those "all types" of solid-state phase transitions are not different types, because all of them are of the *nucleation and growth* (and, therefore, of the first order) type. Thereby most of the content of the Rao and Rao's review was effectively undermined. Nucleation is inseparable from growth, and growth means phase coexistence and rearrangement at interfaces rather than a homogeneous process. In such a case, the *continuous*, *lambda*, *cooperative*, *soft-mode*, *second-order*, *critical*, *displacive*, *distortive*, etc., types of phase transitions turn out to be at variance with the reality. Yet, these nucleation-free types of phase transitions constitute the content of Rao and Rao's book. Maintaining that the statistical-mechanical analysis of critical behavior "provides a basis for understanding the general features of phase transitions," or that "another unifying concept which makes possible a microscopic understanding of structural phase transitions in solids is the soft mode," these authors contradict their own acknowledgment that all those "cooperative" phase transitions occur by nucleation and growth. There cannot be a microscopic understanding of a nucleation-growth process if it is assumed not to involve nucleation and growth.

Yet, the literature on nucleation in a solid state existed, mainly owing to needs of physical metallurgy [66,74]. The theory of nucleation in a liquid phase was slightly modified to also cover nucleation in a solid state and is considered the classical theory on the subject [66,102]. Metals and alloys, however, are a difficult object for collecting basic experimental facts. The theory in question is therefore based rather on conjectures than on hard facts. Besides, there is no guaranty that the nucleation theory in liquids is quite correct. No one previously verified the basic theoretical assumptions by using the simplest and most informative objects, which are good quality small transparent single crystals. The results of this re-examination will now be described. The nucleation actually discovered was quite opposite to that assumed by the conventional theory. Although the conventional theory is not outlined here, its main points will be later compared with the data presented in the section that follows.

2.5.2 Nucleation in small single crystals

The present data on nucleation were accumulated and analyzed for a simple case of temperature-induced phase transitions in tiny, well-bounded, single-component, good quality, transparent single crystals under controlled temperatures. The investigated substances are listed in Table 2.1, PDB being the main object. The experimental facts and their meaning are summarized below. Part of the data has been collected during the years of studying different aspects of crystal phase transitions, and the rest was acquired by special experiments, using optical microscopy and X-rays. It is useful to remember that for PDB, used as an illustration, T_o = 30.8 °C and T_m = 53.2 °C. T_n is the temperature at which nucleation occurs.

● *Nucleation requires overheating / overcooling*. Nucleation never occurs at T_o or in a certain finite vicinity of it. A conservative estimate indicates the "prohibited" range for PDB to be at least 28 °C to 32 °C. Upon slow heating, nucleation in most PDB crystals will not occur until the temperature exceeds 38 °C. Often L phase melts at T_m without transition into H. This event is not a matter of kinetics: storage for any reasonably long time near T_m would not result in phase transition.

● *Nucleation is a rare event*. Slow heating (*e.g.*, 1 °C to 10 °C per hour) does not produce many nuclei. Usually there are only a few units, or only one, or no nucleation sites at all. This observation is at variance with the notion "rate of nucleation" and the statistical approach to nucleation - at least as applied to a single-crystal medium.

● *Exact temperature of nucleation T_n is unknown a priori*. Formation of a nucleus upon slow heating of a PDB single crystal will occur somewhere between 34 °C and T_m = 53.2 °C, and in some cases it will not occur at all. The T_n vary in different crystals of the same substance. There is no way to predict the exact T_n in a particular crystal.

● *Only crystal defects serve as the nucleation sites*. In other words, the nucleation is always heterogeneous as opposed to the *homogeneous* nucleation assumed to occur in an ideal crystal lattice. The location of a nucleus can be foretold with a good probability when the crystal has a visible defect. Another fact is that nucleation in sufficiently overheated / overcooled crystals can be initiated by introducing an "artificial defect" (a slight prick with a glass string or needle). In such a case, a nucleus appears at the place of the local damage. A frequently observed event in cyclic phase transitions L → H → L → H... is that a nucleus appears several times at the same location. There is evidence that homogeneous nucleation does not exist at all (see below).

● *The higher the crystal perfection, the greater overheating or overcooling* $\pm\Delta T_n = T_n - T_o$. In brief, better crystals exhibit wider *hysteresis* -- because $\pm\Delta T_n$ *is* hysteresis (Sec. 2.6.6). The correlation between the degree of crystal perfection and ΔT_n is confirmed in several ways. One, illustrated with the qualitative plot in Fig. 2.23, is the ΔT_n dependence on the estimated quality of the single crystals. Different grades were assigned to sets of crystals depending on their quality, higher grades corresponding to higher crystal perfection. The grades were given on the basis of crystal appearance (visually estimated perfection of the faces and bulk uniformity) and the way the crystals were grown (from solution, from vapor, rate of growth, temperature stability upon growing, etc.). For the solution-grown PDB crystals the graduation exhibited good correlation with the levels of ΔT_n within the grades 1 to 4. Two findings will be noted regarding ΔT_n in these crystals:

1. $\Delta T_n < 1.9$ °C was not found even in the worst (grade 1) crystals, indicating the existence of a minimum (threshold) overheating required for nucleation to occur.

2. The most perfect among the solution-grown crystals (grade 4) were incapable of phase changing L→H, so it was L that melted. The inability to nucleate in the absence of a proper crystal defect follows from the fact that an "artificial defect," created by a slight prick at the temperature near T_m, immediately initiated the phase transition from the damaged spot. Storage alone of the "grade 4" crystals at this temperature, even for hours, did not result in phase transition. An inescapable logical deduction from this simple observation is: dislocation lines, individual vacancies and foreign molecules, as well as the natural crystal faces, do not serve as nucleation sites. There is no doubt of the availability of such defects in those crystals, even though the crystals were of a good quality and prepared from pure material.

T_n

Temperature, °C

T_m

$\Delta T_{n,min}$

T_0

1   2   3.   4   5

Crystal quality grade, n

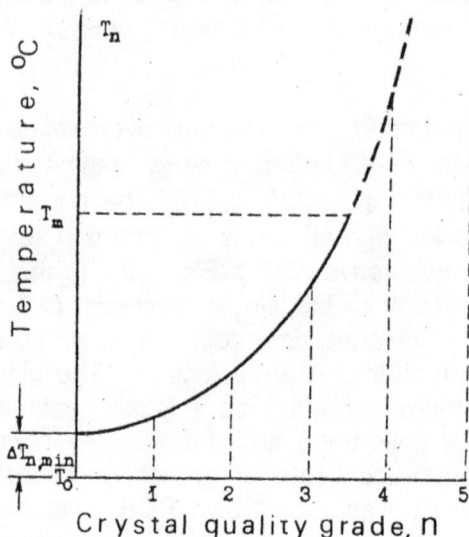

Fig. 2.23 The character of dependence of nucleation temperature T_n on quality of single crystals; (L→ H). The quality is represented by number **n**: the higher estimated crystal perfection, the higher number **n**. No formation of a nucleus is possible until overheating exceeds some threshold value $\Delta T_{n,min}$. Overheating $\Delta T_n = T_n - T_0$ increases with **n**. Nucleation in "grade 4" crystals does not occur spontaneously, no matter how long they are stored just below melting point T_m. Yet, "grade 4" can be coerced into phase transition by a slight prick. No way has been found to induce a phase transition in "grade 5" crystals.

● *Temperature of nucleation T_n is pre-coded in the crystal defect acting as the nucleation site.* Fig. 2.24 shows the results of the experiments involving microscopic observation of PDB (L) single crystals upon slow heating. Temperature T_n was recorded as soon as nucleation of the H phase was noticed. Growth of the H crystal was stopped as quickly as possible with the manual temperature control and the specimen was then returned to the L phase. The cycle was repeated with the same crystal many times. Then the whole procedure was repeated with another crystal.

The experiments revealed three realities of a primary significance: (a) a nucleus appears every time at exactly the same location, (b) the temperature T_n repeats itself as well, and (c) a particular T_n is associated only with the particular nucleation site; a different T_n is found in another crystal, also associated with the defect acting as the nucleation site L. Thus, every crystal defect acting as a dormant nucleation site contains T_n information encoded in its structure. If a crystal has more than one dormant nucleation site and is subjected to very slow heating, only one site with the lowest T_n will be activated. Upon faster heating, the second nucleus of the second lowest T_n may have time to be activated before growth from the first nucleus spreads over its location, and so on.

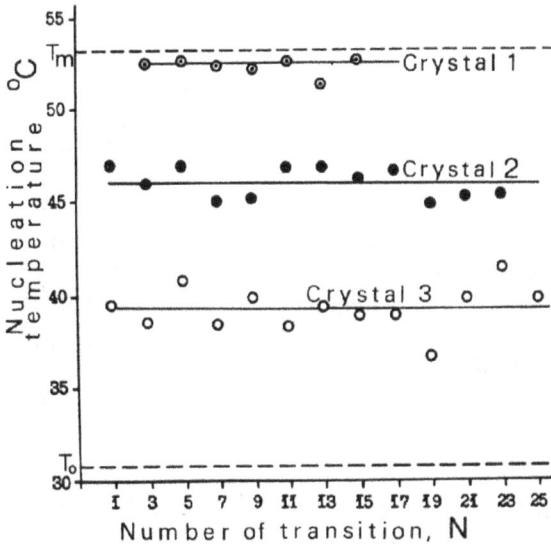

Fig. 2.24 Reiterative formation of a nucleus in small PDB single crystals. Reiteration of nucleation temperature T_n when the nucleus repeatedly forms at the same site in the cyclic process. The procedure was as follows. Temperature of a microscope's hot stage with a small PDB single crystal was slowly raised; as soon as a nucleus of H phase became visible, T_n was measured and the phase transition was immediately reversed. In the subsequent cycles the nucleus formed at the same site (as a rule, at a visible defect) and at the same T_n. The measurements for three single crystals are shown. Every particular site was associated with its particular T_n. N is ordinal number of a phase transition.

● *Orientation of the resultant crystal is pre-coded in the crystal defect acting as the nucleation site.* As has been shown, the 'r' crystals in PDB phase transitions grow in random orientations in the 'i' lattice. One may believe the random orientations to be a result of random fluctuations at the nucleation stage. The real cause turns out to be different, if not the opposite. It was found in the experiments described in (6) that random orientations of the 'r' crystals have a static, rather than dynamic origin. As long as a nucleus forms at the same lattice defect, the 3-D orientation and even the shape of the growing 'r' crystal repeats itself (Fig. 2.25). Another nucleation site will produce a different orientation inherent exclusively to it. Thus, information on the orientation of an 'r' crystal is pre -coded and stored in the structure of the crystal defect serving as the nucleation site. This information, however, remains dormant unless this defect is activated (by heating or cooling the 'i' crystal to the T_n encoded in the defect). This brings about the explanation of the *orientation memory* phenomenon found in the X-ray study described in Sec. 2.2.4.

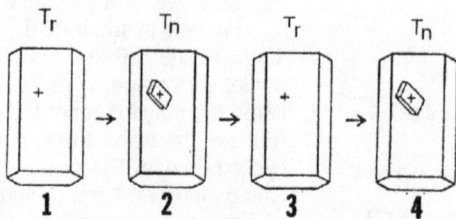

Fig. 2.25 Reiteration of orientation and shape of a nucleus when it forms at the same site. (1) Initial crystal (L) with a local defect (+). (2) Crystal H appears. (3) Complete return to L. (4) H appears at the same site, the same temperature T_n and in the same shape and orientation.

● *T_n in the same crystal can be changed.* Information encoded in the heterogeneous nucleation site can be "erased" from the crystal. One way to decrease T_n is to create an "artificial defect," as mentioned earlier. The T_n also goes down with every transition in the cyclic process $L \rightarrow H \rightarrow L \rightarrow H$... if no precaution is taken to change the temperature slowly (Fig. 2.26). As the number of cycles N increases, the crystal deteriorates. It is no longer a single crystal, but a set of differently oriented blocks filling up the frame of the initial single crystal. Each successive transition originates from a new defect with a lower encoded T_n. The plot is tangent to a horizontal line placed above the T_0, thereby revealing existence of a threshold $\Delta T_{n,min}$. It is worth noting that T_n can also be increased by enhancing crystal perfection with annealing (storage at the temperature slightly lower than T_n).

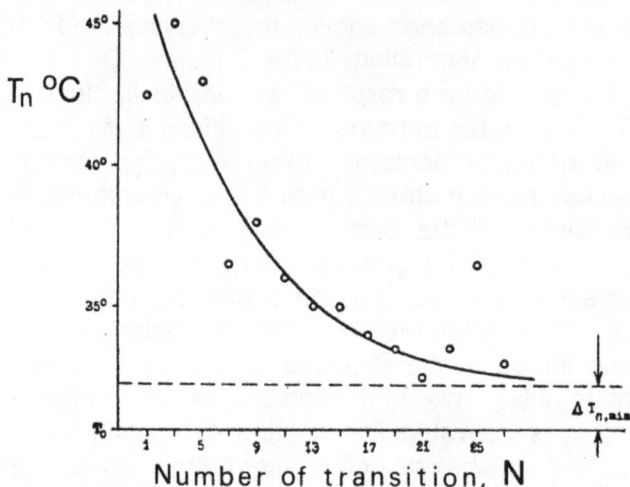

Fig. 2.26 Hysteresis T_n as a function of the number N of transitions

$L \rightarrow H \rightarrow L \rightarrow H \rightarrow L \rightarrow$
$N = 1 \quad 2 \quad 3 \quad 4 \quad 5$

in one of the investigated single crystals.

● *A nucleation site is stable, although only to some extent.* A crystal can store a nucleation site, together with all the encoded nucleation information, for a long time. The site can also withstand some influences such as moderate annealing, internal strains, and passage of an interface. Suppose, the nucleation site is located at some point *A*. At

- 94 -

the encoded temperature $T_{n,A} > T_o$ the site turns into a nucleus. If the transition 'i' → 'r' is completed and reversed, the interface in the 'r' → 'i' transition will pass through the point A. It was found that this may leave the nucleation site A unaffected in a few successive cycles, so it will remain as such in the 'i' → 'r' runs. Ultimately it will be destroyed and the transitions 'i' → 'r' will not originate at the point A any more. These observations account for reappearance of the same X-ray Laue pattern several times (Fig. 2.12) and are useful, as will be seen later, in identifying the type of crystal defects acting as nucleation sites.

● ΔT_n *in a solid→solid phase transition is smaller than in a liquid→solid phase transition.* At least, it was so in our experiments reproduced several times with different crystals. First, $\Delta T_n = |T_n - T_o|$ was measured in a PDB single crystal upon L → H and H → L transitions and was found to be in the 7 to 10 °C range in both cases. Then the crystal was melted on the microscopic hot stage and permitted to cool down to room temperature. No crystallization occurred, thus demonstrating that ΔT_n (liq. → sol.) > 32 °C. It should be noticed that, for a given crystal, both evaluations were related to the same mass and chemical identity, including the chemical and mechanical impurities.

● *Neither nucleation nor growth is possible in "too perfect" single crystals.* In other words, crystal defects (of a certain kind, Sec. 2.5.4) are a necessary constituent of the mechanism of a solid-solid phase transition. Direct evidence of that was discovered when some vapor-grown PDB single crystals were found completely incapable of changing from L to H phase. These crystals (grade 5 in Fig. 2.23) were of two kinds: very thin rod-like ones, prepared by evaporation into a glass tube, and extremely uniform and brilliant crystals found on the internal walls of the jar where PDB was stored. Not only did they not change into the H phase upon heating, but it was also impossible to induce this change by means of the "artificial nucleation." Even after the surface of some of the crystals was badly damaged by multiple pricks with a needle, they still "refused" to change into the H phase at the temperature as high as 53 °C, only 0.2 °C below T_m. Although all conditions for nucleation were thus provided, the nuclei could not grow: some additional condition was absent. We shall return to this problem in Sec. 2.9.2. Anyhow, infeasibility of a homogeneous nucleation manifested itself unambiguously.

2.5.3 Formation of a nucleus: a predetermined act, rather than a successful random fluctuation

The conventional theory of nucleation in solids considered nucleation of 'r' phase to be the consequence of random heterophase fluctuations that give rise to formation of clusters large enough to become stable. The only change made in the formulae over nucleation in liquids and gases is an extra term representing the idea that nucleation barrier in solids is higher due to arising strains. The theory in question originally considered nucleation to be homogeneous (assuming equal probability for nuclei to appear at any point in the crystal), but had later to acknowledge that heterogeneous, *i.e.*, localized at crystal defects, nucleation prevails. Therefore the theory was slightly modified to take this into account. Heterogeneous nucleation is believed to occur at dislocation lines, foreign molecules and vacancies being present in real crystals in great numbers. In other respects the statistical-fluctuation approach has been left intact.

The general principles of nucleation in a crystal, stemming from the findings presented in Sec. 2.5.2, are quite the opposite. Formation of a nucleus is a rare and reproducible act bound to a predetermined location. Homogeneous nucleation is impossible. As for heterogeneous nucleation, the formation of nuclei at dislocation lines or vacancies has to be ruled out as inconsistent with evidence. There are plenty of these defects in every real crystal and they change their position under the action of even the slightest internal strains, to say nothing of the strains caused by moving interfaces. Dislocation lines, vacancies, and foreign molecules cannot account for the observed rarity, stability, and reproducibility of a nucleus. They cannot provide the required structural diversity in order to store individual information both on T_n and orientation of the resultant crystal. The observed predetermined phenomenon has nothing to do with "successful" random fluctuations.

In fact, never before had the common theory a chance to be seriously examined. On one hand, it was based on untested assumptions. On the other hand, its conclusions have not been verified either due to, as Russel [66] pointed out, "the absence of ancillary data to compare theory and experiment." Now we see that the whole set of theoretical notions are not applicable to such a simple system as small single crystal. These are: heterophase fluctuations, homogeneous nucleation, time dependent nucleation, rate of nucleation, induction period, strain energy (of nucleus formation), nucleation on dislocations, etc. For example:

(a) An isothermal *rate of nucleation,* a standard value of interest, is not applicable to nucleation in a crystal, because there every T_n is "pre-coded." Only growth, and not nucleation, can proceed isothermally.

(b) The experimental fact, ΔT_n (solid → solid) < ΔT_n (liquid → solid), presented in Sec. 2.5.2, item 10, runs counter to the principal theoretical premise that the activation energy of nucleation in solids, E_a', is higher due to the internal strains associated with the act. The particular structure of the crystal defects acting as nucleation sites will explain why E_a' is relatively low.

2.5.4 The structure of a nucleation site

Finding the particular structure of a crystal defect serving as a nucleation site thus turns out to be the key problem of nucleation in a solid. The solution is proposed below. It inevitably involves an element of speculation, but only to link all elements of comprehensive evidence into a self-consistent and clear picture. Our method is *logical analysis.* It is qualitative, but more reliable than a detailed mathematical description of an idea that has not been verified.

Let us sum up the properties of these defects. There are few such defects in a good single crystal; they reside at permanent locations; they do not form spontaneously over long-term storage at any level of overheating / overcooling; they are stable enough to withstand rather strong influences; they are not quite as stable as macroscopic defects; they possess a memory large enough to contain individual information both on T_n and nucleus orientation; they are capable of activating the stored information repeatedly; their structure permits nucleation without development of prohibitive strains.

Microcavities (conglomerations of vacancies) of some optimum size (Fig. 2.27) are, perhaps, the only type of crystal defect that meets all these requirements. An optimum microcavity (OM) eliminates the problem of the great strains that, possibly, prevent nucleation in a defect-free crystal medium. OM consists of many individual vacancies and is therefore relatively stable and bound to a permanent location. There can only be a few such large-sized defects in a good single crystal, or even none.   Yet, the defect in question is far from a macroscopic size, so it can be affected in some way to lose its nucleation function. Variations of its size and shape account for its capability of storing individual information on nucleation.   (One additional vacancy can be attached to OM in many different ways, thus adding many new variants of "encoded" information).

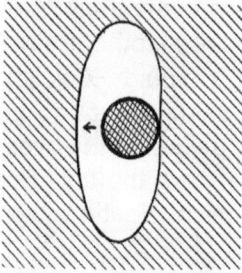

Fig. 2.27   Formation of a nucleus in an optimum microcavity. Not only there are no accompanied strains, but the activation energy of molecular relocation is especially low owing to attractive action of the opposite wall.

The relatively large size of OM has already been discussed. But OM also cannot be too large. Limited stability of OM is one argument in favor of such a conclusion. Now let us return to Fig. 2.27. It is true that a very large cavity eliminates the associated strains in the same way a smaller one does. In both cases a nucleus can grow freely on a cavity wall. The fact is, however, that a large cavity is equivalent to an external surface where no nucleation strains would be involved. But crystal faces do not facilitate nucleation, as it follows from the observation that nucleation in a "too perfect" crystal fails to occur both in the bulk and on the faces. Nucleation on the crystal faces obviously lacks some additional condition.

The idea on the nature of the lacking condition comes from the mode of molecular rearrangement at the *contact* interface (Sec. 2.4). There, molecular relocation from one side of interface to the opposite one proceeded under the attractive action of the latter, and this circumstance lowered the activation energy. Similarly, a strain-free nucleation with low activation energy is offered by a microcavity that is sufficiently narrow to facilitate the molecular relocation by an attractive action of the opposite wall. As we already know, the gap must be of a molecular dimension. The particular width and configuration of this gap in a given OM may well be that same parameter which determines the encoded individual T_n. The detailed shape of OM may then be responsible for the encoded orientation of the nucleus. Thus, there is an intrinsic alliance between the molecular mechanisms of nucleation and growth. Both are based on the principle "relocation under attraction." When nucleus formation is completed, crystal growth takes over so naturally that these two stages of phase transition merge into a single unified process.

From this point on we shall use the terms *contact phase transition* and *contact mechanism* to represent the combination of nucleation and rearrangement at interfaces as described.

2.6 Temperature of phase transition

2.6.1 Analogy with crystallization from melt

The nature of solid-state phase transitions cannot be understood without the realization of what a transition temperature T_{tr} is. The terms usually used as synonyms are: *critical* (or *Curie*) *temperature* (or *point*) and *transition temperature (or point)*. One routinely uses the latter term with no definition, believing that the temperature at which a phase transition *occurs* is (or in an ideal case must be) its physical characteristic. This has been a major source of misinterpretation of solid-state phase transitions.

For illustration, let us compare crystal $\rightarrow$ crystal and melt $\rightarrow$ crystal phase transitions. Suppose, a number of independent workers are recording the same melt $\rightarrow$ solid phase transition with different techniques, such as calorimetry, X-rays, spectroscopy, *etc.*, everything except a visual observation. As well known (but is assumed unknown to our workers), overcooling is imminent in such phase transitions. The reports by different workers on the transition temperature will (a) be systematically lower than T_0 (melting temperature T_m) and (b) differ from one another. (A curious fact is that sloppy workers will report T_{tr} closer to T_0). Further extending the analogy, a probable development could be that some workers will try to improve the reported T_{tr}. They will use highly purified specimens and change temperature very slowly, not realizing that are going in the wrong direction. The recorded T_{tr} will further depart from T_0 and may well be dozens degrees lower. (Overcooling of liquid PDB exceeding 33 °C was mentioned in Sec. 2.5.2).

The described hypothetic situation is real in the field of solid-state phase transitions. A parallel visual monitoring of the phase transitions is impossible with opaque materials and is a difficult experimental task when they are transparent. The indirect techniques that are utilized to study solid-solid transitions would not usually reveal their *nucleation and growth* nature. They are, however, nucleation and growth in both H $\rightarrow$ L and L $\rightarrow$ H directions, so overheating is as inevitable as overcooling. The common view that a phase transition must normally occur at a fixed T_{tr} is an invalid assumption. It is inapplicable to liquid crystallization, neither is it applicable to crystallization in a solid medium, which phase transitions in solids are.

2.6.2 The meaning of a recorded transition temperature T_{tr}

A recorded temperature of phase transition is neither a *critical point*, nor a *Curie point*; nor is it an exact physical characteristic of the phase transition. It is not quite reproducible. It always differs from T_0, which is the only temperature characteristic of the process.

Let us trace where the conventional interpretation of T_{tr} failed. Typically, a solid-solid phase transition is described as follows. As temperature approaches the *critical point*, the system becomes increasingly unstable while *pre-transition* processes prepare it for the forthcoming phase transition; clusters of the new phase emerge and dissolve ("heterophase fluctuations"), increasing in size until, at the critical point, one would comprise the system as a whole, thus completing the transition.

The theoretical idea on heterophase fluctuations is erroneous: the phenomenon does not exist. The particles in a crystal substance are too highly organized and tightly bound to allow even homogeneous nucleation, to say nothing about the fluctuation clusters of the alternative phase. If the theoretical physicists regarded solid-state phase transitions as crystallization, as we do, they might notice that even melt crystallization does not occur by heterophase fluctuations. The statistical-mechanical idea on the fluctuation nature of phase transitions in a crystal matter had for its experimental basis only the false "anomalies" (or "singularities") frequently observed in the temperature range of transitions. They will be discounted in full in Chapter 3.

There is no such phenomenon in crystal phase transitions as the absolute instability of a crystal. There are always two competing phases. If the alternative phase becomes more energetically advantageous at some temperature, the initial phase will not cease to exist until a nucleus can form. The latter will not form if a special condition (an appropriate microcavity) is absent, in which case the initial phase will continue exist in the "foreign" area of the p - T phase diagram. The "pre-transition" phenomena, as they are imagined, do not exist either. There are only the usual temperature-dependent changes involving thermal expansion and intensity of molecular vibrations. These changes do affect the energy balance between the phases, but will not cause phase transition in the absence of nucleation.

In effect, a crystal phase remains quite stable not only as T_o is approached, but even after T_o has been passed by a threshold (sometimes small, but finite) $\Delta T_{n,min}$; ΔT_n can be greater than $\Delta T_{n,min}$ by any value. As soon as a nucleus emerged, its growth can proceed at lower levels of overheating / overcooling, $\Delta T_{tr} < \Delta T_n$. An additional fact of importance is that there must also be some overheating (or overcooling) exceeding a threshold $\Delta T_{tr,min}$ in order to make or keep the interface moving.

To summarize, when dealing with a phase transition, three temperatures have to be distinguished:

- **T_o**, the temperature at which the free energies of the two phases are equal. No thermodynamic reason for the phase transition exists. Being the only invariable temperature characteristic of a given phase transition, it might be called "temperature of phase transition," were it is not for the fact that *T_o is the only temperature at which (and even in a certain vicinity of which) the phase transition is unconditionally impossible.*
- **T_n**, the temperature of nucleus formation. Its value is predetermined in the crystal undergoing the phase transition. It varies from crystal to crystal and related to T_o by the inequalities $T_n > T_o$ upon heating and $T_n < T_o$ upon cooling.
- **T_{tr}**, the temperature at which a phase transition proceeds by interface propagation after the nucleus has been formed. ΔT_{tr} can be controlled and set much lower than ΔT_n, but not too close to T_o without the transition being halted.

2.6.3 The source of typical ambiguities: discrepancies in the reported temperatures, range of transition, rounding, and hysteresis of critical points

As known, experimental determination of the "temperatures of phase transitions" was marked with great ambiguities. Now we are in the position to clarify them all.

◆ *Poor reproducibility; divergence between the data reported by different authors.* It is usually attributed to traces of impurities in the specimens, and sometimes to crystal imperfections. In so doing, one assumes that more purified and defect-free crystals will show a more exact transition temperature.

Clarification. Only T_o is the characteristic of a phase transition. Its exact value cannot be found directly, but only by the measurements both upon heating and cooling, then minimizing the interpolation interval and taking the midpoint as T_o. The reported transition temperatures, however, are those at which changes of the physical properties were detected, *i.e.*, T_{tr}, subject to variation from crystal to crystal. Because the reported data are usually taken from measurements in one direction (say, upon heating, as in adiabatic calorimetry), the systematic error $\Delta T_n(+)$ is added to the above scattered data. Matters are further aggravated if some reported data are taken upon heating, and other data upon cooling: then the divergence $[\Delta T_n (+) + \Delta T_n (-)]$ and the above scatter add up.

◆ *Temperature range of transition* ("sigmoid" curve in Fig. 1b) instead of a point. Sometimes these transitions are called *diffuse*. Nevertheless, the inflection point is usually taken as the transition temperature T_{tr} or even as the critical point T_c. No reasonable explanation existed to why they are stretched over a temperature range.

Clarification. The "diffuseness" is not the manifestation of a specific transition mechanism. It is a consequence of the non-simultaneous nucleation in different particles, or parts, of the specimen. Any transition in a powder or polycrystalline specimen is "diffuse," the variation being only how much. The origin of the *range of transition* is explained in Fig. 2.28. Suppose, a sample of many small-sized crystal particles numbered 1,2...k...K is slowly heated up. The phase transition is controlled by nucleation: as soon as the "pre-coded" T_n in the particle # k is attained, the particle changes from L to H (propagation of the interfaces takes short time). Different transition temperatures T1, T2, T3...Tk....TK are "pre-coded" in the particles (Fig. 2.28a, left). At intermediate temperatures T' the sample is two-phase (Fig. 2.28a, right). The phase transition will never be completed without further heating, for there still are particles with $T_n > T'$ (Fig. 2.28b). The integral effect upon heating over T1... TK will be a sigmoid plot $m_H = f(T)$ where m_H is the quantity (mass) of the H phase (Fig. 2.28c).

Summarizing:
- range of *transition* is range of *nucleation*;
- range of transition lies entirely above T_o (L → H) or below T_o (H → L), never including T_o;

- inflection point of the sigmoid plot corresponds to the T_n "encoded" in the maximum number of the constituent particles;
- any physical property that is proportional to the mass fraction of a phase in the two-phase specimen (such as electric resistance or dilatometric density) will exhibit a sigmoid curve as a function of temperature.

◆ *"Rounding" of the λ-peaks.* The origin of these peaks will be clarified later in all detail; at present, our interest in them is limited by their relation to the temperature of phase transition. A λ-peak, *e.g.* one emerging in measurements of specific heat, spreads over a temperature range like the "sigmoid" curves do. According to the theory, such a peak manifests a critical behavior (second-order phase transition), and its top must be quite sharp, locating the critical point T_c. The problem is, however, that these peaks are not sharp. Schwartz [103] described it as follows: "It is well known that most second-order phase transitions exhibit deviations from the theory called *rounding* which tend to limit the height of the heat-capacity peak and broaden it out." In fact, λ-peaks free of "rounding" were not found. Rounding is usually attributed to impurities and crystal imperfections, but experiment shows something quite opposite, that impure and imperfect crystals have sharper peaks [103,104]. Various authors treated the rounding as a "continuous distribution of critical points," which is a manifestation of the erosion of the critical point concept. No reasonable explanation of the "rounding" existed.

Clarification. The λ-peaks have nothing to do with critical behavior. As in the case of the sigmoid curves, they are a consequence of the T_n distribution in the specimen (refer to Chapter 3). Calorimetric "C_p"-peaks are the latent heat of transition. Like all exothermic and endothermic peaks, they have a rounded shape. Position of the peak top corresponds to the maximum number of the same T_n "encoded" in the particles or parts of the sample. The temperature range of nucleation, which the λ-peak rests on, does not include T_o.

◆ *Two "critical points": one for heating, the other for cooling.* In many cases one remains unaware of a hysteresis of the "critical point," relying on measurements in one direction, usually upon heating. If the "critical points" of "second-order" phase transitions are carefully measured in both directions, the "critical point" upon heating is found at a higher temperature than upon cooling. No reasonable explanation existed to the effect. Although hysteresis is compatible neither with a critical point concept, nor with the idea of second-order phase

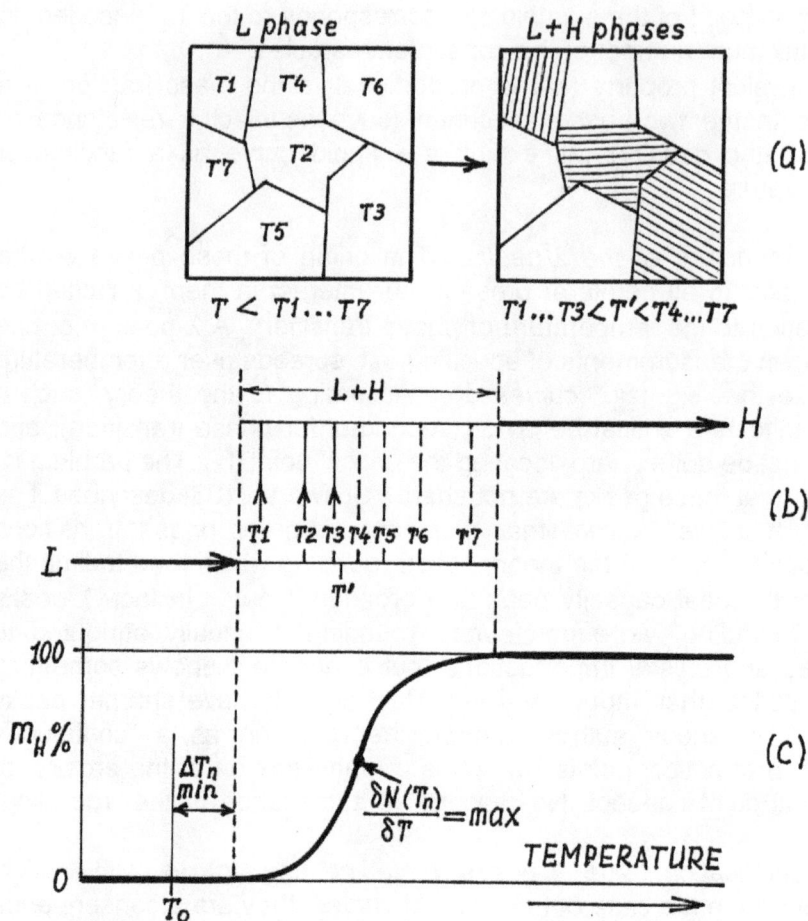

Fig. 2.28 Temperature range of a phase transition.

(a) Phase transition upon heating of a powder or polycrystalline specimen is predetermined by the nucleation temperatures T_n = T1, T2, T3,... pre-coded in the crystalline particles or grains. At T' there are two coexisting phases (shown as shaded and non-shaded). The phase transition cannot be completed without raising the temperature over T'.

(b) A diagram where L and H phases are represented by two different levels and the L → H phase transition of every crystalline particle shown in 'a' is marked on temperature scale. T' is an intermediate temperature. The dashed vertical lines symbolize the temperatures encoded in crystals 4 to 7 at which the phase transitions would be activated upon heating above T'.

(c) The integral effect of phase transition in a fine-particle system. Quantity of H phase vs. temperature, $m_H = f(T)$, is represented by a sigmoid curve. Its inflection point corresponds to $\delta N(T_n) / \delta T$ = max, where $N(T_n)$ is the number of particles with the nucleation temperature T_n encoded in them.

- 104 -

transitions, it is usually ignored even when noticed. For example, in order to find the "critical exponents," one chooses the "critical point" upon heating and performs measurements as close to it as small fractions of a degree, even though the "critical point" in the reverse direction is known to be whole degrees apart.

Clarification. This is the hysteresis of nucleation $\Delta T_n(+)$ and ΔT_n (-), and not of critical points. The ranges of nucleation upon heating and cooling do not overlap, and neither do the λ-peaks erected over these ranges, T_o being in between (Fig. 2.29). The hysteresis in question is unavoidable; recording the peak in one direction signals existence of another on the opposite side of T_o.

A major goal of quite a few experimental studies has been to trace a physical property as a function of temperature as close to the "critical point" as possible. They were aimed at finding the *critical exponents*, major parameters of the theory of critical phenomena. It should be obvious from the foregoing that these efforts were misdirected. The temperatures that one considered to be *critical points* were neither critical, nor points, nor were they exactly reproducible, nor were they the same upon heating and cooling, nor can they ever coincide with T_o.

2.6.4   Misplacement of transition temperature – inevitable misinterpretation of the phase transition phenomenon

Arbitrary placement of the "transition point" at the middle of a λ-peak or at the inflection point of a sigmoid curve turns phase transitions into an extremely complicated, even mysterious phenomenon. They appear to start well before the "transition point" and to complete well after it. The pre-transition and post-transition processes in a crystal medium became the dominant problem. Not surprisingly, development of the theory turned out to be an enormously difficult task: one should not expect simple theoretical solutions of non-existent phenomena.

Realization of the fact that a phase transition starts *after* T_o has passed eliminates the mystery. One example is the "critical phenomenon" called *critical opalescence*, the observed sharp increase of light scattering in the temperature region of some solid-state phase transitions. The phenomenon is believed to be caused by pre-transitional phase fluctuations as the temperature approaches T_c. The idea of pre-transitional unstable clusters of the new phase in a crystal lattice is highly unrealistic and has nothing in common with the evidence presented here. There was the arbitrary assumption that the center of the observed peak of opalescence marks T_c. Now the phenomenon can get a simple explanation. The peaks of light scattering form *after* T_0 has been passed; a sharp increase in the light scattering is caused by the mass nucleation of the new phase. Hysteresis of the peaks of light scattering was indeed found in the experiments. (For more on "critical opalescence" see Sec. 3.5).

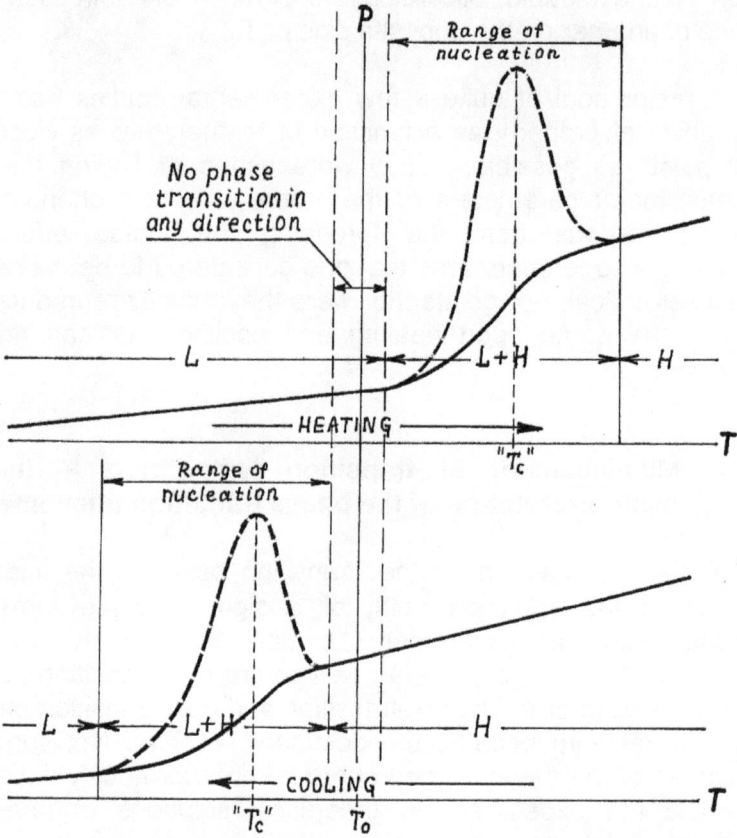

Fig. 2.29 Any "anomalies" of a physical property P, be they sigmoid curves or λ-peaks, reside in the ranges of transition (actually, ranges of nucleation). The "critical point T_c" at the λ-peak top (the common choice) will be a subject of hysteresis, for there are *two* non-overlapped transition ranges, one above T_0 - for heating, and the other below T_0 - for cooling.

2.6.5 More on hysteresis of solid-state phase transitions

Many years ago Landau [34] stated that temperature hysteresis in second-order phase transitions must be zero. This feature followed from the idea itself of a second-order phase transition and has never been disputed. Indeed, hysteresis attributed to a second-order phase transition is nonsense. It follows that if temperature hysteresis (TH) is found (*e.g.*, as is the case in Ni, Sec. 2.6.8) the phase transition is of first-order rather than second-order. No excuses based on the argument that TH is "very small" are acceptable. Our human sensation of what is "small" in the evaluation of a physical value is subjective and depends on the current sensitivity of measuring instruments. Zero TH in second-order phase transitions is a matter of principle. Any *finite* TH, however small, identifies the first-order phase transition, and, consequently, the nucleation and growth process. The above statement on hysteresis by Landau was rigorous: no such words as "usually," "often," "possibly," "occasionally," "may," and alike were added (which nowadays are used too frequently, making clear identification of phenomena and notions impossible). Landau did not anticipate that he thus provided a powerful tool that eventually would help to prove the nonexistence of second-order phase transitions.

TH is an inalienable feature of solid-solid phase transitions. That is why it is drawn in our considerations prior to becoming a special topic itself. As shown above, TH is caused by nucleation lags relative to T_0. The same phenomenon is responsible for overcooling upon melt crystallization. However, nucleation lags in the latter case are observed in one direction only (owing to specifics discussed in [36]), and this prevented them from being termed hysteresis. The difference between overcooling of a melt and TH is that TH is nucleation lags not only in cooling but also in heating.

To illustrate how much TH deserves a serious consideration, it suffices to say that TH is an underlying factor in the formation of all so called phase transition anomalies, or singularities (Chapter 3) and it sheds light on the nature of ferromagnetic / ferroelectric hysteresis (Chapter 4). There are many questions waiting to be answered:

- Is TH a feature of all solid-state phase transitions, or only some of them?
- If found in a particular case, can it be avoided under different conditions?

- Is the magnitude of TH a physical constant for a given phase transition?
- Are nucleation lags the only cause of TH?
- Why, and how, are the notorious loops of TH formed, and what are the factors responsible for their width and shape?
- How is the TH of λ-anomalies explained, and what is the interrelation, if any, between the two?
- What is the relation, if any, between TH and ferromagnetic (as well as ferroelectric) hysteresis?

All these and other related questions will now be answered. The solution of the TH problem was proposed by us in 1979 [105] and was not challenged since. A common perception of TH can be best illustrated with the following casually selected quotations (italicized by us). "The high-temperature phase *often may be* undercooled, and *occasionally* the low-temperature phase *may* be superheated" [46]; (First-order transitions in solids) "occur at *precise* transition temperatures; or, rather, the free energies for the two forms become equal at a definite transition temperature, although the transition itself *may* or *may not* occur" [44]; "A transition from one phase to the other occurs *always at a strictly definite* temperature" [106] "Superheatings and undercoolings are *possible* at first-order phase transitions" [31]; "Transition points...are *often definite* temperatures...It is *quite possible* to superheat a polymorph without transformation taking place" [97].

What can one learn from these wordings? That, as a rule, solid-state phase transitions occur at strictly definite temperatures; that occasionally, for unspecified reasons, overcooling or overheating occurs; that, probably, there are circumstances when only undercooling occurs, while overheating does not (or even *vice versa*?); that whatever the reason is, these phenomena distort the correct way phase transitions have to proceed. No one of the quoted authors identified the phenomenon as TH.

Nothing of the above is correct except that a span of the overcooling or overheating (*i.e.*, TH) is not always considered to be specific to a given phase transition. Two early attempts we know about that account for hysteresis assume the opposite, that the width of hysteresis is characteristic of particular phase transitions. One of the attempts [107,108] is not in accord with thermodynamics. The second [109] (also [54], pp.37-39) correctly relates hysteresis to nucleation. Unfortunately, the authors were bound by the ideas of the classical nucleation theory that, as shown in Sec. 2.5, is not valid. According to that nucleation theory, the emerging nucleus will be in a state of compression or tension due to the difference Δv between the specific volumes of the phases; the hysteresis $\pm \Delta T$ was expected to be wider when Δv is greater.

That the cause of TH was not well understood is seen from the way the TH problem was summarized by Rao and Rao [42]. "Such a situation could arise through two possible causes. One is the formation of a hybrid single crystal ... [the above-mentioned first hypothesis by Ubbelohde was described] ... Hysteresis could also occur for kinetic reasons [the second mentioned hypothesis was described]...Pressure hysteresis appears to be less understood than the thermal hysteresis."

As a result of our own investigation, we cannot now find anything unexplained in the TH phenomenon. The TH is exclusively and totally determined by the specifics of nucleation in solid state presented in Sec. 2.5. The width of TH is not a fixed value. The overheating $\Delta T = T_n - T_o$ in PDB single crystals of the same chemical grade ranged from about 1.9 to 22.4 °C, the latter number being limited only by the melting point 53.2 °C. The TH depended only on the presence or absence of suitable lattice defects. It was irreproducible even in the same crystal (Fig. 2.24). Taking into account the negative arm of the TH (22.4 °C was only its positive arm), the total TH in PDB exceeded 50 °C. Such a wide TH was observed in the transition noted for its small difference in the specific volumes of the phases. The correlation between Δv and TH was nonexistent.

It should be emphasized that optical microscopic observations of nucleation lags ΔT_n in single crystals, described in Sec. 2.5, are the most direct way to establish the cause of TH and investigate its features. Therefore, the source of TH is not a subject of speculations and hypotheses any more. There is no indication of any additional source of TH. The nucleation-growth mechanism and TH are inseparable. *All solid-state phase transitions exhibit TH.*

As for the nature of a pressure hysteresis, common sense prompts that it cannot in principle differ from TH. Pressure hysteresis is as inevitable as TH. The molecular mechanism of a phase transition cannot depend on which controlling variable, temperature T or pressure p, was used to change the energy balance between the phases. As soon as the alternative phase became energetically preferable, the only additional condition for the phase transition to occur is the availability of a nucleus. Phase transitions induced by change of pressure also occur by nucleation-growth. In practice, application of pressure is never quite hydrostatic and, therefore, more or less harmful to the crystal lattice. This leads to two mutually opposite effects. Increased crystal imperfection tends to narrow the hysteresis (but not to eliminate it) owing to the increased availability of the nucleation sites. On the other hand, the disturbance and dismemberment of crystal continuum tends to widen it by stopping crystal growth at the newly emerged boundaries. The resultant hysteresis depends on the prevailing effect.

2.6.6 Resume and final notes on temperature and hysteresis of solid-state phase transitions

Now we have a self-consistent picture accounting for all phenomena associated with measurements of solid-state phase transition temperatures. Solid-state phase transitions are no more or less than nucleation and crystal growth in a solid medium. The only sound temperature characteristic of a transition is T_0 at which the free energies F_L and F_H of the L and H phases are equal. Because neither phase has an energetic advantage at T_0, the transition cannot occur at T_0. At T_0 the system is in the state of *absolute stability,* which is the opposite of the theoretical notion of a "critical point T_c." After T_0 is passed, the phase transition still will not start in a certain vicinity of T_0 until the difference between the free energies ΔF exceeds ΔF_{min} required to overcome a certain nucleation barrier. Some overheating in an L $\rightarrow$ H (overcooling in H $\rightarrow$ L) phase transition is a necessary condition for the nucleation to occur. These nucleation lags are the only cause of the *temperature hysteresis* of phase transitions. The *range of stability* around T_0 where nucleation does not occur (Fig. 2.30a) can be as large as many degrees, or as small as a minute fraction of a degree, but it is always a finite value. Nucleation occurs outside this range at different and unpredictable temperatures. These temperatures T_n are pre-coded in suitable defects serving as dormant nucleation sites. It is T_n, and not T_0, that is actually found in a small single crystal. The T_n in small single crystals of the same material do not exactly coincide. As a result, nucleation in powder and polycrystalline samples spreads over two separate non-overlapping *ranges of transition* AB (in L $\rightarrow$ H) and CD (in H $\rightarrow$ L). The following points will complete the picture.

◆ **Range of transition.** The width of a range of transition is not a fixed value, because it is a characteristic of crystal imperfection of the particular sample, rather than an inherent property of the crystal substance. For instance, the range for a single crystal and the fine crystal powder made from this crystal will be different. The latter range would probably be much wider.

◆ **Sigmoid curves.** A sample in the range of transition is two-phase. As temperature is changing, the quantity of the new phase gradually increases at the expense of the initial phase. The AE and GD curves in Fig. 2.30a represent mass fraction of the H phase m_H in the sample. A "sigmoid" shape of the plots is indicative of two factors acting in opposite directions as the temperature changes. They are:

    (1) Increase in the number of suitable nucleation sites.

    (2) Decrease in the amount of the original phase.

The former factor dominates in the initial, and the latter in the final stage of the phase transition.

◆ **Hysteresis loop.** Finding one sigmoid plot (*e.g.*, AE in Fig. 2.30a) necessarily means that the second one (GD) would be found in the reverse run. Together they form a *hysteresis loop*. No temperature overlap of the two sigmoid plots is possible. An experimental hysteresis loop is shown in Fig. 2.30b.

2.6.7   Example No. 1: The temperature of phase transition in *p*-diiodobenzene (PDI)

Soltzberg *et al.* [111] (SBPAC) have reexamined the PDI phase transition, which had previously been described as one exemplifying a second-order, or "nearly second-order," or "displacive" phase transition and was the subject of a search for a soft lattice mode. The previously reported temperature of this phase transition was between 50 °C and 53.1 °C.

SBPAC used a highly purified substance and vapor-grown single crystals. Emphasis was placed on optical microscopic study involving measurements of optical path difference and visual observations of the process. They found the results of their study to be surprising in light of the previous descriptions of this phase transition. A resume of their results follows.

The phase transition was found to proceed by propagation of interfaces that sharply divided the crystal into two coexisting phases. Wide hysteresis, up to 30 °C, between the forward and reverse transition temperatures $T_{tr}(+)$ and $T_{tr}(-)$ was found. The temperatures substantially varied depending on the history of the crystal: between 45 °C and 57 °C for $T_{tr}(+)$, and between 24° and 31°C for $T_{tr}(-)$. Actually observed T_{tr} was concluded to be inadequate as a measure of the phase equilibrium temperature. Stopping the interface motion and bracketing indicated T_o to be in the range 35 °C to 40 °C, that is, about 14° C lower than the previously reported transition temperature that was believed to be the equilibrium. The large error in the previously reported transition point was evidently caused by "one way" measurements, namely, upon heating.

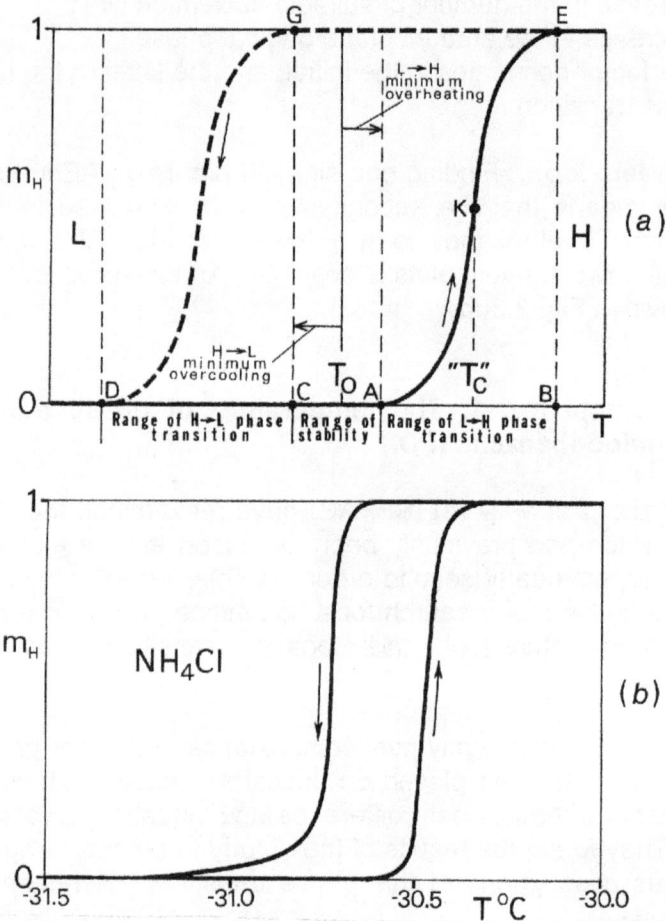

Fig. 2.30 Hysteresis loop of a solid-state phase transition (m_H is mass fraction of H phase).
(a) (Schematic) "Sigmoid" curves AE and GD, representing the m_H in the heterophase (L+H) temperature range of transition. Together they form a hysteresis loop DAEGD. The range of stability around T_o is indicated. 'K' is the inflection point erroneously assigned to be the "critical point" (or "Curie point") in phase transition upon heating.
(b) (Experimental) The hysteresis loop $m_H(T)$ in NH_4Cl (Dinichert [110]).

The SBPAC's work corrected two common mistakes: acceptance of an actually observed transition temperature as the equilibrium temperature and unawareness of the hysteresis. The work also exemplified a reexamination of a "second-order" ("soft-mode," "displacive," etc.) phase transition, resulting in its re-qualification to the *nucleation-and-growth* type.

2.6.8   Example No.2:   the "Curie point" in Ni

The theory of ferromagnetism considers a paramagnetic-ferromagnetic transition temperature to be a "Curie point" (T_c). This is so essential that the theory in its current form would not exist without it (Sec. 4.4). As compared to other ferromagnetic materials, nickel is regarded as the exemplary subject for investigation of the basics of ferromagnetism. While ferromagnetism in general is the topic of Chapter 4, the problem of the Curie point, as it appears in the case of Ni, is treated here in the context of a phase transition temperature. The Curie point had to be a strictly definite characteristic temperature at which spontaneous magnetization abruptly appears or disappears. The magnetization curves shown in textbooks precipitously approach the temperature axis (Fig. 2.31a), thus being seemingly in accord with such a representation. A claim that the experimental curves follow the theoretical course assigned by the Curie-Weiss law completes the apparent harmony.

The harmony vanishes as soon as the real state of affairs is examined closer. The disconcerting facts, however, are scattered over specialized monographs and original articles. When collected, they reveal that curves of the Fig. 2.31a type are misrepresented; the conformance of the Curie-Weiss law with the experiment is poor; no one knows how to determine the Curie point exactly; the variance between the reported locations of the Curie point in Ni greatly exceeds reasonable limits; the Curie point upon heating is not the same as upon cooling; and more. It should be stressed that all this relates to the classic ferromagnet Ni, considered one of the simplest and normal regarding its magnetic behavior as a function of temperature. These discrepancies are listed below point by point.

(1)   Rather than a fixed point, a Curie point turns out to be a very ambiguous quantity. The authorities in the field put it as follows: "It even remains unsettled what we should take to be the Curie temperature, and how to determine it" [64]; "The change from the

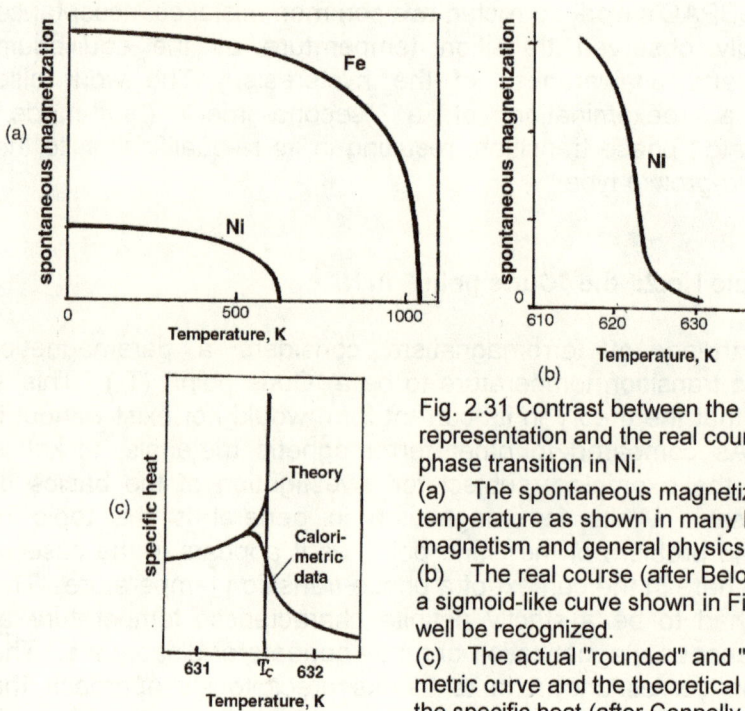

Fig. 2.31 Contrast between the "idealized" representation and the real course of the phase transition in Ni.
(a) The spontaneous magnetization *vs.* temperature as shown in many books on magnetism and general physics.
(b) The real course (after Belov [64]) in which a sigmoid-like curve shown in Fig. 2.28c may well be recognized.
(c) The actual "rounded" and "tailed" calorimetric curve and the theoretical sharp peak of the specific heat (after Connelly *et al.* [104]).

ferromagnetic to the paramagnetic state is not perfectly sharp, and it is difficult to find and determine the Curie point exactly" [65]. Such statements are concerned not only with Ni, but with all ferromagnetics in general.

(2)   All possible experimental errors and variation due to impurities can hardly account for the scatter as wide as 35 °K (from 619.7 to 655.2 °K) in the position of the Curie point reported by different authors for Ni (Table 2.4). Such a variance in itself is an important experimental fact that has not yet been explained. Nor it could be until the *range of nucleation* and *temperature hysteresis*, as presented in this book, are taken into consideration.

Table 2.4   The Curie point in NI reported by different authors.

Weiss & Forrer [112]	Bloch [in 113]	Bozorth [65]	Belov [64]	Conelly et al. [104]	Petrovic &Napijalo [114]	Vygovskii & Ergin [115]	Kittel [116]
631.2	655.2	631.2 – 636.2	619.7 – 621.4	631.58	Heating 633.2; cooling 631.7	625.85	627.2

(3)   In contrast to the "idealized" plot shown in Fig. 2.31a, the real magnetization does not drop precipitously towards zero. Rather, it has a "tail" of about 7 °K to 10 °K long (Fig. 2.31b). Belov has discussed the issue at length [64], stating that all real ferromagnetic phase transitions are "smeared out" and have "tails" of spontaneous magnetization. Suggesting a particular way of extracting the Curie point, which he believed to be better than others, he reasoned: "At this temperature an overwhelmingly large part of the volume of the specimen is in the paramagnetic state, and there remain in the ferromagnetic state only very small regions, the 'remnants' of the spontaneous magnetization." The fact described in this statement is of importance, for it means coexistence of the ferromagnetic and paramagnetic phases in a temperature range, a unique attribute of the nucleation and growth phase transition. Belov, however, was far from drawing this conclusion, because it would undermine his basic assumption that "the transitions between ferromagnetism and paramagnetism are phase transitions of the second kind" (*i.e.*, second-order). He attributed this phenomenon to the inhomogeneous state of the material, which results in "not a single Curie point, but a whole set of somewhat different Curie points." (The theory would rather retract its basic premise than recognize its own failure.)

A mere substitution of "different nucleation temperatures T_n" for "a whole set of somewhat different Curie points" accounts for all the seemingly contradictory facts in a simple and consistent manner. Once all evidence is considered both here and in Chapter 4, the fact of a fundamental significance becomes obvious that a ferromagnetic phase transition is nucleation and growth, and its temperature does not have the meaning assumed by the theory of ferromagnetism. It is indicative that quite a number of ferromagnetic phase transitions have already been classified as of the first-order (see Sec. 4.2.2) - even though in a formal manner, without realizing that the "first order" is equivalent to "nucleation and growth."

(4)   Evidence of a range of *phase coexistence* in the Ni phase transition has been reported (neutron depolarization study by Drabkin *et al.* [117]). Bartis [118], in discordance with the standard treatment of this transition, suggested that "the transition may look continuous for the crystal as a whole, even though it is discontinuous at some point for every crystallite." This was basically the correct idea, although neither the cause of the phenomenon was specified (different nucleation temperatures encoded in the crystal defects, refer to Sec. 2.5.2), nor it

was made clear that "discontinuous" means *nucleation and growth*. The *range of nucleation* exhibits itself as a *range of transition* where the two phases coexist. The concept of a Curie point is inapplicable to such transition (Sec. 4.4).

(5)   The specific heat measurements reveal a "rounded" peak in the region of the Curie point, rather than a sharp one predicted by the molecular-field theory (Fig. 2.31c). The phenomenon has been studied time and again, the work by Connelly *at al.* [104] (CLM) being especially meticulous. It was aimed at giving "a more detailed picture of the form of the singularity in $C_p(T)$" in the vicinity of T_c. Small Ni single crystals of very high purity were used. The authors rightfully claimed their calorimetric technique to be "unusually" precise. The temperature resolution, for instance, was as high as 0.01 °K. The measurements, however, were carried out upon heating only, which, as will be seen, significantly degraded the advantage of the high precision.

The work by CLM proved to be devoted in the most part to the phenomenon of *rounding* (Sec. 2.6.3). Some of their comments are instructive: "This rounded appearance of the maximum...is one of the most consistently reproducible features... Rounding must be treated as a real effect...Unfortunately, the criteria for defining T_c from rounded experimental data are somewhat subjective and imprecise...This ambiguous situation... Smeared out discontinuity instead of the sharp step...." CLM accepted the view of other authors that the rounding is the "result of a continuous distribution of Curie points [which] may be regarded as arising from small differences in the critical temperatures of microscopic regions within the specimen." With a special theoretical treatment to circumvent the predicament, CLM found, as they believed, the exact number for T_c equal to 631.58 °K. Unfortunately, this number was of no value for several reasons. *First*, even according to CLM, it was not a unique Curie point for Ni, but, rather, an average taken over the "range of Curie points." *Second*, as seen from the data by CLM, the position of T_c changed upon annealing, *i.e.*, was sensitive to even a small content of crystal defects. (Here the reader may recall that crystal defects are a necessary condition for nucleation, Sec. 2.5.2). *Third*, the T_c found by CLM exceeded that of the preceding (also very accurate) work [119] by as much as 2 °K with, it seemed, no apparent reason. CLM conceded the "discrepancy between the temperature scales" used in the two investigations to be "mildly surprising since both used the same model of commercially available calibrated platinum resistance thermometer for standardization." Indeed, the discrepancy between the temperature scales, exceeding hundreds times the precision of the calibrated thermometers, was improbable. Simply the overheating $\Delta T_n = T_n - T_0$ in the work by CLM was by 2 °K higher owing to a higher crystal quality of their samples. This immediately suggests that T_0 in Ni phase transition is at least 2 °K lower than the "Curie point" specified by CLM; probably, it is lower than 629 °K. *Fourth*, the "one way" technique upon cooling would produce "T_c" lower than upon heating (see below).

Other observations in the CLM work were also inconsistent with the common interpretation of the ferromagnetic transition in Ni. Specifically, (a) the impure crystals had a sharper transition, and (b) contribution of chemical impurity to rounding was found less important than physical imperfection. CLM were unable to explain either these or other effects, such as why the maximum in the preceding study by Maher and McCormick [119] was not only situated at a lower temperature, but was also sharper;* nor could they account for the source of the rounding. They concluded: "Farther work is required to identify the physical variables responsible for rounding of the specific-heat maximum." A farther work is not required any more. All the above-mentioned results, observations and seeming discrepancies are in a complete accord with the *range of nucleation*.

(6)   Hysteresis of Curie points in ferromagnetics is a phenomenon rarely mentioned, but well-known to experts. No one, it seems, was willing to consider the problem in any detail. Kirenskii *et al.* [120] have written a monograph on the phenomenon of temperature magnetic hysteresis. They designated hysteresis of Curie points as one of two types of the temperature magnetic hysteresis, in which "Curie point detected upon heating turns out to be somewhat higher than upon cooling." This statement was referred, in particular, to cobalt and some ferromagnetic alloys. Structural transformations in the heterogeneous ferromagnetic materials were mentioned as the cause of the hysteresis. This brief note was given only to devote the monograph to the alternative (not relevant) type of hysteresis. The hysteresis of Curie points is a disturbing and unexplained reality in the theories of ferromagnetism and phase transitions.

Hysteresis of the Curie point in Ni has long been presumed non-existent. The phenomenon was reported in 1978 by Petrovic and Napijalo [114]: the Curie points determined from the heating curves were higher than from the cooling curves (see Table 2.4). They

---

* The same phenomena were observed in the NH_4Cl phase transition by Schwartz [103] who noted that they were opposite to what had to be expected. See Sec. 3.3.9 where the phenomena are explained in terms of nucleation and growth.

proposed the *mean* value to be designated as T_c. That would imply, however, that the "critical phenomenon" - ferromagnetic transition - does not occur at the "critical point," but only after some overheating or overcooling. This would not help T_c to regain its physical significance to the theory of ferromagnetism. The mean value is, evidently, T_o rather than T_c.*

The "Curie point" concept cannot withstand a close examination in the case of Ni ferromagnetic transition, nor should one expect it to be valid in other cases. That the Curie point was a "weak point" of the orthodox theory was apparently realized by some authors. Otherwise Vonsovskii [121] would not need to defend the "point" nature of the Curie point (pp. 466-467). He conceded that it is difficult to establish some definite temperature at which the ferromagnet -- paramagnet transition takes place and that it is only possible to specify some *transition temperature range*. He attributed this, however, to specific problems of magnetic measurements. He found a more direct proof of the "point" nature of the magnetic transition in the sharp peaks of nonmagnetic properties. Considering the real facts ("rounding," "tails," "distribution of Curie points"), this assumption was incorrect, not to mention that the position of these rounded and tailed peaks depended on whether they are recorded upon heating or cooling. Another authority in the field, Belov [64], warned earlier: "There is no basis for taking the position of the maximum of the anomaly of a 'nonmagnetic' property as the true Curie point..."

Thus, we have demonstrated that in the classic case of the Ni ferromagnetic transition the concept of a critical (Curie) point is a total failure. We actually deal with the nucleation-growth process and its temperature characteristic T_o.

2.7 Order-disorder phase transitions as crystal growth

2.7.1 The deceptive "common sense"

Orientation-disordered crystals (ODC) are a specific solid state in which the constituent particles are engaged in a thermal hindered rotation, while retaining a 3-D translation order. A concise description of the phase transitions from a crystal state to ODC (C $\rightarrow$ ODC) was given,

---

* The T_c = 632.3 °K from the data by Petrovic and Napijalo is higher than T_o < 629 °K suggested above, which may have resulted from an insufficiently adjusted absolute temperature scale in their work.

e.g., by Rao and Rao [42], and in more detail by Parsonage and Staveley [54]. However, even the latter source will leave a reader in the dark about the actual physical process of the phase rearrangement. A common attitude towards all solid-solid phase transitions, and the order-disorder in particular, is that they can be generally understood by comparing the two crystal phases. This is implied in almost every analysis, consideration, or discussion, being assumed as a matter of common sense. However, this apparent "common sense" is deceptive, for it involves an invalid idea that a phase transition is only a partial rearrangement of the initial structure, so that the resultant phase is a mutated initial phase. As applied to the order-disorder transitions, this subconscious idea gave rise to the conviction that the function of such a transition is to execute a "disordering" of the initial phase, leaving it otherwise basically unchanged. (*cf.* "...Some weaker bonds are broken while the stronger bonds remain" [44]). However, it is shown in Sec. 1.4 that common sense plus the principle of minimum energy require the two phases to be quite independent of each other. The nucleation-growth *contact* mechanism executes the total rearrangement.

When one discusses a C → ODC transition, switching from the comparison of terminal phases to its mechanism is so subtle that the logical slip is left unnoticed. Possible similarities between the phases have nothing to do with the mechanism of the phase transition (Sec. 1.4). For instance, Parsonage and Staveley [54, p.311] discuss the III-II transition in ammonium chloride (which, they point out, is one of the most studied of all order-disorder phase transitions in a crystal). They start from a detailed consideration of the structure and properties of the "ordered" form III and "disordered" form II, noting their similarities and dissimilarities. The conclusion is that all ions in III have a parallel alignment, while in II the ions are distributed at random between two orientations. This difference between two terminal states is called "interpretation of transition." The subsequent discussion confirms that, indeed, the case in point is a *mode* of the transition: "Basically, therefore, the nature of the transition is very simple... Nevertheless, a number of rather subtle features prove to be involved, which are still being actively investigated... Careful studies ... have shown that while the transition begins gradually, it is undoubtedly completed isothermally, so that it is partly gradual and partly first-order."

So, the comparative approach sheds no light on the actual behavior of the system during phase transition. Neither a model for the phase rearrangement has been proposed, nor was an explanation suggested to such a "hermaphrodite" nature of the phase transition.

Later we will clarify the matter. At this point it suffices to say that there is nothing "gradual" in this phase transition other than gradual change in the mass ratio of the two distinct phases over a temperature range of their coexistence (Secs. 3.3.6 to 3.3.11). The transition proceeds according to the general mechanisms of nucleation and growth described in the previous sections. These mechanisms are valid irrespective of the particular crystal structures and physical properties of the phases, including order / disorder status.

The contemporary views on order-disorder phase transitions mainly rest on two observations: (1) one of the phases exhibits some kind of orientational or positional disorder, and (2) a "heat capacity" λ-peak is usually found. Clearly, this is insufficient for a molecular mechanism of these phase transitions to be conclusively inferred. Setting aside any conjectures based on phase comparison, little specificity can actually be found in these phase transitions. They cannot be tied exclusively to second-order phase transitions as one initially believed. Rao and Rao summarized [42, p.122]: "An order-disorder transition can take place discontinuously as in a first-order transition or continuously as in higher-order cooperative transitions and behave like a λ-transition." Even the λ-peaks are not their exclusive characteristic since these peaks are found in "order-order" phase transitions as well. The capricious behavior of different order-disorder phase transitions mentioned by Rao and Rao will prove to be harmonious: λ-peaks are shown in Chapter 3 to be a characteristic of first-order (and, consequently, nucleation and growth) phase transitions.

In pertinent literature, when it comes to interpreting *rotational* order-disorder transitions, all attention is paid to the alleged "cooperative" process. Their heat capacity λ-peaks are considered a consequence of the cooperative mechanism of these phase transitions [42,122,123]. It is explained that at some temperature upon heating the molecules start rotating in the crystal; the rotation of any molecules makes it easier for its neighbors to rotate, so the additional amount of heat absorbed per degree that sets the molecules in motion increases rapidly until, at the peak's maximum, all the molecules are more or less free to rotate. This typical interpretation, however, raises more questions than gives answers. In such a case, why do the λ-peaks have a gradual descending part as well? Does it mean that molecular rotation increases when the phase transition is approached by cooling? And how are the heat capacity λ-peaks in phase transitions to be interpreted where no rotation is involved? The answer to these and other disconcerting questions is rooted in the fact that the λ-peaks do not represent a heat capacity. They are a side effect of the nucleation-growth mechanism (see Chapter 3).

2.7.2 Direct microscopic observations of C -- ODC transitions

A variety of experimental methods were earlier employed in studying order-disorder transitions. Rao and Rao listed them [42]: thermodynamic measurements such as those of heat capacities; scattering techniques; X-ray diffraction, electron diffraction and neutron diffraction; infrared, Raman, neutron and NMR spectroscopy; measurements of dielectric constants, magnetic susceptibilities, resistivity, thermoelectric power, Hall coefficient, and more. All these technique are indirect, so interpretation of the results inevitably involves unproven assumptions. To make up for this shortcoming the below described investigation based on direct microscopic observations was carried out.

The molecules in an ODC are lacking a long-range orientation order due to their rotation in the crystal lattice. Frenkel [123] named the C $\rightarrow$ ODC phase transitions "orientation melting," thus prejudging their mechanism as an onset of molecular rotation in the original crystal lattice. This view became prevalent. Buerger [37,43] classified them "Transformations of Disorder: Rotational; rapid, barrierless, second-order." At that, disregarded were (a) discontinuous change in the specific volume, (b) temperature hysteresis reported in some cases, and (c) statement by Landau and Lifshitz [36] that these phase transitions are not of second order, but first.

Yet, some rotational phase transitions seemed to exhibit no overheating, while the fact of overcooling was beyond doubt. Two such instances, CBr_4 and C_2Cl_6 (I-II), were chosen for our examination. The only other reason to select them was that their temperatures of phase transitions, $T_0(CBr_4) = 46.9\ °C$ and $T_0(C_2Cl_6) = 71.7\ °C$, were convenient for studies with a microscope.

What should one expect to see in a microscope if a phase transition would take place exactly at T_0? This would mean a transition mechanism very different from nucleation-growth. In the ideal case of quite uniform heating the whole crystal will change into the resultant phase simultaneously over the bulk, indicating a *cooperative* process. In reality such uniform heating is hardly possible, therefore a very unstable shapeless pseudo-interface representing an isotherm separating the crystal parts with $T>T_0$ and $T<T_0$ would be seen. Conversely, $\Delta T \neq 0$ is a necessary condition for a crystal growth. A mere observation of the growth exhibiting natural faces will prove the existence of some hysteresis $\pm\Delta T \neq 0$.

No measurements of hysteresis in CBr_4 and C_2Cl_6 phase transitions $C \rightarrow ODC$ were carried out previously. The work by Marshall *et al.* [124] left the impression that $\Delta T=0$ in the CBr_4 case. Therefore, the first task was to examine this point. Since higher ΔT_n was expected in smaller and more perfect crystals, such crystals were grown in quantity. The crystals, one by one, were placed on a microscopic hot stage and heated as slowly as $2°$ C per hour. The relative precision of the temperature measurements was $\pm 0.05°C$. Temperature readings were taken at the moment when first sign of the phase transition was noticed. A typical result for the crystals listed in Table 2.5, column 1, is given in column 2. For every crystal the temperature of its $ODC \rightarrow C$ transition was also recorded (column 3). While the numbers in column 2 relate to the crystals as they were grown from a solution, the data in column 3 relate to the crystals already damaged by the $C \rightarrow ODC$ transition. In order to somewhat equalize the initial conditions for the transitions in both directions, the crystals listed in column 4 were grown from the melt; their transition temperatures upon cooling are shown in column 5.

Table 2.5 shows that (a) there is hysteresis ΔT_n in both $C \rightarrow ODC$ and $ODC \rightarrow C$ directions; (b) the overheatings are rather small, which was the reason why they were previously overlooked; (c) $\Delta T_n(ODC \rightarrow C) > \Delta T_n(C \rightarrow ODC)$; (d) the magnitude of hysteresis, being dependent on crystal quality, is not a fixed characteristic of the phase transition. The visual observation of the phase transitions made it obvious that the hysteresis was caused by nucleation lags. Also, an instructive result comes from the comparison of the data in columns 3 and 5. The crystals in the latter case were of a better quality and therefore exhibited hysteresis 14 times as large (16.4 °C to 1.19 °C).

Existence of overheating ΔT_n and its dependence on crystal quality have also been confirmed in another way. Forty similar single crystals of CBr_4 were divided into two equal groups, A and B, with the only difference that A were, by visual estimation, more perfect than B. From them 20 pairs of 1A+1B were made. Every pair was slowly heated together on a microscope hot stage and viewed in the microscope. In 19 pairs of 20 it was found that $\Delta T_n(A) > \Delta T_n(B)$.

The next task was to observe the C → ODC phase transitions in single crystals with an optical microscope in the same manner as was done in the C → C transitions. The first fact to note was that the rotational-disordering phase transitions also start locally and proceed by

Table 2.5 Hysteresis ΔT_n in CBr_4 transitions C ↔ ODC; $T_o = 46.9°C$

1	2		3	4	5
Crystal	T_n, °C			Crystal	T_n, °C
(*)	C → ODC	(C→) ODC → C		(**)	ODC → C
No.1	47.1	45.9		No.10	29.0
No.2	47.15	46.4		No.11	31.1
No.3	47.45			No.12	29.9
No.4	47.1	46.0		No.13	29.8
No.5	47.0	44.6		No.14	29.9
No.6	47.1	46.1		No.15	31.9
No.7	47.0	45.2		No.16	30.9
No.8	47.0	45.4		No.17	29.9
No.9	46.9	46.1		No.18	31.7
				No.19	30.7
Average	$T_n = 47.09$	$T_n = 45.71$		$T_n = 30.5$	
Average	$\Delta T_n = 0.19$	$\Delta T_n = -1.19$		$\Delta T_n = -16.4$	

(*) Crystals 1 to 9 were grown from a solution at room temperature. They were quite transparent and had good natural faces.
(**) Crystals 10 to 19 were grown from the melt. They consisted of several blocks of ODC and did not have good natural faces.

propagation of interfaces. This suggested that, once again, we deal with a crystal growth. In order to make it clear, an interface had to be made to move extremely slowly and uniformly. With this in mind, an interface had to be stopped and sustained at a very stable temperature level under a fraction of the 0.2° overheating. With the available techniques this condition could be met only approximately. Nevertheless, as the photographs in Fig. 2.32 show, the order-disorder transition *is* crystal growth of the rotational-disordered phase in the "ordered" crystal medium; it is *not* a cooperative phenomenon. The ODC could not be as perfect as the 'r' crystals in the C → C transitions shown in Figs. 2.2 to 2.10. It is known that rotational-disordered crystals do not grow from the melt so well as the usual crystals do; there is every reason to extend this observation to the growth of such crystals in a solid medium as well. The relatively large increase in the specific volume, typical for the transitions into ODC, is another adverse contributor to the crystal growth conditions.

While only the results for CBr$_4$ were described above, C$_2$Cl$_6$ has also been studied – with the similar outcome. In Fig. 2.33 the X-ray Laue patterns prior to and after a C $\rightarrow$ ODC $\rightarrow$ C cycle are reproduced for both cases. It is seen that single crystals turn into polycrystas, a fact attesting to both multiple nucleation and a lack of fixed orientation relationships in these phase transitions.

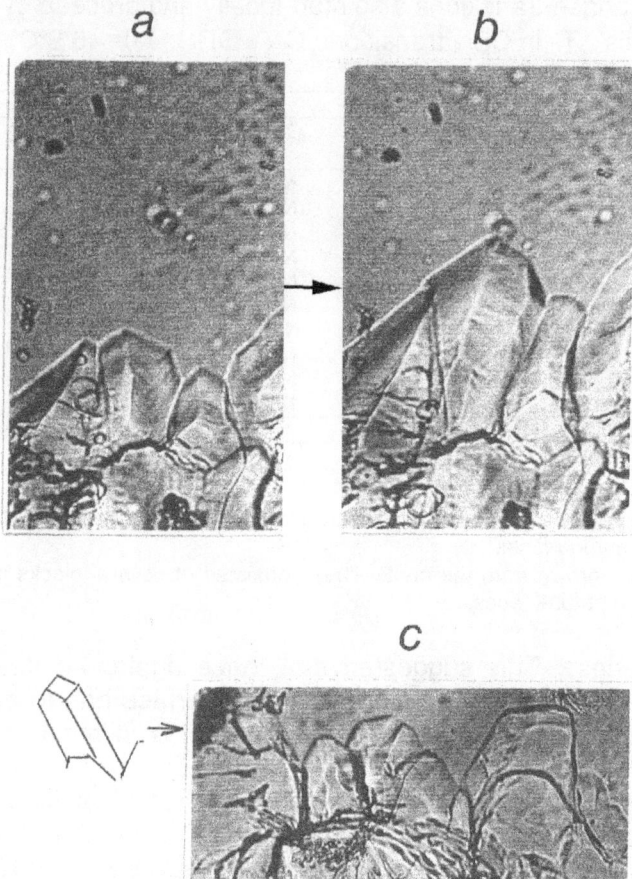

Fig. 2.32 Rotational order-disorder L $\rightarrow$ H phase transition in CBr$_4$.
(a,b) Growth of a conglomeration of single crystals (two successive stages). The growing ODC crystals are not well shaped, but the natural facing is evident. Note that the phase transition is not cooperative in the bulk. The rotational phase (below the interface) and the non-rotational phase (above the interface) merely coexist while all phase rearrangement occurs at the interfaces.
(c) Another conglomeration of growing ODC crystals. Note the ODC crystal reproduced in drawing. X 100.

Rotational order-disorder transitions are no different from the transitions between ordered phases: hysteresis caused by nucleation lags, its dependence on the crystal defects, phase coexistence in a temperature range, crystal growth displaying natural faces, the lack of fixed orientation relationships. All that exhibits a typical *contact* phase transition as was specified previously. The phase transition in itself is not "rotational," even though the resultant phase is "rotational." Any solid-state phase transition brings about changes in the physical properties, and the commencement of molecular rotation is one such change. Buerger classified these transitions as "rapid" (Sec. 1.5.2), but they are not. They can be "rapid" or "sluggish" depending on ΔT_{tr}, or they can be stopped at any intermediate stage by setting $\Delta T_{tr} = 0$. The phase transitions into a crystal state with rotating molecules again demonstrate the generality of the crystal growth principle.

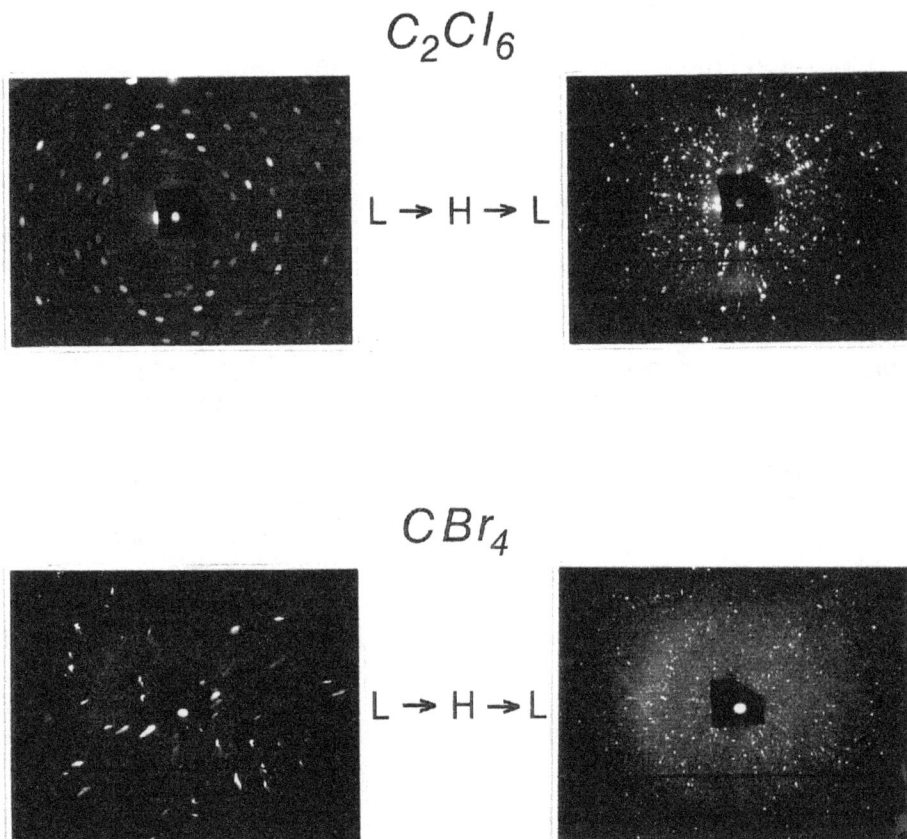

$$C_2Cl_6$$

$$L \to H \to L$$

$$CBr_4$$

$$L \to H \to L$$

Fig. 2.33   X-Ray Laue patterns taken from CBr_4 and C_2Cl_6 before and after a phase transition cycle C → ODC → C. The initial specimens were single crystals. Polycrystal patterns after the transitions are indicative of multiple nucleation and growth in random orientations.

2.7.3   Order-disorder phase transition in CH_4: no ambiguities

This phase transition has attracted a good deal of attention; ample experimental literature is available, but the results were ambiguous. According to Aston [122], "there is a transition, presumably of the second order (λ-point), at 20.4 °K." Parsonage and Staveley, however, called it "partly gradual" and noted that there was a discussion about the order of the transition and that the issue had not been resolved [54, p.572].

A review of the literature on this phase transition can be found, *e.g.*, in [54]. All the "disturbing" points can be best illustrated by the following quotes from just one detailed experimental work (Colwell, Gill and Morrison [125]). They merely represent the features of *nucleation-growth contact mechanism*: "The transitions in the methanes are often referred to a being of the 'lambda' type or of a nonisothermal type, but these are approximate descriptions." "There is some ambiguity in the order or type of transition specified by the experimental results." "No unambiguous assignment of order can be made [for the transitions in methanes] because of the apparent dependence of their detailed behavior upon thermal history." "Detailed shape of the transition depends upon thermal history." "The calorimetric studies of the transitions are complicated by the occurrence of hysteresis." "Since calorimetric measurements are usually made by adding energy, the results will normally refer to... ascending... curves." "The transition can proceed isothermally... over part of the transition region, but the factor controlling the width of the region cannot be defined." "The greater the strain, the narrower is the transition region." "Impurities...are without measurable effect on the transition." "A small volume change (<0.2%) has been measured." "The X-ray measurements also show the presence of extra lines (or reflections) in the transition region and their occurrence has not been explained." "The characteristics displayed by III-II transition are of a kind which might be expected of a crystallographic change." "The fundamental causes of the transition can still not be established." "In spite of the rather large amount of both experimental and theoretical work ...it is still not possible to draw definite conclusions about the mechanisms of the transitions in the solid methanes. This is also true, of course, of nearly all molecular crystals which display transitions."

It remains to add only few remarks. The admission by Colwell *et al.* in the last excerpt still holds true nowadays regarding not only molecular crystals, but all solid-state phase transitions as well, for the scientific community at large. The above-listed unexplained phenomena are in accord with the *contact* nucleation-growth concept. But the notions "nucleation" and "growth" are not present in the quoted article. Another classical example of an order-disorder phase transition is that in NH_4Cl. Later it will be analyzed and shown that it matches our unified picture of solid-state phase transitions as being nucleation and growth.

2.8 Epitaxial phase transitions

2.8.1 Oriented and non-oriented crystal growth in crystal medium

The phenomenon of epitaxy is generally understood as oriented crystal growth of one substance on a crystal substrate, such as a natural crystal face, of another substance. To be oriented by the crystal field of the substrate there must be sufficient closeness between the crystal lattice parameters of the substrate and deposit. Namely, at least two lattice translations in the deposit, as regard to their length and the angle they form, must not differ by more than ~ 10% from those on the substrate surface. The orientation can be exact or approximate. Other conditions being equal, the closer the crystal parameters, the stronger the orientation effect, that is, epitaxy. A substance can be deposited on a substrate from a liquid phase (liquid epitaxy) or vapor phase (vapor epitaxy). The present point of interest is a new kind of epitaxy, *solid epitaxy*, which we encountered in the course of experimental investigation of those phase transitions that exhibit a rigorous orientational relationship (OR) between the polymorphs.

For a long time strict OR between the phases was presumed (and therefore not verified) in almost every interpretation of crystal phase transitions. If OR appeared not to be strict, it was regarded as a deviation of the real system from the ideal. Even random OR did not attract due attention. Physical metallurgists took strict OR as a characteristic of martensitic transformations, theorists as a characteristic of second-order transitions, followers of the Buerger's classification - as an indication of a displacive transformation. Most experimental techniques, including all those using powder specimens, are not suited to dealing with OR. When OR was not known, a strict OR was implied as a matter of course.

If solid-state phase transitions are nucleation and growth, orientation of the 'r' phase is determined by the orientation of the nucleus. Then the cause of a rigorous OR, when it takes place, is *oriented nucleation*. The theory of nucleation in a uniform crystal medium regards the nucleation to be *homogeneous* and *coherent*, that is, strictly oriented. As shown in Sec. 2.2, it is not the case. Generally, any orientation is possible. It is not unusual to encounter a case of a quite strict OR. When and under which conditions does a strict OR occur? In the subsequent sections a detailed analysis of particular examples of such phase transitions will show that the conditions in question are those required for oriented nucleation (epitaxy).

The *epitaxial* phase transitions have a specific theoretical and practical importance. It is this kind that is most frequently mistaken for one or another "cooperative" transition: second-order, displacive, martensitic... Moreover, the epitaxial nucleation offers a straightforward explanation to the formation of ferromagnetic and ferroelectric domain structures, and through this to the origin and properties of ferromagnetics and ferroelectrics.

2.8.2   Epitaxial transition in hexamethyl benzene (HMB)

We shall consider the transition in HMB (see Table 2.1) near 110 °C. Previously it was reported [126] to occur "instantaneously" without changing the direction of light extinction when it was observed under crossed polarizers.

The crystal structures of L and H are compared in Table 2.6. To facilitate the comparison, the *a'b'* pseudocell, similar to the *ab* unit cell of L, was chosen to represent H. Both structures consist of molecular layers parallel to *(001)* with the flat benzene rings lying in the layers. Because the molecules have a circular shape, their close packing in a layer is quite, or almost, hexagonal. This accounts for the fact that the parameters *a* and *a'*, *b* and *b'* of the *(001)* layers are so similar. It can be seen from row 2 that the divergence of the specific volumes Δv is ~10%, but this is for the most part due to the difference in the temperatures at which the parameters were measured. The exact parameters of L just below the transition temperature are unknown. But we can roughly estimate them, because the specific volume changes in the phase transition by $\Delta v/v = 2.4\%$ [127]. The interlayer spacing d_{001} differs by 2.2% (row 3), but the divergence again relates to the phases at different temperatures. There is no doubt that d_{001} in L will be greater at 110 °C due to normal temperature expansion, probably approaching that in H. Suppose, $\Delta d_{001} = 0$ in the phase transition. Then the whole 2.4% change in the specific volume will be attributed to a change in the layer parameters. Now, using the data of row 5, it can be found that *a* and *b* are smaller than *a'* and *b'* by only 1%. This estimate is conservative: the difference between *a* and *a'*, *b* and *b'* can only be exaggerated. The molecular layers in the two polymorphs are almost identical: the main difference between the polymorphs is in the mode of layer stacking.

Table 2.6. Comparison of hexamethyl benzene polymorphs.

Row	L, stable below ~110 °C	H, stable above ~110 °C
1.	Triclinic System $a = 8.92$ Å $\alpha = 44°27'$ $b = 8.86$ $\beta = 116°43'$ $c = 5.30$ $\gamma = 119° 34'$ At room temperature T_r,	Rhombic System $a = 9.13$ Å $b = 16.12$ $Z = 4$ $c = 7.52$ At temperature 115°C
2.	Volume per molecule	
	252 Å^3 (at T_r)	277 Å^3 (at 115°C)
3.	Distance d between molecular layers (001)	
	368 Å (at T_r)	376 Å (at 115°C)
4.	Orientation of molecules (planes of the benzene ring)	
	Coincides with (001) within 1°	Coincides with (001)
5.	Area per molecule in the (001) layer	
	68.4 Å^2 (at T_r)	73.7 Å^2 (at 115°C)
6.	Arrangement of molecules in the (001) layer	
	(at T_r,)	(at 115°C)
	 $a = 8.92$ Å, $b = 8.85$ Å $\gamma = 119° 34'$ (at T_r)	 $a' = 9.25$ Å, $b' = 9.13$ Å $\gamma' = 119° 32'$ (at 115°C)
7.	Relative shift of (001) layers in stacking	

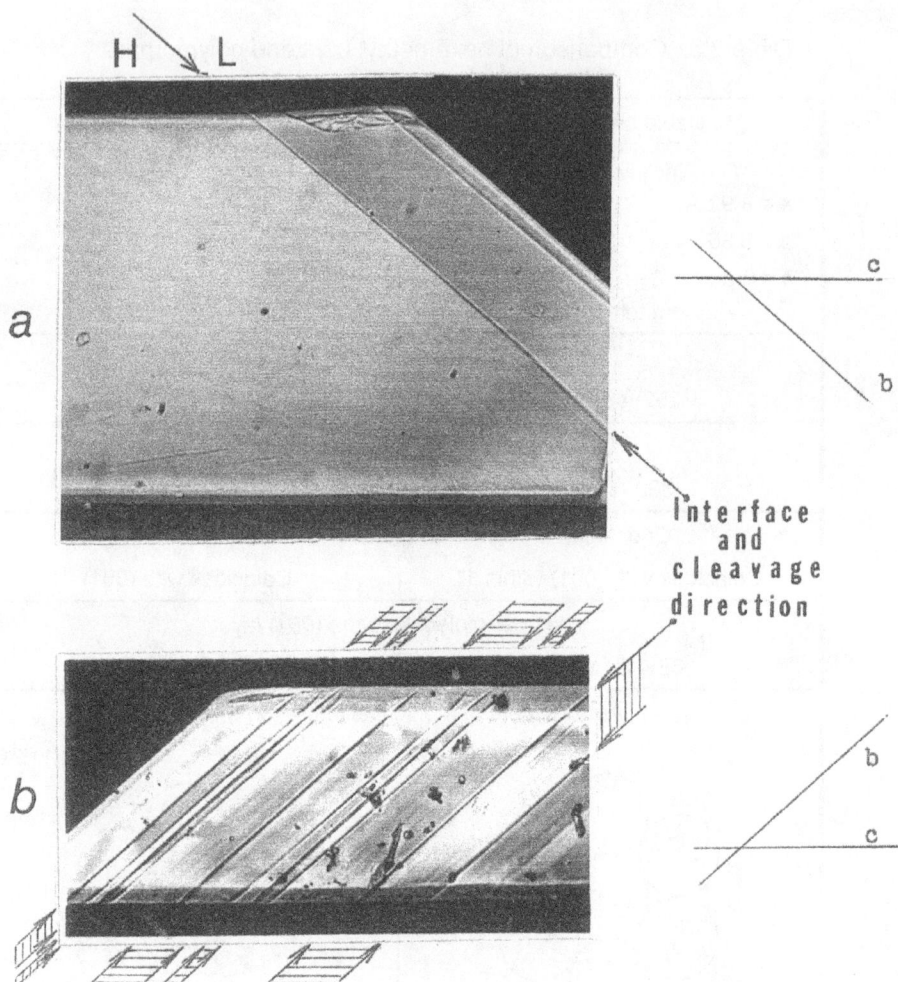

Fig. 2.34   The interfaces (arrows) during phase transitions in HMB crystals as observed in a microscope. Crossed polarizers. Upon rotation of the microscope stage both phases were extinguished simultaneously.

(a)   The first transition (L → H) in a fairly perfect single crystal (a rather rare case of a single interface moving through the crystal). The interface remained parallel to itself and the cleavage planes.

(b)   The last in the chain of cyclic phase transitions (L → H → L → H →) L → H.
The transition was initiated at several points on the edges, continued by formation of thin H strips parallel to the cleavage, and then proceeded by a gradual width increase of the H bands denoted by shading between the arrows. The frontal advancement of the interfaces visible on the photographs was not, however, truly gradual; it rather proceeded by lateral strokes of very small steps along the interface lines (edgewise mechanism, Sec. 2.3), as a closer examination has revealed.

It would seem, HMB is an ideal example of a *displacive-martensitic* transformation in which entire molecular layers slip over one another. However, the data described below refute such a mechanism. First of all, the transition does not occur "instantaneously"; it occurs by nucleation and gradual growth of the 'r' phase. The nucleation takes place at crystal defects as seen from the following: (a) direct observation with a microscope suggests this to be the case, (b) ΔT_{tr} is greater in more perfect crystals, (c) nucleation can be initiated at any point by a touch with a needle. The interface is observable (Fig. 2.34). When overheating ΔT_{tr} is carefully adjusted, the transition proceeds as a smooth interface movement. This movement can be halted by lowering ΔT_{tr} to zero. Even small increases in ΔT_{tr} sharply accelerate interface motion (Fig. 2.35). At $\Delta T_{tr} > 2.5°$ the interface advances at the rate > 2 mm/sec. If nucleation lags are of this order or higher (a realistic assumption), the phase transition in single crystals or grains of 0.2 mm size will be completed within 0.1 sec and appear to the eye as instantaneous. Such an "instantaneous" transition is still 10^5 -10^6 times slower than the velocity of elastic wave in a solid medium.

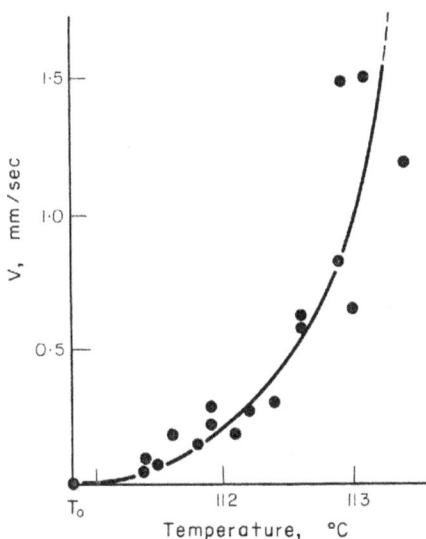

Fig. 2.35 The temperature dependence of velocity V of interface motion in HMB phase transition (viewing a *bc* face). $T_o = 110.8 °C$.

The natural faces of the solution-grown L crystals shown in Fig. 2.34 have their edges directed along the unit cell axes. Unlike the phase transitions considered so far, interfaces in HMB always have the same direction, which is parallel to *b* on *bc* faces and *a* on *ac* faces. Therefore, the interfaces are planes parallel to (*001*). On the other hand, (*001*) is a pronounced cleavage direction. A prick with a needle causes lamination of the crystal along (*001*) planes (Fig. 2.36a). This layered structure is directly related to the epitaxial character of the phase transition. Fig. 2.36b explains why HMB has this layered structure: the particular shape of its molecules prevents sufficiently close packing between the molecular layers (*001*).

A closer microscopic examination of interface motion, as viewed from the *bc* side, reveals small steps, or ledges, running along the interface, that is, in the *b* direction. Nucleation usually occurs at the *c* edges and initially gives rise to formation of narrow wedges of the H phase with the numerous growth steps along a wedge generating line (Fig. 2.37a). The wedges then penetrate through the crystal to form bands of the H phase seen in Fig. 2.34b. An important feature of this growth morphology is the edgewise mechanism similar to that described in Sec. 2.3. In the present case the observed edgewise growth means that the molecular layers (*001*) do not slip as a whole over one another during a phase transition. Rather, every layer (*001*) is subjected to a complete rearrangement when the interface separating L and H parts in the layer runs along from one edge *a* to the opposite one (Fig. 2.37b). These quick runs along the individual crystal layers were also observed from the *ab* side where the interfaces appeared as lines of a color separation.

A set of additional test experiments was performed. It was confirmed with the X-ray Laue patterns that a strict OR was always preserved as a result of the phase transitions. Another test was the optical study with crossed polarizers. Measurements of the extinction angle relative to the crystal axes *a*, *b*, and *c* prior to and after phase transition have shown that the orientation of the (*001*) layers remained unchanged, even though a small change in the angle between the crystal edges *b* and *c* by ~2° occurred (Fig. 2.37c).

Thus, the two polymorphs of HMB consist of almost identical molecular layers packed into weakly bounded stacks. The only major difference in their structure is the mode of layer stacking. On the basis of a structural comparison alone, this would be an ideal example of a "displacive" phase transition. But such a conclusion would be in error: no displacement of the layers takes place. The phase transition is actually a complete structural reconstruction, molecule by molecule, at the interfaces moving along every layer and from one layer to the next. The end result - the new structure - consists of almost (but not exactly) the same layers of a different stacking mode. This process is *oriented (epitaxial) crystallization*.

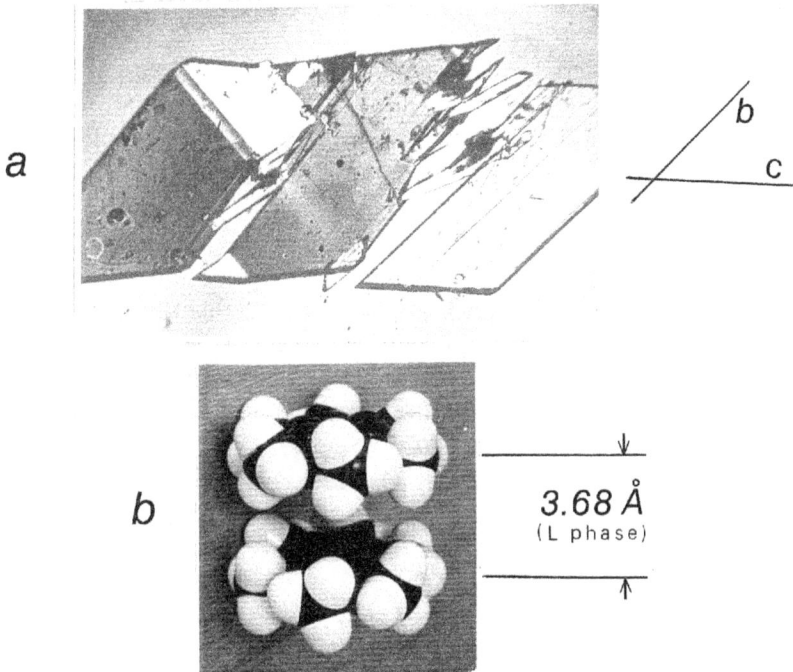

Fig. 2.36 A layered structure of HMB crystals.
(a) Lamination of a crystal along (*001*) when it is pricked with a needle.
(b) The minimum distance between the carbon atoms of the benzene rings in adjacent (*001*) molecular layers is larger than the sum 3.40 Å of carbon Van der Waals' radii due to repulsion of the CH₃ groups. As a result, the interaction between the layers is weakened.

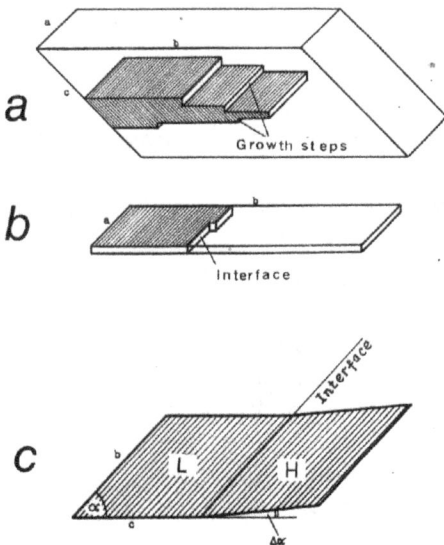

Fig. 2.37 The morphology of phase transition in HMB (schematic). Development of a wedge-like resultant crystal in the bulk. Its faces are parallel to the cleavage planes (*001*), and its growth proceeds according to the edgewise mechanism (Sec. 2.3). Very small steps shown in the figure run along the interface, adding every time a thin layer to the new phase. Interface motion in the thin layer parallel to (*001*). The direction of the molecular layers parallel to (*001*) is retained after the transition, while the angle α changes a little.

2.8.3   Epitaxial transition in DL-norleucine (DL-N)

DL-N is a short-chain aliphatic substance (see Table 2.1). It is a synthetic amino acid, a close analogue of naturally occurring methionine. It was found that DL-N has a phase transition at ~117 °C. Its L crystal structure has been solved by Mathieson [128]. It is of a layered type (Fig. 2.38) with the parameters a = 9.84 Å, b = 4.74 Å, c = 16.56 Å. The molecular layers are parallel to the (001) plane. The c axis makes a 14.5° angle with the layer normal, hence the layer spacing d_{001} = 16.03 Å. The molecular axes are aligned roughly along the c axis. Each layer is bimolecular: the CNCOO groups of the molecules are pointed toward the center of the layer where they form a network of hydrogen bonds N-H... O. This central "skeleton" turns the bimolecular layer into a rather firm structural unit. The interlayer interaction is of a purely Van der Waals' type, so the layer stackings are governed by the principle of close packing in the same way as in the n-paraffins [129].

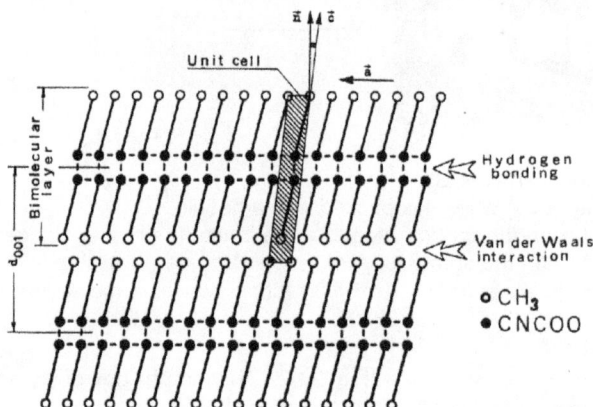

Fig. 2.38
Characteristic features of the DL-norleucine (DL-N) crystal structure.

The initial microscopic observations of the phase transition revealed a vaguely defined running polychromatic wave that easily reversed its direction when the temperature was slightly changed up and down near 117 °C. Therefore, the first task was to establish whether there was a hysteresis caused by nucleation lags. (The same problem as in order-disorder transitions, Sec. 2.7.2). To this end, smaller and more perfect single crystals were prepared. They were 0.5 to 2 mm size rhombus- or trapezium-shaped plates as thin as 0.02 to 0.1 mm. With these crystals and temperature control better than 0.1 °C the nucleation hysteresis was detected, although it was rather small, comparable to that in C → ODC phase transitions (Sec. 2.7). Fig. 2.39 attests that all L → H transitions start at $T > T_o$, while H → L at $T < T_o$. For the first two transitions $|\Delta T|$ is about 0.8°, then decreases gradually to stay at 0.2° level. This is the same kind of behavior as in Fig. 2.26 (Sec. 2.5.2). In another experiment, ΔT_n was compared in pairs of one visually perfect and one less perfect single crystals of equal size. For each of the 10 pairs selected for the examination it turned out that $|\Delta T_n|_{perf} > |\Delta T_n|_{imperf}$ by 0.3° to 0.8°. Finally, introduction of a mechanical defect initiated transition at a lower ΔT. All the observations indicated nucleation at the crystal imperfections.

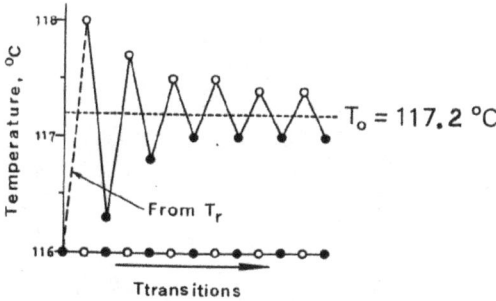

Fig. 2.39
Temperature hysteresis ΔT upon cyclic phase transitions in a DL-N crystal. The hysteresis is relatively small, not a constant and decreases with the number of the transitions to a low, but not zero, level.

When a (001) face was viewed, the direction of maximum extinction of the crystal under crossed polarizes changed by less than by 2° as a result of phase transition. This change was quantitatively irreproducible, the circumstance pointing at internal strains as its probable cause. These observations suggested that there was a rigorous OR between the phases.

The X-ray investigation involved the following techniques: rotating-crystal photography to identify and verify the unit cell axes; Laue photography to study OR; powder photography to study the parameters in a layer plane as a function of temperature through the transition; small-angle technique with a live projection of the X-ray patterns on the screen to study how the interlayer spacing d_{001} changes through the transition.

Laue photographs taken before and after phase transition were almost indistinguishable (Fig. 2.40a), which proves both the rigorous OR and a strong similarity between the two structures. Then a set of the powder photographs as a function of temperature was taken and used to find the intralayer spacings d_{200} and d_{020} vs. temperature. As Fig. 2.40b shows, there is a quite noticeable (~1%) change in these spacings, which means that the layers in the two phases are only similar, and not identical. Finally, the long spacing d_{001}, which is a characteristic of a layer stacking mode, was observed on a screen and photographed (Fig. 2.41) upon heating over the transition range. *Two* distinct phases of a different layer stacking mode coexisted in a temperature range. The L phase was represented by the interlayer spacing d_{001} = 16.5Å. Upon heating, the second line representing the interlayer spacing d_{001} = 17.2 Å of the H phase appeared. Its intensity gradually increased from zero to a maximum at the expense of the intensity of the L line until the latter was completely extinguished.

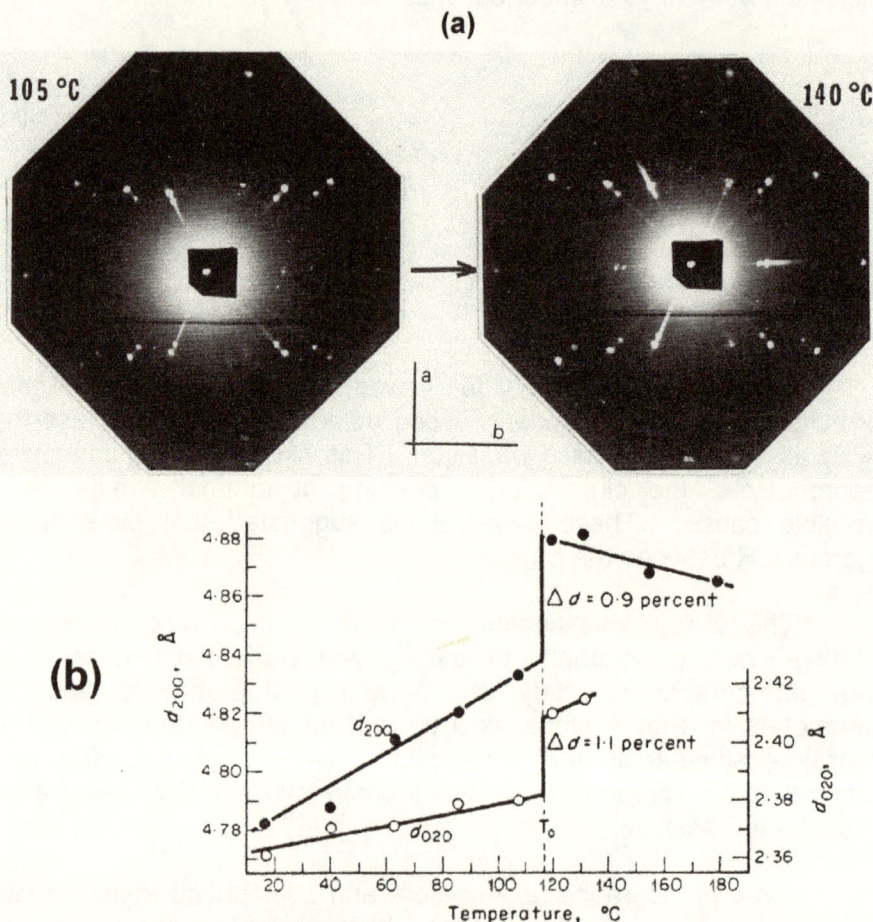

Fig. 2.40   Large-angle X-ray examination of the DL-N phases.

(a)   Laue diffraction patterns of a single crystal before (105 °C) and after (140 °C) the transition. A single-crystal character is retained; there is a strict OR, as well as almost complete identity of the patterns.

(b)   The temperature dependence of the intralayer spacings d_{200} and d_{020} (indexed in the L lattice) through the transition. The plot has been derived from the sequence of powder patterns taken with a 114.6 mm diameter camera. A discontinuous change in the above spacings by ~1% is seen.

The morphology of the phase transition can be described as follows. The real layered crystal can be viewed as a stack of thin-plate single-crystal blocks of irregular thickness. These blocks with (001) surface will be referred to as lamellae. The phase transition proceeds by motion of interfaces separately in every lamella, dividing them into L and H parts. Considering that in an imperfect single crystal the lamellae are very thin and weakly bound to one another, the nucleation and propagation of the individual interfaces in them occurs rather independently and is poorly coordinated. While in theory an individual interface may be seen in a microscope, in practice there are too many of them. Such a phase transition will be seen by an observer as a pattern of numerous interfaces moving at different levels in the individual lamellae, the picture being at best a diffuse "wave." In order to see individual sharp interfaces as those in Fig. 2.42, two conditions must be satisfied: sufficiently perfect thin crystal consisting of only few lamellae and carefully regulated temperature to insure a slow interface motion at very small ΔT_{tr}.

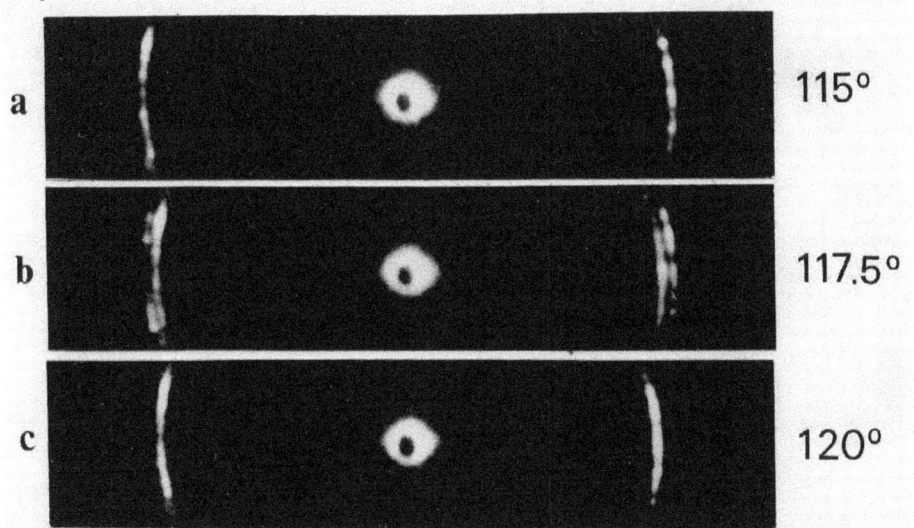

Fig. 2.41   Change in the interlayer spacing d_{001} upon the L $\rightarrow$ H phase transition in a DL-N powder specimen. The photographs were taken from the screen of a device for direct viewing small-angle X-ray patterns.
(a)   Before the transition. d_{001} = 16.5 Å.
(b)   During the transition. Two separate lines coexist, each representing one of the phases. It was visually observed that the intensity of the H line was gradually increasing at the expense of the L line. Thus, the H *quantity* in the heterophase specimen was increasing, and L decreasing over a temperature range. This experiment visualizes how apparent "continuous" and "displacive" phase transitions actually proceed.
(c)   After the transition, d_{001} has increased by 4.1%.

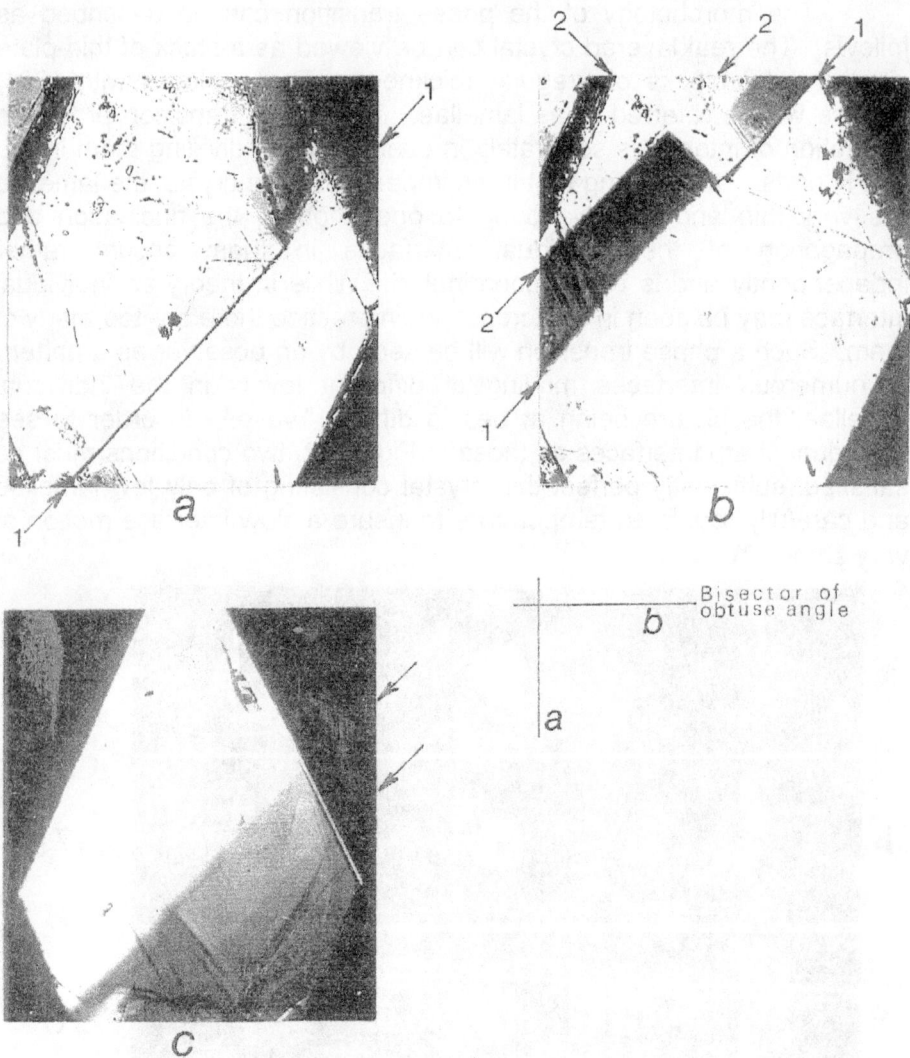

Fig. 2.42 The interfaces (arrows) during phase transitions in DL-N crystals as observed in a microscope under crossed polarizers. Upon rotation of the microscope stage, both phases extinguish simultaneously. The phase transition proceeds by growth of new (001) layers over the previous ones (the layers are parallel to the plane of picture). The rectilinear interfaces (2,1, ℓ)indicated by the arrows represent steps that new growing layers make with the previous. If a growing layer is very thin (of a micron order), as 2 - 2 is, a color separation of the layers is observed owing to interference of light (hence the black-white contrast on the photographs).

(a,b) Two stages of phase transition.

(c) Simultaneous growth of several very thin (001) layers forming steps with one another.

- 138 -

The phase transition in DL-N is a variation of crystal growth, its oriented (epitaxial) version. Several facts in line with this conclusion will be added. Interfaces at T_0 are at rest, but set in motion with accelerating speed as ΔT_{tr} increases - as expected from crystal growth. An edgewise advancement of the interfaces (Sec. 2.3) in the individual layers was seen repeatedly, although because of lack of optical contrast* no attempt was undertaken to record this phenomenon under maximum optical resolution. The interfaces were rectilinear, of one and the same crystallographic direction $(2,1,\ell)$ in the L lattice. In contrast to the non-epitaxial phase transitions described previously, no change of the interface was noticed when the transition was reversed, although the interface already became a growing crystal face of the H rather than L phase. This is not surprising, considering the expected close structural identity of the (001) layers in the two phases and the rigorous OR.

2.8.4 Significance of layered structures. Epitaxial nucleation in the interlayer microcracks

The experimental data presented in two preceding sections on phase transitions exhibiting strict OR are sufficient to account for their specificity. These transitions are consistent with the concept of solid-state crystal growth. HMB and DL-N differ in their properties, molecular shape, and crystal structure. HMB is an aromatic substance with flat circular "coin-like" molecules close-packed into pseudo-hexagonal molecular layers, a molecular plane coinciding with the layer plane (Table 2.6). All intermolecular bonding, both in the layers and between the layers, is of Van der Waals' type. DL-N, on the other hand, is an aliphatic substance with a layered crystal structure typical of chain molecules, where the molecular axes are quite or almost perpendicular to the layer plane. The layers in DL-N are bimolecular with a firm "skeleton" of hydrogen bonds in the middle (Fig. 2.38). In spite of this dissimilarity, the two exhibit similar features of their phase transitions, not found in those described previously. *This is due to the layered crystal structure.*

A layered structure has strongly bounded, energetically advantageous two-dimensional units, molecular layers, while the interlayer interaction is weak. Since the layer stacking contributes

---

* Owing to simultaneous extinction of the two phases under crossed polarizers, a level of transparency could not be a source of optical contrast between them. There were two weaker sources of the contrast: color separation due to light interference in thin lamellae and polarization of light as a result of reflection from the interfaces.

relatively little to the total lattice energy, the difference in the free energies of the structural variants with different layer stacking is small. This is a prerequisite for polymorphism in layered crystals. Change from one polymorph to the other mainly involves the mode of layer stacking. The layers themselves change only slightly under influence of different layer stacking.

Speaking figuratively, nature does not take advantage of HMB and DL-N layered structure to make their phase transitions "displacive" by slipping whole molecular layers over one another into the new mode of layer stacking. The transitions proved to be crystal growth on every account: nucleation, moving interfaces, phase coexistence in a temperature range, hysteresis, and large discontinuous change (2.4% and 6%) in the specific volume. Every molecular layer undergoes a reconstruction, molecule by molecule, to build up a new layer of almost the same structure, but now in a different layer stacking. Even the cementing action of hydrogen bonding inside the DL-N molecular layers does not prevent them from 'molecule by molecule' reconstruction. The universal *contact* mechanism fits that purpose: in "contact" rearrangements only *relative* strength of bonding on both sides of the interface matters, and not the strength *per se*. A specificity of this kind of crystal growth is (a) rigorous OR, (b) a uniform direction of the interfaces, and (c) a relatively low level of the hysteresis ΔT_n. These features can be understood in terms of the *contact* mechanism combined with the phenomenon of epitaxy. For HMB and DL-N, the conditions for epitaxial nucleation and growth of the 'r' phase are guaranteed by their layered structure and proximity within ~1% of the structural parameters of the layers in the two polymorphs.

The *contact* mechanism involves nucleation of the 'r' phase at particular crystal defects - optimum-sized microcavities (Sec. 2.5.4). Real layered crystals always have numerous defects that result from imprecise layer stacking. Most of these defects are minute wedge-like interlayer cracks at the crystal faces. In such a microcavity there is always a point where the gap has the optimum width for nucleation. There the molecular relocation from one wall to the other occurs with no steric hindrance and, at the same time, with the aid of attraction from the opposite wall. In view of the close structural similarity of the layers in the two polymorphs, this nucleation will be *epitaxial*. In accord with this description, the nucleation was indeed found to occur at the crystal faces.

Fig. 2.43 shows schematically the initial stage of phase transition in a layered crystal. The flat shape of the embryo enhances the surface interaction and hence the orienting effect by the substrate. In fact, both sides of the microcrack, and not one, provide the orienting action ensuring an even more rigorous OR than in the traditional "single-side" epitaxy.

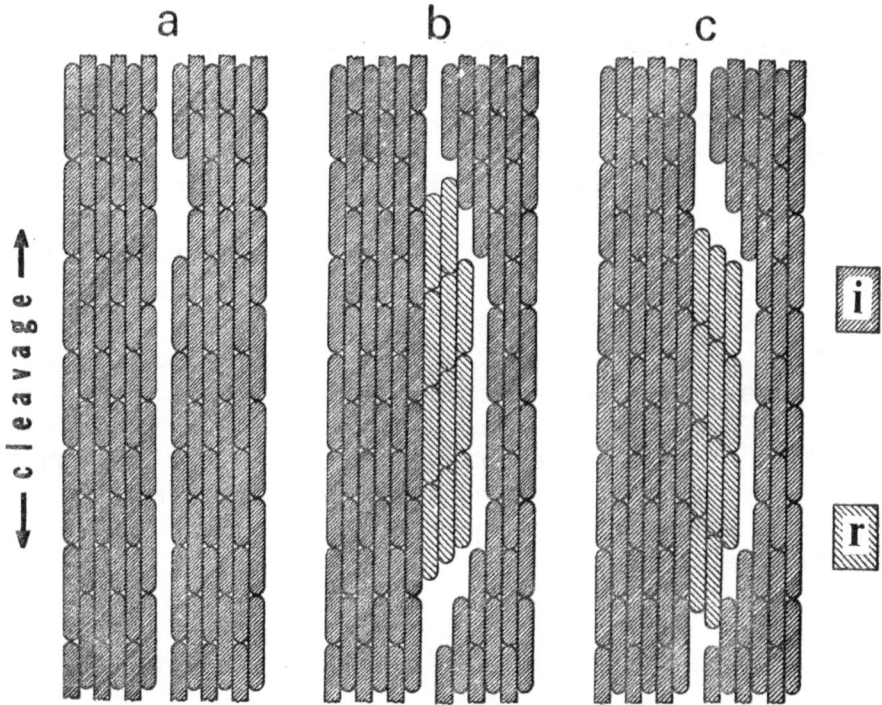

Fig. 2.43   Initial stage of an epitaxial phase transition.
(a) A lattice defect in the form of a submicroscopic crack parallel to the cleavage.
(b) Oriented embryo of the resultant phase, formed by consecutive transfer of molecules from one side of the flat microcavity to the other.
(c) Equally probable embryo in the "twin" orientation relative to its counterpart; it can form if the resultant lattice has a lower symmetry.

Once the conditions for a strict OR have been formulated, it is useful to look back at the phase transitions where OR was more or less random. Those were instances where the unit cells of the polymorphs were not sufficiently similar and/or there were no pronounced layered structures. The PDB, exhibiting practically random OR (Sec. 2.2.2), may satisfy one of the two conditions. Its unit cell parameters below and above T_o = 30.8 °C, are close [130,131], the translation *a* in L being twice as large:

(L) *a*=14.80, *b*=5.78, *c*=3.99 Å, (α=90°), β=113°, (γ=90°), Z=2;
(H) *a*=7.32, *b*=5.95, *c*=3.98 Å, α=93°10', β=113°35', γ=93°30', Z=1;

But the PDB crystal structures do not meet the condition to be layered structures. There are no flat microcracks between the appropriate crystal planes to offer substrates for the oriented nucleation. Thus, a layered character of a structure is a necessary condition for the *epitaxial* phase transitions.* Moreover, a layered structure alone is almost a sufficient condition, because, as pointed out above, it is usually combined with structural proximity of the layers in the two phases.

Finally, there is a simple answer to why the temperature hysteresis ΔT_n in epitaxial phase transitions is small (see Fig. 2.39) as compared with non-epitaxial transitions. Due to the abundance of wedge-like microcracks, there is no shortage in the nucleation sites; at that, the presence of a substrate of an almost identical surface structure acts like a "seed." As a result, only small overheating or overcooling is required in order to initiate phase transition. Without a scrupulous experimental verification, the phase transitions in question may seem "instantaneous," without a hysteresis, "cooperative," "displacive," "second-order," etc.

2.8.5 Three examples in support of epitaxial crystal growth

The following three examples that were earlier classified in literature as "displacive," "not of first order," "λ-type," "higher-order," will be considered and explained in terms of epitaxial growth on the cleavage planes of the original phase. In each case a strict OR was definitely established. The relevant data about two of them, 1,2,4,5-tetrachlorobenzene (TCB) and aniline hydrobromide (AHB), are listed in Table 2.7.

TCB, on cooling from room temperature T_r, changes at -85 °C from a stable monoclinic (H) phase to triclinic (L). The unit cell parameters for H and L are very close, so an important condition for epitaxial transition is satisfied. The H crystal structure is similar to HMB in two respects: (a) the molecules form layers parallel to (*001*), the

---

* Yet, see p.195.

Table 2.7   Data on the polymorphs of 1,2,4,5-tetrachlorobenzene (TCB) and aniline hydrobromide (AHB).

		TCB		AHB	
		$C_6H_2Cl_4$		$C_6H_5NH_3Br$	
T_0, °C		-60		+22	
Phase		L	H	L	H
References		[132]	[133]	[134,135]	[136]
Symmetry		Triclinic	Monoclinic	Monoclinic	Rhombic
Unit cell (Å)	a	9.60	9.73	16.725	16.77
	b	10.59	10.63	5.95	6.05
	c	3.76	3.86	6.81	6.86
	α	95°	(90°)		
	β	102.5°	103.5°	91°22'	(90°)
	γ	92.5°	(90°)		
Z molec/cell		2	2	4	4
Meas. at T°K		150	300	163	343
Shape of crystals			Rod-like along [001]	Lath-like along [010]	
Faces			{110}	{001}, {100}	
Cleavage		May break up during phase transition		Main is (100); Also (001)	

benzene rings being in the layer planes; (b) the distance between the benzene rings in adjacent molecular layers is enlarged by a repulsion of the Cl atoms, thus weakening the interlayer Van der Waals' interaction. The interlayer spacings d_{001} are 3.76 Å (H, T_r) and 3,67 Å (L, -120 °C) - almost as in HMB (Table 2.7 and Fig. 2.36b). Inevitably, TCB crystals must be *layered* along (001) for the same reason as HMB; the second condition of epitaxial phase transition is met as well.

The reported OR slightly deviated, however, from that expected for epitaxial crystal growth.   The latter requires the orientation of molecular layers (001) to be unchanged upon transition, which in the present case means a ~5° change in the axis c direction. But the X-ray determination showed an unchanged c direction, but the (001) orientation changed by the same ~5°. This discrepancy is cleared up after we take into account one particular experimental detail.   The crystals were rod-like along the c axis. They were placed in a capillary mounted on an X-ray goniometric head. The capillary prevented the c direction from changing during phase transition, forcing the (001) planes to turn inside the capillary as shown in Fig. 2.44.

In a similar manner, an examination of the experimental data on AHB revealed the presence of every characteristic of epitaxial phase

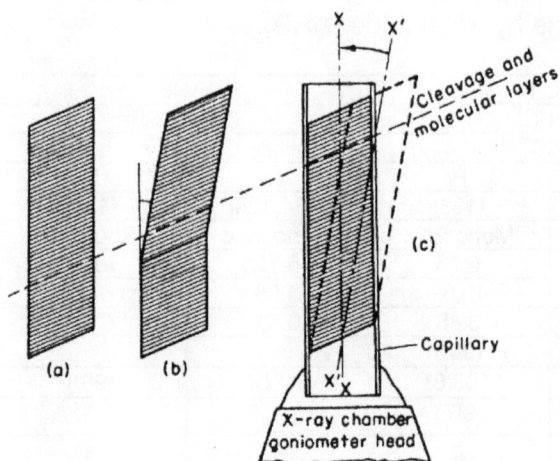

Fig. 2.44 To the X-ray determination of orientation relationship of the two phases in epitaxial phase transitions. The drawing illustrates a problem resulting from change in crystal shape.
(a) A free crystal before the transition.
(b) During the transition.
(c) The crystal placed into a capillary mounted on an X-ray goniometric holder for OR determination. The capillary prevents the crystal axis X---X from deflection to the X'----X' position, thus forcing the molecular layers (shaded lines) to change their orientation.

transition: pronounced layered structures with the layers parallel to (*100*), close proximity of the layer parameters, preservation of the cleavage direction (*001*), small hysteresis (but not zero, as the authors of the experimental study suggested).

A fixed OR was also reported [144] in the phase transition of s-triazine:

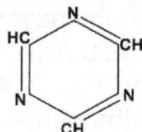

Its rhombohedric crystals, stable at T_r (P3c; a = 9.647 Å, c = 7.281 Å in the hexagonal cell), change at -60 °C into a monoclinic phase. In H, the plane molecular rings are parallel to *ab*. The rings are placed over one another, $c/2$ = 3.64 Å apart, suggesting a layered structure with a cleavage along *ab*. The X-ray data on OR allow to conclude that the axes *c* of the phases have the same direction, while there is an 6.2° angle between the *ab* planes. That is, the OR that was actually found deviates from that expected in epitaxial transition by 6.2°. Again, we attribute the discrepancy to the effect of the capillary (Fig. 2.44) where the crystals were encapsulated.

Thus, the available data in those three phase transitions are in accord with the concept of epitaxial crystal growth. The epitaxial nucleation-and-growth mechanism also found support in the experimental study [138] in which the above-described results on HMB and DL-N (initially published in [139]) were positively referred.

2.8.6 The origin of domain structures

A peculiarity common to all three phase transitions discussed in Sec. 2.8.5 was reported by their investigators. In TCB and AHB, the 'r' phase in the H → L transitions appeared in two equivalent orientations, approximately in equal quantities. The phenomenon was interpreted as the consequence of a *displacive* mechanism acting in different directions in different parts of the crystal. However, these phase transitions occur rather by *epitaxial crystal growth.* It will be shown now that this process carries with it the explanation as to why and how the lamellar-domain structures, such as found in ferroelectrics, come into existence. The fact is that the occurrence of domain structures has nothing to do with the energy of magnetic or dipole interaction. This reverses the causes and effects in the whole interpretation of ferromagnetism and ferroelectricity (Chapter 4).

Formation of the lamellar-domain structures is almost inevitable in epitaxial transitions if the emerging 'r' phase has a lower symmetry. The phenomenon is observed only in H → L transitions because, with very few exceptions, it is L that has a lower symmetry. As illustrated in Fig. 2.43, an oriented L nucleus can appear with equal probability under two orientations related to one another as crystallographic twins. Fig. 2.45 shows this in more detail. In Fig. 2.45a, two orientations of the 'r' monoclinic lattice would form with equal probability in the initial rectangular lattice whose cleavage is (*001*). These orientations differ in the direction of the c axis. The two 'r' crystals can be brought into coincidence with a two-fold axis perpendicular to (*001*). The initial unit cell is a_i, b_i, c_i. Growth of a nucleus gives rise to the formation of a *laminar domain* of one or the other orientation.

In practice, phase transitions in layered crystals are always multinuclear. Even a seemingly perfect single crystal of a layered structure contains multiple submicroscopic wedge-like chinks parallel to the cleavage planes. These tiny chinks are located at all natural faces except the ones parallel to the cleavage. They act as the nucleation sites from which the lamellar domains of the 'r' phase grow. Approximately one half of the lamellar domains assume the orientation No.1, and the rest assume No.2. When the sides of the adjacent domains of the same orientation meet, they merge into a single domain. The lamellar-domain structure with the strict alternation of No.1 and No.2 orientations, sketched in Fig. 2.45b, emerges. Two features of the resultant lamellar-domain structure will be noted: (1) any two adjacent domains are crystallographic twins and (2) the thickness of different domains varies.

A domain structure, as in Fig. 2.45b, represents the simplest case. For example, two or three equivalent cleavage directions, rather than one, can exist. Or a domain structure can consist of the domains of three rather than two orientations. The latter version has been found in s-striazine (Fig. 2.45c). There the resultant monoclinic phase emerges from the initial rhombohedral phase in three orientations that can be made coincident by rotation about a 3-fold axis. An attempt to explain this phenomenon by a "3-fold displacement" would be even more involved than in the case of two domain orientations. However, it is easy to account for the phenomenon in terms of epitaxial growth on the cleavage planes of H, for the (001) substrate has a 3-fold symmetry. All three orientations of the monoclinic phase shown in Fig. 2.45c must be equally probable for nucleation. A lamellar-domain structure of the three crystallographic orientations in a random sequence appears after multinuclear transition.

2.8.7 On classification of solid-state phase transitions

Many phase transitions were classified in the past on the basis of observations that were either too limited, insufficiently precise, erroneously interpreted, or a combination of the above. The assertions such as "continuous change in physical property," "no change in crystal structure," "λ-transition," "no latent heat," "no hysteresis," "occurs instantaneously," "second-order," "higher-order" were typical. If studied more attentively, it can always be proven that these phenomena do not exist. Phase transitions in solid state do not occur "cooperatively." They do not proceed by "displacement," "turn," "rotation," "dilatation," "shear," "expansion," "deformation," "distortion," "catastrophe," or the like. The only real molecular mechanism is *crystal growth at a contact interface*. It manifests itself in two modes:

(1) Oriented (epitaxial) crystal growth.
(2) Non-oriented crystal growth.

This is the complete classification of *all* solid-state phase transitions, including ferromagnetic, ferroelectric and superconducting.

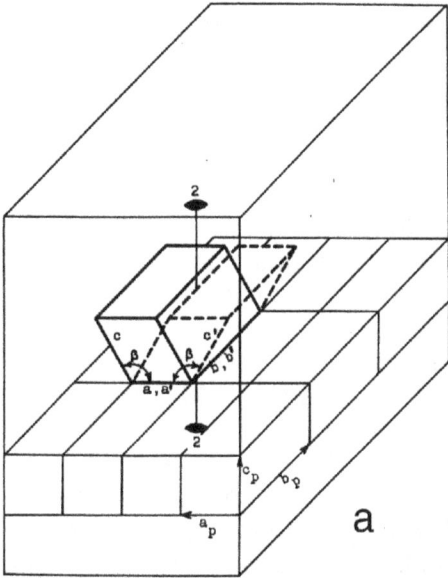

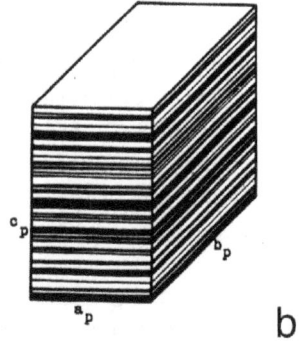

Fig. 2.45   Formation of "twin" domain structure in epitaxial phase transitions as a result of oriented multinucleation of a lower-symmetry phase. The nuclei assume two or more equally probable orientations.
(a) Two equally probable orientations of a monoclinic lattice of the resultant phase in the initial crystal characterized by a rectangular lattice and cleavage (*001*).
(b) Growth of each nucleus leads to formation of a lamina of one particular orientation of the two possible. The initial single crystal turns into a lamellar structure of the domains of two alternating fixed orientations shown as black and white. (Two neighboring laminas of the same orientation merge into one.)
(c) A three-orientation case represented by *s*-triazine. A transition from the H rhombohedral lattice to the L monoclinic gives rise to three, rather than two, equally probable domain orientations of the latter (one is shaded, two others are shown by dashed lines). The subscripts *m* and *h* are related to the monoclinic and hexagonal unit sells respectively. The lamellar structure forms of the domains of three different orientations alternating at random.

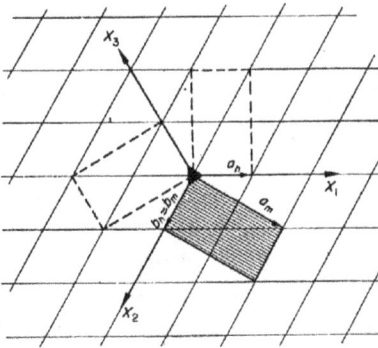

2.9   Kinetics of solid-state phase transitions

2.9.1   Definition; current status; bulk and interface kinetics

The term *kinetics* will be assigned the following meaning: any relationships between the macroscopic rate of a phase transition and conditions or parameters that it depends on.   There were attempts to measure rates of solid-state phase transitions aimed at finding their mechanism.   They were never productive for the reason that proper measurements of the kinetics already require a basic understanding of the mechanism.   This point will be illustrated later in this section.

The current status of the problem in question is best represented in the "Rates of Phase Transformations" by Doremus [140].   The stated goal of that work was "to summarize the present state of our knowledge of kinetics of phase transformations."   Since the term "phase transformations" is usually reserved for solid-state phase transitions, some readers would be surprised to find them not included in the book.   In the 12 lines of an introductory table all known phase transitions were listed: vapor-liquid, vapor-solid, liquid-solid, etc., including solid-state recrystallization and solid-state phase transitions.   Of them, 11 were considered in the following chapters, but not solid-state phase transitions.   This was a recognition, although without a direct statement, that the present state of our knowledge does not include kinetics of solid-state phase transitions.   To fill the blank, the foundations of kinetics of solid-state phase transitions are now put forward.

Let us consider why kinetics of solid-state phase transitions does not exist as an established scientific topic even in the form of a plausible hypothesis.   If phase transitions are treated as critical phenomena, the problem of their kinetics must not exist.   Such transitions must occur at its critical temperature T_c and comprise all the matter at once.   There is no kinetics: the rate of transition is a function neither of time t, nor temperature T. The concepts "kinetics" and "critical phenomenon" are incompatible: any attempts to circumvent this fact (and there were some) effectively nullify the notion itself of a critical phenomenon.

It should be unambiguously stated that the notion *kinetics* implies *phase coexistence* and, consequently, a first-order phase transition. There were experimental studies where the change in the mass fraction 'm' of one phase in a two-phase specimen was the value of interest. This value was measured as a function of time t by recording changes of a physical property or intensity of a signal from one of the phases (such as a characteristic X-ray reflection or an optical spectral line). This type of kinetics is a *bulk kinetics*. Powder or polycrystalline specimens were usually used. Specimens of L phase at temperature T were placed in a thermostat of a higher temperature T_{th} where the transition L $\rightarrow$ H occurs. The measured rate dm/dt was attributed to the bulk rate of phase transition under the isothermal conditions T_{th} = const.

The described technique is fruitless, illustrating that a certain prior knowledge of the phase transition mechanism was necessary in order to design an experiment meaningfully. The fact is that 'm' would not change isothermally in a powder specimen, for it is controlled by the distribution of nucleation temperatures pre-coded in the individual microcrystals (Sec. 2.5). Review of some experimental works where $dm/dt \neq 0$ was found revealed that isothermal conditions were not attained there: 'm' actually changed while the sample was warming towards T_{th} and passing the temperature range of nucleation. The bulk rate dm/dt was proportional to the rate of warming over this temperature range. The latter is described by the equation $dT/dt = k (T_{th} - T)$, which caused the illusion that the equation of chemical kinetics for a first-order reaction (no relation to a first-order phase transition) $dm/dt = K (1 - m)$ was in control. The "reaction equilibrium constant K" found from the latter equation was delusive. Calculating the activation energy E_a of phase transitions on the basis of these experiments made no more sense. The equation of chemical kinetics for first-order reactions is applicable to homogeneous reactions only. It is not suited for a heterogeneous process, which a solid-state phase transition is. For the same reason it cannot describe interface motion. Suppose, there is a rod-like single crystal in which a phase transition is proceeding as a uniform movement of an interface from one end to the other. The equation, however, would require the velocity of the interface motion to be proportional to the remainder of the initial phase, *i.e.*, to rapidly decelerate.

The rate of a phase transition in a bulk kinetics depends on both nucleation and growth in unknown proportion, different in every particular case. Nucleation critically depends on the presence and distribution of specific lattice defects. It is quite different in a perfect and imperfect single crystal, in a big and small crystal, in a single crystal and polycrystal, in a fine-grain and coarse-grain polycrystal or powder. Growth also depends on many parameters discussed later. The parameters on which the nucleation and growth depend are not only irreproducible and uncontrollable, there is also no way, when they act together, to theoretically separate their contributions in order to calculate the total bulk rate. In this context, the theoretical approach called "formal kinetics" should be mentioned [141] where an attempt to separate them was undertaken. One of the main assumptions there - isothermal rate of nucleation - was invalid due to "pre-coded" character of both nucleation sites and the nucleation temperatures.

Bulk kinetics can shed no light upon the physics of phase transitions. As a minimum condition, the nucleation and interface motion contributions must be experimentally separated. The best way to do this is a visual observation of motion of a flat interface in an optically transparent single crystal. This is the method of *interface kinetics*. Now we can return to the issue as to why solid-state phase transitions were omitted in the work "Rates of Phase Transformations" which was intended to be comprehensive. One can conclude from this book that its author probably believed, correctly, that only "interface kinetics" is physically sound. But actual manifestations of the interface motion had no plausible explanations. Its velocities can be by orders of magnitude higher than diffusion could account for (solid-state recrystallization was not excluded from the book: grain boundaries migrate slowly, so diffusion considerations could still be plausible). On the other hand, the phase interface moves by orders of magnitude slower than it is assumed by the martensitic mechanism. Above all that, the interface motion is utterly irreproducible. Thus, the conventional theory did not have a clue to the interface kinetics. The clue is the *contact crystal growth*.

As mentioned, there is a reproducibility problem. The absolute velocity V of interface motion is not reproduced well even when observed again in the same single crystal. Another part of the problem stems from the fact that flat interfaces are the natural faces of the growing 'r' crystal. All faces grow with their own velocities that can differ by orders of magnitude. There are a number of other uncertainties. Fortunately, valuable information can be obtained from *relative* V changes. This type of study is described below. As will be shown, it helps not only to verify and substantiate the *contact* mechanism, but also to penetrate deeper into its details. It will be demonstrated that the universal *contact* mechanism easily accounts for all the complexity, versatility, and poor reproducibility of the kinetics of solid-solid phase transitions. In other words, this simple molecular mechanism and the complicated manifestations of kinetics will be shown to agree well with each other.

2.9.2 A more detailed representation of interface motion

In Secs. 2.4.2 and 2.4.3 the *contact* molecular model of an interface was put forward. It was outlined in general terms how the interface moves forward by a "molecule-by-molecule" relocation at the molecular steps. The availability of some extra space at the steps to provide sufficient steric freedom for the relocation was noted as an important condition. This extra space eventually comes out on the surface. A new space must be found for the process to continue. There is no such space in an ideal crystal. Consequently, a phase transition can take place only in a *real* crystal with a sufficient concentration of vacancies and/or their clusters. Our experimental study of interface kinetics offered strong support to the above basic concept. Moreover, it made it possible to gain insight into the intimate details of interface motion. Now this more detailed *contact mechanism of phase transitions* will be presented. It should be emphasized that it was logically inferred from the whole set of the collected observations. Then these observations are described and, in turn, explained in terms of that more detailed picture.

An interface between two polymorphs is a natural face of the 'r' crystal. Its motion is *growth* of this crystal face. It takes an *edgewise* course (Sec. 2.3), which means a molecule-by-molecule packing of the molecular rows in a molecular layer and the construction of one layer after another in succession. The flat shape of the interface means that $V_\tau \gg V_n$ (Fig. 2.14b). The normal velocity of the interface motion V_n is controlled by the *two-dimensional nucleation*: formation of 2-D nuclei on the surface of the underlying molecular layers (refer to Fig. 2.15). In a crystal-crystal phase transition we deal with two major types of nucleation: 3-D to initiate a phase transition, as described in Sec. 2.5, and 2-D to initiate a new layer to advance the interface in the direction of its normal.

The 2-D nuclei form only heterogeneously, as the 3-D do. There is a significant difference in their function. While only one specific crystal defect (OM, optimal microcavity, Sec. 2.5.4) is needed to start a transition, a sufficient concentration of appropriate defects is required to keep the interface moving. These defects are also microcavities, but smaller than OM, although not just individual vacancies. They will be called *vacancy aggregations* (VAs). One VA can act as a site for a 2-D nucleus only once; it disappears after the running ledge delivers it to the crystal surface. The VAs may differ by the number of vacancies they consist of, as well as by how close to the interface they are.

In the process of its motion the interface intersects the positions of VAs, which is equivalent to a flow of VAs onto the interface. The intensity of the flow depends on the concentration of VAs and is affected by VAs migration. Not all VAs of the flow can be effective, but only those with the activation energy of nucleation lower than a particular level. The latter is determined by the overheating / overcooling $\Delta T_{2-D} = |T_{tr} - T_0|$, where T_{tr} is the actual temperature of phase transition. One transition in a crystal "consumes" a certain part of the total number of effective VAs, leaving repetition of the process possible.

Another major phenomenon of interface kinetics resulted from the fact that the interface motion is crystallization, and, consequently, a process of purification. The cause of the purifying effect is obvious: attachment of a proper particle to the growing crystal face is preferable and more probable than of a foreign one. Due to the "repulsion" of foreign particles by a growing crystal surface, crystallization is widely utilized in practice for purification of substances. In this purification process any crystal defects are "foreign particles" as well. In particular, vacancies and VAs form a "cloud" in front of the moving interface, as sketched in Fig. 2.46. This phenomenon gives rise to the following effects:

(a)   intensification of the VAs flow to the interface, resulting in a faster interface motion;

(b)   intensification of coagulation of vacancies into VAs and VAs into larger VAs in the "cloud";

(c)   dissipation of the "cloud" with time due to migration of the defects towards their lower concentration, that is, away from the interface.

Interface

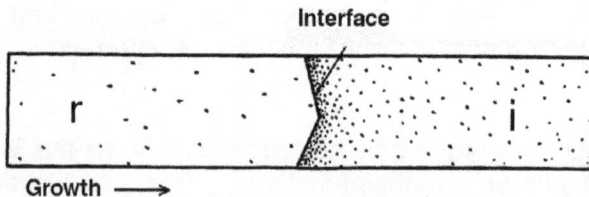

Growth ⟶

Fig. 2.46 Accumulation of crystal imperfections in front of moving interface. The phenomenon resulted from the fact that adding a proper particle to a growing crystal face is always more preferable than an improper. (Crystal defects such as vacancies or linear dislocations can also be considered "improper particles"). The well-known zone refining technique is based on the same principle.

The described mechanism of interface motion is responsible only for the basic phenomena of interface kinetics. These phenomena

- 152 -

clearly exhibit themselves if some precautions are taken: change in the specific volume upon the phase transition of the chosen object is small, the specimens are good small single crystals, the interfaces are flat, and the velocities of the interface motion are low. If these conditions are not met, the "basic kinetics" can be completely obscured by secondary phenomena caused internal strains. The secondary phenomena and their effects will be considered in Sec. 2.9.6.

2.9.3 Ten experimental facts of interface kinetics

2.9.3.1 *No phase transition in defect-free crystal*

This experimental fact, described in Sec. 2.5.2, item 11, is most fundamental. If crystals are "too perfect" they do not change their phase state at any temperature. Even when sites for 3-D nucleation are purposely created, the interface motion cannot proceed, lacking a sufficient concentration of VAs. Obviously, the crystals in which the phenomenon was observed were still far from being ideal, still containing numerous defects such as vacancies and linear dislocations. The defects that these crystals were lacking were VAs. A single VA is required to form a new 2-D nucleus on the interface every time a moving edge-bound ledge exits on the crystal surface together with the previously "consumed" VA. If the VAs are present, but their concentration is lower than required for uninterrupted 2-D nucleation, phase transition would still be impossible.

The fact that "too perfect" crystals are stable at all temperatures, no matter that their free energy is not the lowest, is of a primary importance for the theory of solid-state phase transitions. Crystal defects are the necessary participants in the process. One would be wise to take this fact into account prior to undertaking a theoretical work on phase transitions in ideal crystal medium.

2.9.3.2 *Temperature dependence*

There is a strong dependence of the velocity V of an interface motion on temperature. There is a problem, however, with V(T) measurements due to the V dependence on the availability of VAs. Therefore, the experimental curve in Fig. 2.47 should be considered as exhibiting only general qualitative features of temperature dependence. Every experimental point in the curve is the result of double averaging, first on all the transitions in each crystal, and then on different crystals. The curve shows that

- $V = 0$ at T_o; ($\Delta T_{tr} = T_{tr} - T_o = 0$). While this is self-evident, it may not seem as such if formulated differently: temperature of phase transition is the temperature at which the phase transition is unconditionally impossible (refer to Sec. 2.6).
- $V(T)$ is tangent to T-axis at T_o.
- V increases from zero as T_{tr} moves away from T_o up or down.
- the left-hand part of the $V(T)$ curve exhibits a maximum similar to that found in melt crystallization.

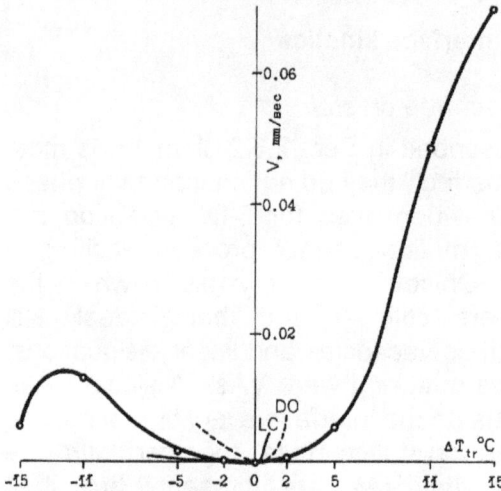

Fig. 2.47 Velocity V of interface motion in PDB against T_{tr}. Each experimental point was the result of averaging. The curves marked with letters were drawn from literature data: LC [Compt. Rend. *248*, 3157 (1959)], and DO [Docl.Acad.Nauk SSSR *73*, 1169 (1960)] to demonstrate poor reproducibility typical for kinetics measurements.

Phenomenologically, two factors shape the curve in Fig. 2.47: driving force and absolute temperature. The former is determined by the difference between the free energies of the phases; V is zero at T_o and increases as T_{tr} is moved away from T_o in any direction. The absolute temperature factor, on the other hand, affects V in one direction: the higher T_{tr}, the higher V. To the right from T_o the two factors act in the same direction, causing progressive V increase; to the left from T_o they act in opposite directions giving rise to the maximum.

2.9.3.3 *Hysteresis of interface motion*

There are two types of temperature hysteresis in solid-state phase transitions.   One, described in Sec. 2.6.6, is that of 3-D nucleation, ΔT_{3-D}, required to initiate phase transition in a crystal. The other is that of 2-D nucleation, ΔT_{2-D}, required to continue the transition, that is, to keep the interfaces moving; as a rule, it is much smaller. In order to observe it, interface motion should first be stopped by setting $T_{tr} = T_o$, and then set in motion again by deviation from T_o. However slow and careful the last procedure is performed, one will find that some finite overheating / overcooling $\Delta T_{tr} \neq 0$ is required in order to resume interface motion. The situation in the vicinity of T_o is shown schematically in Fig. 2.48. (For example, it would be observed if the part of Fig. 2.47 near T_o could be enlarged). Interface can move only if ΔT_{tr} is greater than some threshold value. Thus, a phase transition is intrinsically a non-equilibrium phenomenon. Therefore, when one encounters the title "Non-equilibrium phase transitions," the question "Do equilibrium phase transitions exist?" would be justified.

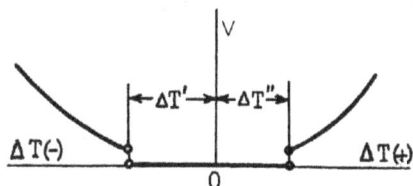

Fig. 2.48 Hysteresis of interface motion. The interface motion requires 2-D nucleation, and the latter requires overcoming some energy barriers. The sketch is to illustrate the experimental fact that ΔT lower of a certain minimum ($\Delta T'$ for cooling and $\Delta T''$ for heating) will not set an interface in motion. The phenomenon is similar to the hysteresis of 3-D nucleation.

2.9.3.4 *Depletion of the reserve of lattice defects*

Interface motion in the small rod-like (0.22 x 5 mm) PDB single crystal shown in Fig. 2.49 was subjected to long investigation on a hot microscopic stage. The crystal, grown from a vapor phase, was of a rather high quality. By temperature control the interface (seen in the photograph) was moved back and forth many times without letting it to reach the crystal ends. After 10 to 15 cycles, the interface was found moving slower in every successive cycle under the action of the same ΔT_{tr}. Additional 10 to 15 cycles stabilized the interface at a fixed position: it became completely insensitive to temperature changes. In this state the specimen consisting of the two phases divided by the interface could be stored for days at room temperature 20 °C, that is, about 11° below than $T_o = 30.8$ °C. After several days the resumed experiments revealed that the dependence $V = f(\Delta T)$ had been partially restored, but the same ΔT_{tr} produced much lower V. The initial V had not been regained even after several weeks.

The schematic in Fig. 2.49 illustrates the cause of the phenomenon. Cyclical movements of the interface over the length ℓ cause "cleaning" effects. While "consuming" some part of the VAs for the 2-D nucleation, the interface pushed other VAs from the ℓ area as a result of the "zone refining" effect. The interface completely stopped when the concentration of VAs, C_{VA} fell below the critical level required for the renewable 2-D nucleation. A partial restoration of the motion capability after the long "rest" was due to VAs migration from the end regions to the "working" region ℓ. This experiment makes the intrinsic irreproducibility of V quite evident. The velocity in the same specimen can differ by orders of magnitude under identical temperature conditions.

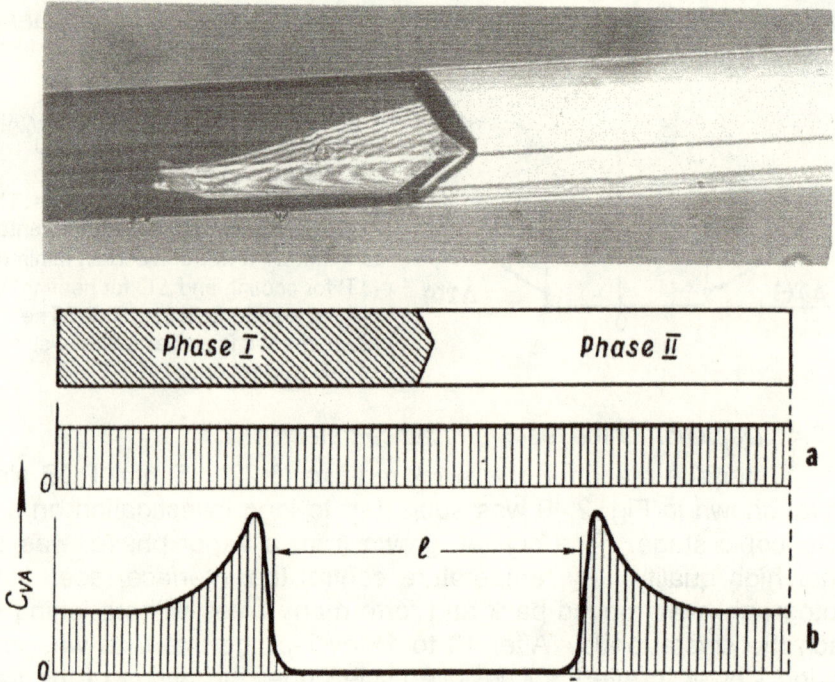

Fig. 2.49 Depletion of the reserve of lattice defects (VAs) needed for 2-D nucleation to keep the interface moving. The photograph shows the thin rod-like PDB single crystal with which the experiments were carried out. The drawings show what happens to the concentration of defects C_{VA}, being initially uniform (plot 'a'), after the interface was moved back and forth many times over the length ℓ (plot 'b').

2.9.3.5 *Velocity V as a function of the number of transitions*

This experiment was similar to that described in the previous section, but with two differences: the crystals were not so perfect and V was measured in every interface run. The specimens were oblong PDB single crystals grown from a solution. Due to significant V scatter the measurements were averaged over 20 crystals. A region of 1 mm long was selected in the middle of a crystal and the time required for the interface to travel this distance was measured. The outside regions played a certain auxiliary role. Two microscopes with hot stages set at $T_1 = T_0 + \Delta T$ and $T_2 = T_0 - \Delta T$ were used in the measurements. The V dependence on the ordinal number N of transitions in the cyclic process was measured for a fixed $|\Delta T|$. The $V = f(N)$ plots are shown in Fig. 2.50. Only qualitative significance should be assigned to them.

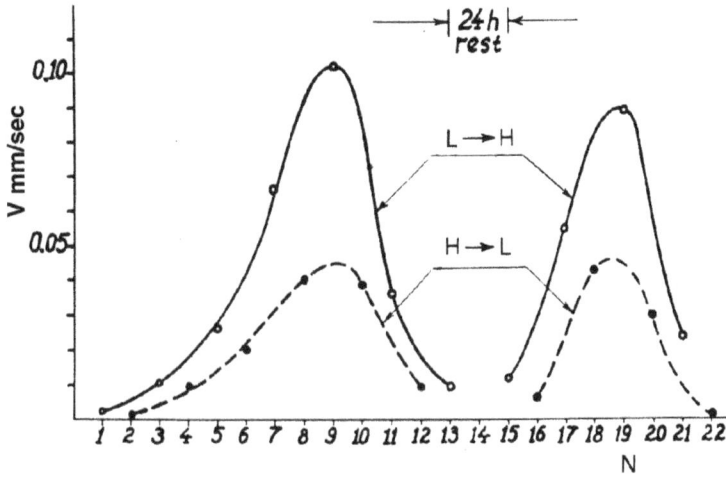

Fig. 2.50   Velocity of interface motion V in PDB single crystals as a function of number N of phase transitions in the cyclic succession L→H→L→H... Two phenomena are revealed: maxima of V and ability to show them again after a sufficiently long "rest."

A new finding is the maxima, and more specifically, their ascending side - because their descending side has been explained earlier. In general terms, a moving interface initially creates more VAs than it consumes, but the tendency is reversed after a number of successive transitions. In more detail, it occurs as follows. A moving interface accumulates a "cloud" of vacancies and VAs in front of it. If density of the vacancies in the "cloud" is sufficiently high, their merging into VAs creates more VAs than is expended on the 2-D nucleation. Considering that the number of vacancies in the region is limited, the consumption eventually prevails and V begins dropping. The recurrence of the whole effect after a long "rest" is due to migration of the vacancies from the end parts of the crystal.

2.9.3.6   Lingering in resting position

If an interface moving with a speed V_1 at a fixed ΔT_{tr} was stopped by setting $\Delta T = 0$, it tends to linger in the resting position once the initial ΔT_{tr} is restored. After some lag the interface leaves the resting position, but under a lower speed $V_2 < V_1$. The longer the resting time is, the lower the V_2. Once resumed, its movement accelerates to approximately the previous steady V_1 level.

The diagram in Fig. 2.51 explains this peculiar behavior. A uniformly moving interface pushes a "cloud" of crystal defects in front of itself. Its speed V_1 under isothermal conditions is controlled by the density of VAs in the cloud. The density, in turn, is controlled by the balance between accumulation and consumption of VAs. Holding the interface in one position allows the cloud to dissipate to the extent depending on the resting duration. In order to start moving again, the interface must now "dig" for VAs from the uniform distribution, leaving behind a "hole." As a result, $V_2 < V_1$, but approaches V_1 as the new "cloud" accumulates.

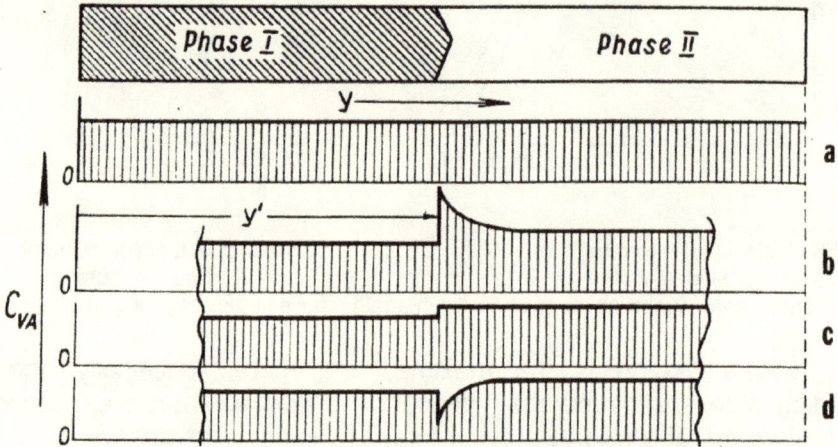

Fig. 2.51   The effects of moving and resting interface on the VAs concentration, C_{VA}.
(a)   The initial uniform distribution.
(b)   In the course of uniform motion of the interface which is in the position y' at the moment.
(c)   After long rest in the position y'.
(d)   After the motion was resumed.

2.9.3.7 *Memory of the previous position*

If after the procedure described in the previous section the reverse run immediately follows, the moving interface "stumbles" (is retarded spontaneously) exactly at the position where it was previously resting. The phenomenon, first reported in [146] without explanation, is almost a visual proof that the "hole" shown in Fig. 2.51d really exists. It provides a compelling support to the concept of interface kinetics based on the flow of VAs on the interface.

2.9.3.8 *Slower start upon repetition*

Using temperature control, it is possible to set up a cyclic process in which a single nucleus of the 'r' phase will appear, grow to a certain small size, and then dissipate back to the 'i' phase. In such a process, growth of the 'r' crystal in every subsequent cycle requires a longer time. The cause: the growing crystal consumes the available surrounding VAs for its 2-D nucleation, while the traveling distance is too short to accumulate a "cloud" of the defects. The concentration of VAs in the area is reduced with every successive cycle, providing a lower V.

2.9.3.9 *Acceleration from start*

Just after its nucleation, an 'r' crystal grows very slowly. It takes some traveling distance for the interface to accelerate (Fig. 2.52) and attain a steady V level. A "cloud" of VAs is initially absent, being accumulated during this "running start." Eventually a kind of equilibrium between the accumulation and consumption is reached, producing (in a uniform crystal medium) a uniform interface motion.

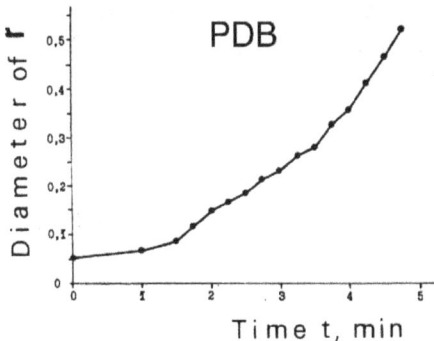

Fig. 2.52. An example of acceleration of crystal growth after onset of phase transition.

2.9.3.10 *Acceleration induced by approaching interface*

When there are several 'r' crystals growing from independent nucleation sites in the same 'i' crystal, it can be easily seen that the rates of their growth vary in a wide range. Considering that the initial crystalline matter and ΔT_{tr} are equal for all the growing 'r' crystals, this fact in itself is instructive in regard to kinetics of solid-state phase transitions: which of these rates does any existing theory account for? There is another phenomenon observed repeatedly: these rates are not quite independent of one another. In one instance, pictured in Fig. 2.53a, the crystal r_1 was almost not growing when a fast-growing face from r_2 began approaching from the opposite end. The latter crystal noticeably activated the growth of the former when the two were still separated by as much as 1.5 mm. As the r_2 was coming closer, growth of r_1 sharply accelerated (Fig. 2.53b).

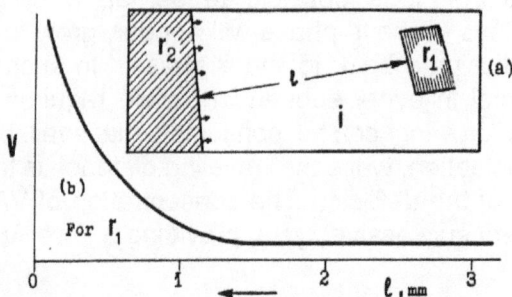

Fig. 2.53 Actuation and acceleration of growth caused by an approaching interface. (a) A sketch picturing a real case when one crystal (r_1) was initially not growing under some overheating ΔT = const, and then was actuated by the approaching interface from r_2. (b) Change in the observed velocity V for r_1 (a qualitative representation, but the distances ℓ between r_1, and r_2 are close to real).

The crystal r_1 was initially not growing due to the lack of VAs in its vicinity. Transport of VAs from r_2 to r_1 has spurred its growth. The growth progressively sped up as the flow of VAs increased from the "cloud" driven by the approaching r_2. A plausible additional cause is the strains spreading from r_2 (faster-moving interfaces produce stronger strains). The strains set in motion the static VAs dwelling at some distance from r_1 and thus start to foster its growth even before it is approached by the "cloud" from r_2.

2.9.4 Revision of a common concept of activation energy

In experimental studies of kinetics of solid-state phase transitions the phase ratio was measured *vs.* time with the objective to find the "activation energy of phase transition E_a." The process of a phase transition was heterogeneous, but E_a was interpreted as the energy barrier to be overcome in the process of a cooperative homogeneous rearrangement of one ideal crystal structure into another. The inconsistence of this approach is conspicuous. Because phase transitions between crystal states occur by 3-D nucleation and subsequent growth, there must be at least two activation energies: one for nucleation, the other for rearrangement at the interfaces. The latter process, in turn, involves two major stages: 2-D nucleation of molecular layers and molecular relocation at interfaces. The three basic activation energies that control the above three major stages of a solid-state phase transition are now considered.

E_a'. Activation energy of the formation of a 3-D nucleus. The E_a' depends on the particular structure (size- and configuration) of the lattice defect (OM) acting as the nucleation site (Sec. 2.5.4). The nucleation temperatures encoded in these sites are different, therefore

E_a' is not a unique characteristic of a particular phase transition. Rather, it can be of any magnitude greater than $E'_{a,min}$ corresponding to the $\Delta T_{tr,min}$. Absence of even a single OM in the crystal is equivalent to $E_a' = \infty$. This leads to the conclusion that attempts to find E_a' that is characteristic of a given phase transition would be physically unsound. This activation energy has nothing to do with interface kinetics. Phase transition in a fine-crystalline powder exemplifies the case when bulk rate of transition under changing temperature is governed exclusively by different E_a' encoded in the individual particles.

E_a''. Activation energy of the formation of 2-D nuclei on a molecular-flat interface. E_a'' is not a fixed number either. It varies owing to structural differences (size and shape) in the lattice defects (VAs) acting as the nucleation sites. The VAs must be present in quantities and located near the interface in order that the latter be able to propagate. If this condition is not met, the phase transition (interface motion) will not be possible, which is equivalent to $E_a'' = \infty$. At moderate concentrations of VAs the interface motion is controlled more by the availability of VAs than the E'' magnitudes. A lower speed of an interface motion at the same temperature is an example of interface kinetics governed by VAs availability. In the case of high VAs concentrations, when only a small part of the available VAs is "consumed" during interface motion, molecular relocation across the interface starts limiting the interface speed.

E_a'''. Activation energy of molecular relocation at the contact interface. As shown in Sec. 2.4.4, the process in question is a "stimulated sublimation." Hartshorne and associates (Sec. 2.4.1) used experimental dependence V(T) to estimate the E_a (E_a''' in the present notation) from the Arrhenius equation and received a meaningful result - even though with 30% error. Their experiments fit the conditions at which E_a''' plays a part in controlling the interface kinetics. The E_a''' was "extracted" from the mutual action of the VAs and E_a''' because absolute temperature affects molecular mobility much more strongly.

2.9.5 Relationships between the controlling parameters

The complications and instabilities of interface kinetics are rooted in feedbacks. An interface needs certain conditions for its motion, but its motion affects these conditions. Flowchart 2 summarizes relationships between the parameters responsible for the interface kinetics controlled by VAs flow. After the foregoing discussion, the flowchart is self-explanatory even if it may seem cumbersome. Connections between the parameters should be traced from the bottom up following solid-line arrows. The feedbacks that turn the process into autocatalytic are shown by broken lines. The temperature effects are of two kinds. One is ΔT_{tr}, which provides energy gradient for phase transition. The other is absolute temperature - the cause of molecular vibrations and other mobilities. The flowchart illustrates that (1) phase transition in an ideal crystal is not possible and (2) the phenomena of kinetics are complex, multiparameter and irreproducible in spite of the simplicity of the *contact* mechanism. Yet, the flowchart represents only the simplest case of slow interface motion in a real single crystal of a good quality and when the accompanying strains are sufficiently small not to create the additional complications described in the next section.

FLOWCHART 2: interface kinetics controlled by VAs flow
(non-epitaxial phase transitions).

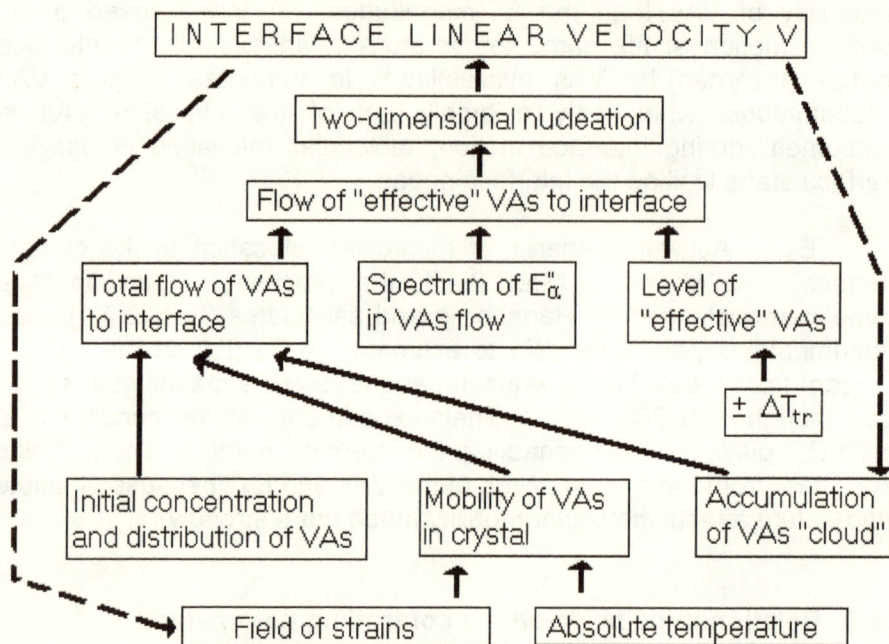

2.9.6 The "truly out of control" kinetics

If the previously described interface kinetics may seem "out of control," it still represents the simplest and most orderly case. A smooth advancement of a flat interface takes place only if certain precautions are taken: the specimen is a good small single crystal and ΔT_{tr} is low; it is also helpful if the specific volumes of the polymorphs are close and the crystal has a plate-like shape. Then the strains arising at the slowly moving interface can dissipate before damaging the 'i' crystal medium. If these conditions are not favorable, the crystal growth loses visually orderly character. This disorderly morphology for a century delayed discovery of the underlying phenomenon presented in Sec. 2.2: growth of naturally-faced crystals. (Another cause of the delay was not using optical microscopy and transparent single crystals.) The phase transition in most instances appears to an observer as a blurred thick "wave" rolling over the crystal and quickly completing the process, leaving behind a less transparent material. What kind of kinetics is that? The X-ray patterns reveal that a single crystal turns into a polycrystal. All facts taken together suggest that the interface *generates* multiple lattice defects, OM, acting as the sites for 3-D nucleation immediately in front of itself. This is caused by the strains originating from the fast-moving interface. Because VAs are generated as well, the new growth proceeds quickly in both directions: toward the interface and out of it, creating new strains. Not having time for relaxation they again damage the adjacent lattice. This kinetics is based on the positive feedback:

INTERFACE MOTION → STRAINS → GENERATION OF NEW NUCLEI JUST IN FRONT OF THE INTERFACE → GROWTH FROM THESE NUCLEI (INTERFACE MOTION) → STRAINS ... and so on.

It should be noted that the term "interface" is used here only conditionally: actually, it is a rather thick heterophase layer. Such an interface can move with a relatively high speed.

Fig. 2.54 shows an instance of a sudden change of orderly crystal growth to the kinetics based on the positive feedback. It exhibits itself as a local "explosion" with the higher speed of interface motion by one order of magnitude. Thus, in 'c' two different kinetics transparently manifest themselves in the same initial crystal.

A peculiar negative feedback causing a "rhythmical" interface motion was observed in elongated plates. There the accumulating strains periodically relaxed by mechanical shear along the cleavage direction, temporary interrupting the interface motion. A pattern of roughly evenly spaced strips parallel to the cleavage emerged.

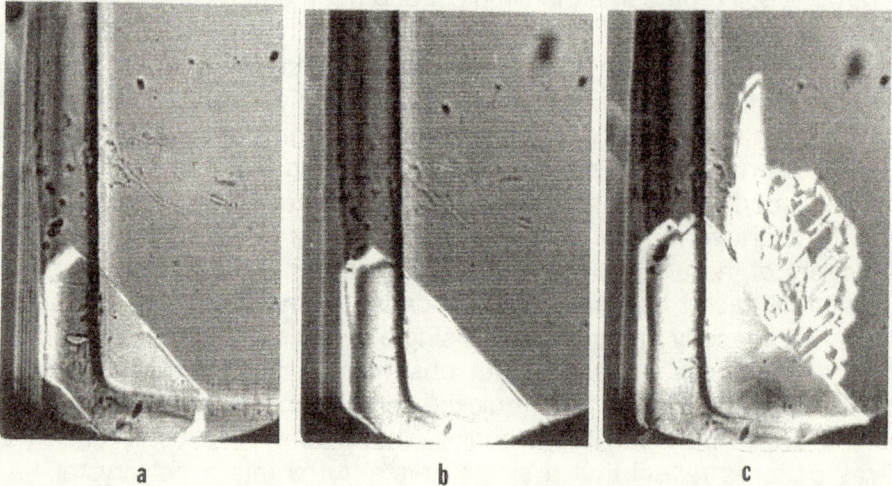

Fig. 2.54 Sharp conversion from a "quiet" interface kinetics based on consumption of available defects ('a' and 'b') to that based on generation of new defects by the strains spreading from the interface ('c'). Only one corner of the 'i' crystal - where the 'r' crystal was growing - is shown. The conversion manifests itself as an "explosion" on the flat surface.

CHAPTER 3. "LAMBDA-ANOMALIES" AND OTHER APPARENT ANOMALIES

3.1 The appearance of different physical properties upon measurement of a heterophase system

The only process which takes place in a specimen passing through the temperature range of transition is a gradual change of the mass fraction m_i of the 'i' phase from 1 to 0, and the mass fraction m_r of the 'r' phase from 0 to 1; $m_r = 1-m_i$. At every instant the two phases are physically separated, each phase possessing its own independent set of properties. If a property P is the point of interest, the physical status of such a system is properly described by $m_r(T)$ (or $m_i(T)$) and *two distinct plots* $P_i(T)$ and $P_r(T)$. However, observation of a *single combined curve* $P_{obs}(T)$ through the range of transition has been the most common way to study phase transitions. *A continuous appearance of these amalgamated curves $P_{obs}(T)$ is due to a gradual increasing m_r in the specimen at the expense of m_i, and not to a continuous changing $P_i(T)$ into $P_r(T)$ as one believed.*

In order to properly interpret the results of any measurements through a transition range, we should notice that all properties and the corresponding methods of measurement of a two-phase entity can be categorized by type regarding the appearance of $P_{obs}(T)$. They are summed up in Table 3.1 and commented below.

TYPE 1. The properties/measurements exhibiting no anomalies. Measurements of some properties can be "tuned" selectively at only one phase in the two-phase mixture. Particularly, this takes place in precision X-ray measurements of lattice spacings and unit cell parameters. The set of X-ray reflections characteristic of one phase does not coincide with that of the other phase. Each set can be observed and treated separately. Such measurements reveal the coexistence of two distinct phases in the transition range, their normal thermal dilatation, and no sigmoid (S-) or λ-anomalies. The X-ray low-angle patterns in Fig. 2.41 are one illustration of such measurements; another is given in Fig. 3.1a. Sometimes, however, the X-ray resolution turns out insufficient to detect the actual two-phase state of the specimen.

Table 3.1 The response upon measurement of a heterophase system (key examples)

	PHYSICAL PROPERTIES / METHODS			
	TYPE 1	**TYPE 2**	**(a) ⇐ TYPE 3 ⇒ (b)**	
Response upon recording	REFLECTING DISTINCT QUALITY OF A POLYMORPH	QUANTITATIVE, FEATURING m_r or m_i	FEATURING dm_r / dT	PROPORTIONAL TO PRESENCE OF NUCLEI AND / OR INTERFACES
OBSERVED "ANOMALY"	**NONE**	**S - CURVE**	**λ - PEAK**	**λ - PEAK**
	X-ray (or another diffraction technique) density	Dilatometric density	Dilatometric coefficient of thermal expansion	Light scattering
	Crystal lattice parameters	Thermal conductivity	Heat capacity as commonly appears in calorimetric measurements	Neutron scattering
	Mössbauer spectra	Dielectric constant (except in epitaxial phase transitions parallel to layers)	Dielectric constant in epitaxial phase transitions - parallel to layers	
	ANY QUALITY UNIQUE TO ONLY ONE OF THE POLYMORPHS (Recording of sufficiently high resolution, not reflecting quantity)	Electrical resistance or conductivity / Magnetization of ferromagnetics / Polarization of ferroelectrics	ANY FUNCTION DEFINED AS THE FIRST DERIVATIVE OF TYPE 2 PROPERTY	

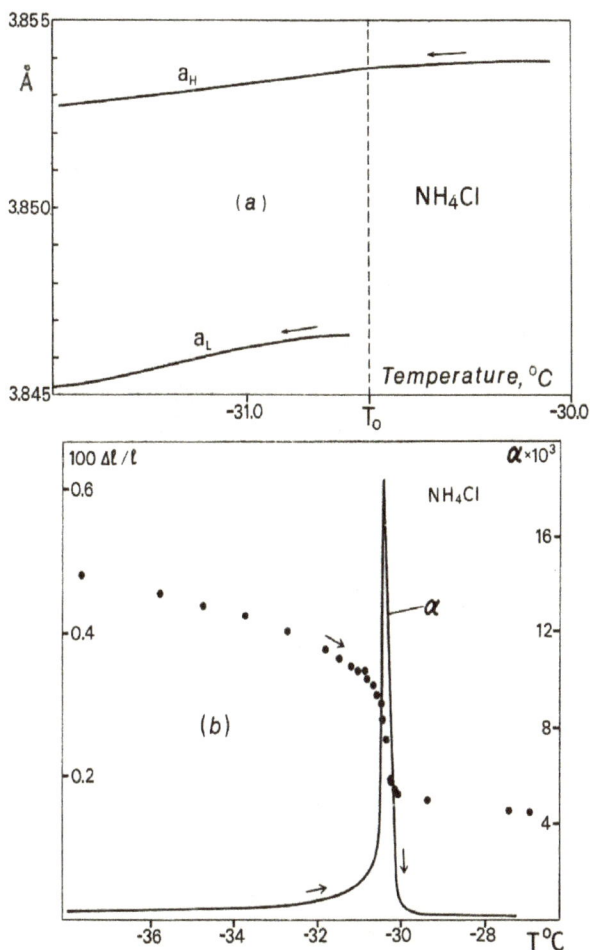

Fig. 3.1 The X-ray unit cell data (Dinichert [110]) and dilatometric data (Lawson [142]) through the temperature range of NH4Cl phase transition. The dilatometric measurements produced a result conflicting with the more informative precision X-ray measurements. The X-ray data should be given a priority as providing "vision" of the process inside the specimen as opposed to a dilatometric method which is "blind."

(a) The lattice parameters a of the cubic unit cells of the H and L phases. The phases independently coexist in a temperature range, showing a normal thermal dilatation with no S-like gradual change from a_H to a_L (and hence the unit cell volume from $(a_H)^3$ to $(a_L)^3$).

(b) The S-like array of experimental points is the dilatometric relative linear thermal expansion $\Delta\ell/\ell$ in NH4Cl. The solid line represents a λ-peak of the volume coefficient of thermal expansion α derived from the dilatometric data. Dilatometry cannot detect the fact of a two-phase state, but the X-ray data clearly reveal that the apparent S- and λ-anomalies are due to a change in the relative content of H and L in the specimen.

- 167 -

Spectroscopy is another example, but only when a sufficient resolving power can be provided, which may not be the case. On the other hand, a Mössbauer spectroscopy with its very high resolution has revealed a coexistence of two distinct phases in Fe and Ni ferromagnetic phase transitions when even X-ray methods failed (refer to Chapter 4).

TYPE 2. The properties exhibiting S-curves. Some other properties can be measured only as a superposition of the responses from the two phases present in the specimen. Such a property is an additive function of the independent contributions from all its constituent parts. Electric resistance R, dilatometry-measured specific volume v, and (with rare exceptions) spontaneous polarization P_s are examples. Given temperature T, the observed property P_{obs} is

$$P_{obs} = P_i m_i + P_r m_r = (P_r - P_i) m_r + P_i$$

Because P_i and P_r are much weaker functions of temperature than m_r this is very closely approximated by

$$P_{obs}(T) = A m_r(T) + B,$$

where A and B are constants, *i.e.* $P_{obs}(T)$ will assume the S-shape of $m_r(T)$. The inverse functions, such as density $\delta(T)$ or electrical conductivity $\rho(T)$, will, of course, assume the S-shape too, although with the inverse inflection.

The classical "second-order" (and later on a "partially second-order") phase transition in NH_4Cl represents a compelling proof of the foregoing account. In Fig. 3.1 two kinds of measurements of specific volume through the transition range are compared: (a) X-ray data and (b) dilatometric data. The dilatometric measurements (array of experimental points), which are of Type 2, exhibit a continuous S-like change, while the X-ray data, which are of Type 1, prove that no such anomaly really exists. The source of the discord is the experimentally established S-like shape of $m_r(T)$ shown in Fig. 2.30b.

TYPE 3. The properties exhibiting λ-peaks. The following general rule is true: *λ-peaks in the $P_{obs}(T)$ plots are characteristic of those properties which are determined by the rate $dm_r(T) / dT$ rather* than by the $m_r(T)$ itself. In the following sections we shall inquire into the specific features of different λ-peaks to show in detail how this rule works.

There is an additional source of "λ-anomalies." As stated repeatedly, two polymorphs have independent sets of physical properties P(T). No restrictions exist regarding disposition of the curves $P_L(T)$ and $P_H(T)$. In principle, there can be four mutual dispositions shown in Fig. 3.2. Two of them, 'c' and 'd', give rise to a figure reminiscent of a λ-peak.

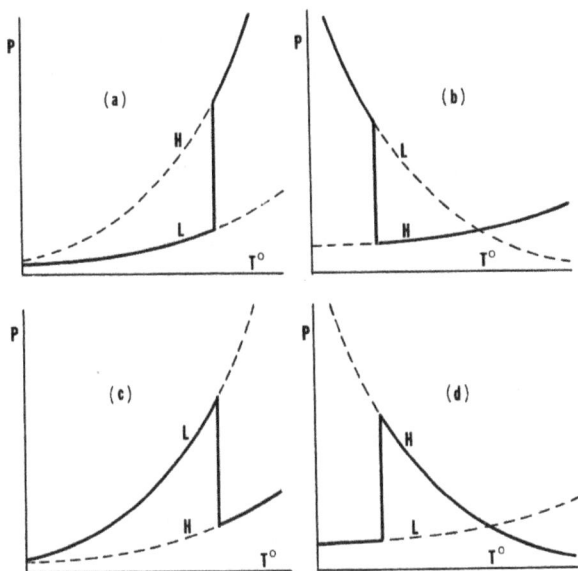

Fig. 3.2 Another reason contributing to the illusion of the λ-anomalies: two of the four possible mutual dispositions, namely 'c' and 'd', of the plots of a physical property P(T) for two participating phases give rise to the resultant plots (solid line) resembling "anomalous peaks." (The sketch does not reflect the existence of a temperature range of phase transitions.)

3.2 Experimental imitation of a "continuous transition" and a "λ-anomaly"

The account for "phase transition anomalies" presented in Table 3.1 will be illustrated experimentally. It will be demonstrated that a nucleation-growth phase transition under typical experimental conditions will have the appearance of a continuous transition and exhibit the "λ-anomaly" proportional to the rate of transition dm_r / dT. Glutaric acid ($T_0 = 64.0$ °C) was chosen. Its phase transition by nucleation and crystal growth was exhibited in Sec. 2.2.3 and Fig. 2.10. The experiment was of two parts: recording the concentration of free radicals with an electron spin resonance (ESR) spectrometer and measurements of T_n using an optical microscope with a hot stage.

ESR experiment. Hundreds of crystalline plates of ~1 mm size were grown from a solution and X-irradiated to produce a concentration of free radicals sufficient to obtain an intensive ESR signal. One hundred crystals of nearly equal size and appearance were set aside to be studied under a microscope; the remainder was used for the ESR study. The effect of heating the specimen through phase transition on the relative number of free radicals was measured. At T_r the content of free radicals did not change for weeks. On heating, outside the transition range it decreased moderately enough to enable the detection of accelerated free radical recombination during the phase transition. The specimen was heated in the ESR spectrometer at the rate of 0.4 °C/min. The spectrum was automatically recorded every three minutes. The amplitude of the most intense maximum was chosen as the indicator of the relative number N of free radicals in the specimen.

The N(T) shown in Fig. 3.3a features a typical "continuous transition" (S-curve) between 70 and 79 °C. This range of transition is obviously shifted up from $T_o = 64$ °C, the S-curve being a right-hand branch of the hysteresis loop.

The curve in Fig. 3.3b (solid line) is obtained by a graphical differentiation of the first curve. It represents a rate of recombination of free radicals dN/dT during the uniform heating of the specimen. The plot is a typical λ-peak. The "critical point" found from the position of the maximum, $T_c = 72.3°$, is 8.3° higher than the actual T_o. Here we deal with the perfect illusion of a "critical phenomenon" when the system starts "preparing" for the phase transition *before* T_c is reached. In fact, the first indications of the transition appear ~6° *after* T_o has been passed (refer to Sec. 2.6.4).

Tn measurements. In each of the 100 selected crystals the nucleation temperature of the L → H transition was separately measured. The crystals were placed in turn on the hot stage and heated at the same rate 0.4 °C/min; the T_n was read out when the first sign of the transition was seen in the microscope. In practice, $T_n = T_{tr}$ in a particular crystal, considering that the transition after nucleation was quickly completed. The temperature scale was divided into intervals $\delta T = 0.5°$ and the number of crystals showing phase transitions in each δT was calculated. The resultant dm_r / dT were marked in the appropriate scale on the plot dN/dT (Fig. 3.3b). The excellent agreement between the ESR and T_n data makes it obvious that the λ-peak merely represents the rate with which the 'r' phase in a heterogeneous system replaces 'i' over the range of transition.

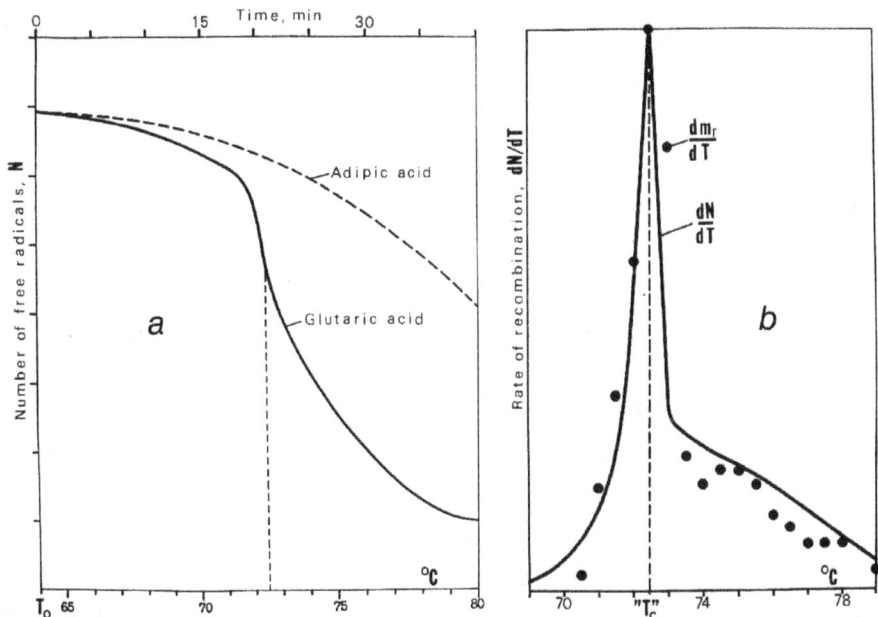

Fig. 3.3 Experimental imitation of a "continuous transition" and a "λ-anomaly" with the quite "discontinuous" (*i.e.*, first-order, nucleation-growth) phase transition in glutaric acid.
(a) The temperature dependence of the number of free radicals N shows a continuous S-like change. The specimen (coarse crystal powder) was heated at 0.4 °C/min rate. The similar plot for adipic acid which does not undergo a phase transition is given for comparison.
(b) The rate of recombination of free radicals dN / dT *vs.* T in the specimen as a whole (solid line) and the rate of nucleation (= phase transition) dm_r / dT in the crystal particles in a parallel experiment (dots). The solid line was obtained by a differentiation of the curve in 'a'. One can see that this curve (a typical λ-peak) represents, in a proper scale, the rate of nucleation.

3.3 "λ-Anomaly" of volume coefficient of thermal expansion

A very typical property that is claimed to exhibit λ-peaks is a volume coefficient of thermal expansion α. An example is given in Fig. 3.1b (solid line). As α(T) is a first temperature derivative of the specific volume v(T) by definition, the correlation between them is trivial: the first derivative of an S-shaped plot is always a peak. One can choose any property of Type 2 (S-curve in Table 3.1), take its temperature derivative (it will be a "λ-anomaly"), give this function a name (such as "temperature coefficient of conductivity," "temperature coefficient of spontaneous polarization") and assert that it is a "critical phenomenon."* A problem with all the reported λ-peaks of α(T) is that these "anomalies" (or "singularities") are not any more real than the dilatometric S-curves of $v(T)$ from which they were derived. So, the source of the λ-peaks in α(T) plots is again the S-like shape of $m_r(T)$. The equation of a dilatometric S-curve is

$$v_{obs} = (v_r - v_i)\, m_r(T) + v_i,$$

and the approximate equation for α(T), assuming v_i, $v_r \neq f\,(T)$ as compared to $m_r(T))$, is

$$\alpha = (1/v_{obs})\, dv_{obs}\,/\,dT = (1/v_{obs})\,(v_r - v_i)\, dm_r\,/\,dT.$$

Thus, "λ-anomalies" of thermal expansion coefficients - a mysterious physical phenomenon - have that simple explanation: they are not anomalies. Rather, it is mysterious how this misconception could survive for such a long time. The 1940 dilatometric study by Lawson [142] revealed the "λ-anomaly" of α(T) in the NH₄Cl phase transition (Fig. 3.1b). But in i942 Dinichert investigated the case with a much more informative technique. His superb X-ray experiments completely clarified the phenomenon [110]. The work made it obvious that there were two coexisting phases in the range of transition, each exhibiting no traces of a singularity regarding the density and thermal expansion (Fig. 3.1a); only the ratio $m_r\,/\,m_i$ was changing along a sigmoid curve over the transition range (Fig. 2.30b).

This complete clarification by Dinichert has either been not understood, or disregarded in subsequent considerations, even though his article was sometimes mentioned. Quite a few studies re-examining the case have been carried out, the 1971 dilatometric by Fredericks [143] and 1970 X-ray powder by Boiko [144] being characteristic. These studies were superfluous. In every respect they fell far behind the Dinichert's work, even though they were carried out 30 years later. Suffice it to say that the heterophase state in the transition range was neither measured, nor detected, nor taken into account, nor even mentioned. An outcome of this backward development was the establishment of a general conviction in the existence of the λ-anomalies of thermal expansion coefficients. The particular case of NH₄Cl played a major role in this matter. Many other λ-anomalies of α(T) were reported. They are all destined to be nullified. A simple way

---

* Example: Rao and Rao [42, p. 295] reproduced a resistance *vs.* temperature plot (S-curve) and its derivative (λ-peak) and supplied them with the caption: "Note the second-order transition indicated in the plot of the derivative of resistance".

to do this is to demonstrate that the S-curves of $v(T)$ exhibit hysteresis, because hysteresis of thermal expansion is unthinkable in the light of basic physics.

3.4 "λ-Anomaly" of heat capacity

3.4.1 Special role of the heat capacity λ-peaks

Specific heat λ-peaks have been reported in hundreds of experimental studies. Since the 1920s, when the phenomenon was found in NH_4Cl, they attracted the attention of both experimentalists and theorists. To cite an example, in the 1979 book by Parsonage and Staveley [54] one can find more than 30 figures with the λ-peaks presented as "Specific Heat C_p of [substance] vs. Temperature T." The most famous was that of NH_4Cl. The specific heat λ-peaks have become the main stimulation to develop the theory of critical phenomena and to be explained by it. Even though the theory was unable to explain them, it was believed that the theory was basically correct. In the meantime, their actual origin was explained in 1979 [105]: the "heat capacity" λ-peaks do not belong to the $C_p(T)$ plots. They actually represent a latent heat of first-order (nucleation and growth) phase transitions. That account was ignored: it was neither challenged, nor accepted.

3.4.2 [Nucleation range] + [Latent heat] = ["λ-Anomaly"]

The λ-peaks are an effect of the nucleation-growth nature of phase transitions. Owing to a temperature range of transition in which the mass fraction m_r gradually changes from 0 to 1, the latent heat Q of the i → r transition is absorbed (in L→ H) or released (in H → L) with the corresponding rate.

Let us consider a typical calorimetric heat capacity experiment that exhibits a λ-peak. A description of the routine procedure can be found in the review by Westrum and McCullough [49]. When passing through the range of transition, they explain, the mode of measurement is on an intermittent basis, allowing time for the establishment of thermal equilibrium after each input of electrical energy into the calorimeter "so that the measurements truly represent equilibrium data." These authors, as usual, believed that a sufficiently long waiting time between the successive measurements could ensure *thermodynamic* equilibrium.

However, as has been shown in the previous sections, a phase transition is a *non-equilibrium* process by its very nature. A prolonged retention of the specimen at a constant temperature within the transition range can provide only the *thermal*, and not thermodynamic equilibrium. The specimen consists of the particles (or blocks, grains, domains) of both phases (refer to Fig. 2.28). Each new input of energy into the calorimeter gives rise to (a) temperature increase by δT and (b) phase transition in the next portion of the particles (blocks, grains, domains) whose "pre-coded" T_n fell in this interval δT. Their latent heat of transition is added to the measured value of heat capacity. Separation of the latent heat from the specific heat is impossible in such calorimetric measurements. As a result, a *false "heat capacity" peak is superimposed the true heat capacity curve.*

Fig. 3.4 illustrates the foregoing. If there was no latent heat of transition, a true C_p would be delineated by the S-curve shown with a dashed line. It can be called "true" only conditionally, though, for it represents an average of two separate contributions from the coexisting

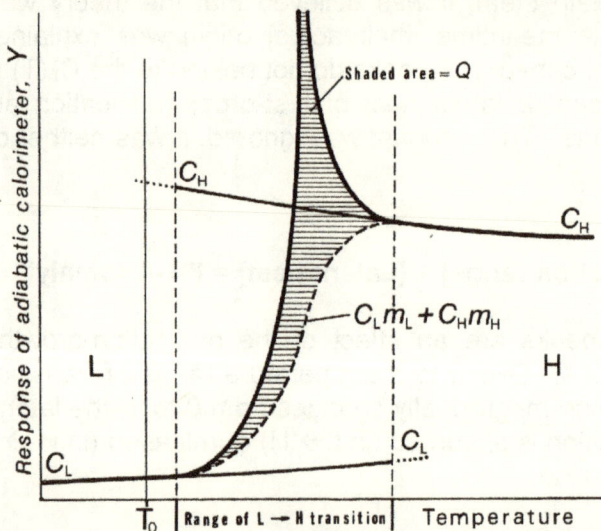

Fig. 3.4   Specific heat C_p in the range of phase transition (schematic). The drawing features the following (the subindex "p" is omitted):

• Two separate overlapping plots $C_L(T)$ and $C_H(T)$ would represent the specific heat most clearly.

• The S-curve (dashed bold) represents amalgamated plot of two independent contributions C_L and C_H from the coexisting phases L and H, taking into account their mass fractions m_L and m_H; $m_L + m_H = 1$.

• The bold-faced curve with a "λ-anomaly," which would be produced by an adiabatic calorimetry. It is composed of both C_L and C_H contributions (dashed curve) and a superstructure of the latent heat Q of the phase transition.

- 174 -

phases. The contribution y of the latent heat is proportional to the rate of phase transition dm_r/dT:

$$y = Q\, dm_r/dT,$$

where Q is specific heat of transition. The equation of the calorimetric response Y is

$$Y = C_p + y = m_r\, C_{p,r} + (1 - m_r)\, C_{p,i} + Q\, dm_r/dT.$$

The latent heat of phase transition is equal to the area of the λ-peak resting upon the dashed line. If the range of transition (actually, range of nucleation) is narrow enough, the peak is narrow and is called "λ-anomaly." At lower temperatures, where the range of nucleation is usually wider, the λ-peak can degenerate simply to a "hump," in which case it is called "Schottky anomaly" after German scientist W.H. Schottky.

3.4.3 Hysteresis of the "heat capacity λ-anomalies"

Thus, the λ-singularity in question does not exist. A "specific heat λ-anomaly" is merely the latent heat of nucleation-growth phase transition, the area of the peak being a quantitative measure of the latent heat. This simple solution to the old problem may seem difficult to accept, considering its grave impact on the critical phenomena concept. Any doubts should be dispelled by the temperature hysteresis of the alleged "heat capacity" peaks, which is irreconcilable with the physical laws.

Let us sum up the picture of a phase transition cycle L → H → L in a powder specimen (Fig. 3.5). The constituents of this picture are all the familiar "anomalies": "continuous transition" in the transition range (S-curve), hysteresis loop, λ-peak. Each crystal particle in the specimen shows hysteresis determined by the ΔT_n "pre-coded" in it (plot 'a'). Because ΔT_n is not the same in different crystal particles, plot 'b' for the whole specimen differs from 'a' by expansion of m_r (T) in each direction into a transition range. Plot 'b' differs from Fig. 3.4 (dashed line) by having the second transition range shifted in the opposite direction from T_0; taken together, the two such dashed curves form the familiar hysteresis loop. Finally (plot 'c'), each S-curve is superimposed by a λ-peak of the latent heat of phase transition in calorimetric measurements.

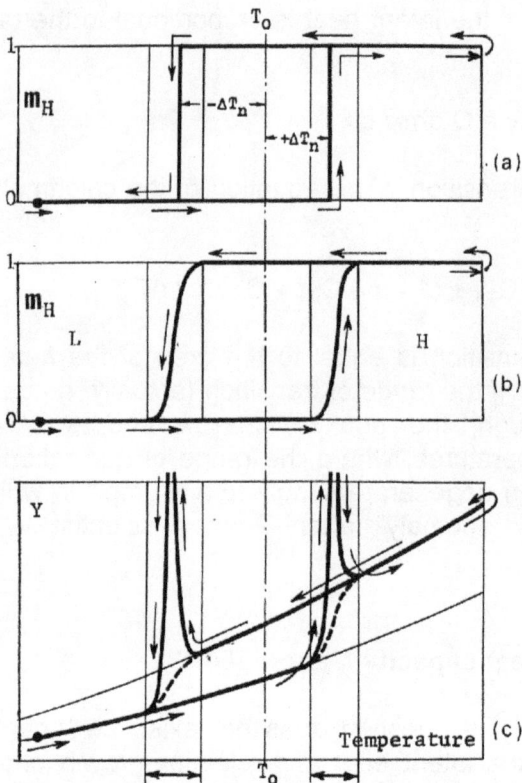

Fig. 3.5 "Continuous" transitions (S-curves), hysteresis loop, and λ-peaks upon a cycle L → H → L. Ranges of transitions are marked by bilateral arrows. There the specimen is two-phase. m_H is mass fraction of H phase.
(a) Hysteresis in a single crystalline particle of a powder specimen.
(b) Spread of each transition over a range for the specimen as a whole owing to different T_n in the constituent particles; two continuous S-curves, one in the forward and one in the reverse direction form a familiar hysteresis loop.
(c) λ-Peaks due to latent heat of phase transition. They are superimposed on the S-curves of $C_p(T)$.

3.4.4 Instructive story of "specific heat λ-anomaly" in NH_4Cl

In this and the following several sections the canonical case of "specific heat λ-anomaly" in NH_4Cl around -30.6 °C will be reexamined. This case is of a special significance. It was the first where a λ-peak in specific heat measurements through a solid-state phase transition was reported; it was in 1922 [145]. Since then the phase transition was the subject of numerous studies by different experimental techniques. Now it is considered the most thoroughly investigated. In every calorimetric work [e.g., 103, 148, 151, 152, 153] a sharp λ-peak in this phase transition was recorded; neither author expressed any doubt in a specific heat nature of the peaks. The transition has been designated as a *cooperative order-disorder phase transition of the lambda type* and used to exemplify this type of phase transitions. Yet, after all that effort no one could claim that the λ-anomaly was well understood. As it has now become clear, there were good reasons for that failure: the specific heat anomaly was nonexistent. If this classical anomaly is dethroned, no credence can be lent to other specific heat λ-anomalies.

The story of the NH_4Cl phase transition and its "specific heat λ-anomaly" is instructive also because NH_4Cl was the only example used by Landau in his original articles on the theory of "continuous" second-order phase transitions - in spite of prior warnings that this could be mistaken. In the 1935 article "On the theory of specific heat anomalies" [146] Landau asserted that his theory "agrees well with the measurements...on NH_4Cl." The case of NH_4Cl was also used by Landau in a 1937 paper to exemplify "continuous" second-order phase transitions. However, as early as 1934, Justi and Laue in several articles [33] argued that second-order phase transitions have no thermodynamic or experimental justification, that phase transitions of the NH_4Cl type have their heat of transition spread over a finite temperature interval, and that overheating / overcooling may occur. In the same year, another author (Euchen [147]) called NH_4Cl-type phase transitions "disguised" first-order transitions. Even though those authors were still far from advancing the nucleation-growth concept, they were closer to the real nature of things than Landau and the whole school of subsequent workers who ignored their publications, as indicated by an absence of references.

In the following few years certain evidence was obtained undermining both a heat capacity origin of the λ-peak and a cooperative character of this phase transition. It was found by Extermann and Weigle [148] that the "anomalies of heat capacity," as they called the λ-peaks, exhibit temperature hysteresis that did not result from kinetics of the transition: the λ-peak appeared upon heating at a higher temperature than upon cooling. Because heat capacity hysteresis runs counter to the basics of thermodynamics, this should have raised doubts about the heat capacity nature of the λ-peaks. Any attempt to account for a heat capacity hysteresis requires making an implausible assumption (as Ubbelohde did, see below) that thermodynamics does not always hold true. Another piece of evidence is the Dinichert's X-ray data [110] touched upon earlier (Figs. 2.30b and 3.1a). They revealed that the NH_4Cl specimen remained two-phase over a temperature range of transition, while only the mass ratio between the two crystal phases was changing. Although both cubic crystal lattices seemed to be close, their difference was distinct at all temperatures of the range and was accountable for the 0.45% divergence in the crystal densities. These findings were inconsistent with the continuous cooperative mechanism assumed for this phase transition, but have not led to a clarification of its mechanism in the minds of workers in the field. The "specific heat λ-anomaly" remained intact. To illustrate the state of affairs during the ensuing years, a few examples will be reviewed.

Ubbelohde [149, 150], in divergence with the mainstream, criticized the contemporary theory for disregarding the structural mechanism of phase transitions. He noticed and discussed the phase coexistence discovered by Dinichert, but failed to recognize the independence of the phases and need to represent them with two separate plots. The λ-peaks in the amalgamated calorimetric plots were still taken for a true $C_p(T)$. Thermodynamics was declared inadequate to account for phase transitions. A "hybrid crystal" concept was introduced [107,109]. Inconsistence of that concept with thermodynamics is reason enough not to discuss it further.

Voronel and Garber [151] concluded the NH_4Cl phase transition to be of the first order in terms of finding some latent heat. The conclusion was not new because phase coexistence found by Dinichert 25 years earlier was even clearer evidence. As usual, it was unnoticed that the λ-peak represented a two-phase rather than monophase system. As a result, the notorious "heat capacity λ-peak" was observed. This was contradictory to the established belief that λ-peaks are an exclusive characteristic of second-order transitions. The transition was termed "close to second-order" and treated as quite second-order (ordering parameter, fluctuations, etc.). Nucleation and growth were not mentioned and apparently were not recognized as the characteristics of a first-order transition. The latent heat was determined as being significantly lower than the "total"(?) transition heat $Q_{total} = 84 \pm 10$ cal/mole.

Several years later Schwartz [103] again undertook a reexamination of the "heat capacity" λ-peak in NH_4Cl. While carrying out the experiments, one specific problem persisted. It was a two-phase state of the specimens. The author was determined to exclude the two-phase state from consideration and ultimately (mistakenly) presented his "heat capacity" peaks as relating to a homogeneous specimen. The Dinichert's results were mentioned, but misinterpreted. For example, Schwartz believed that the transition occurs at two temperature points, one on heating, the other on cooling. But the Dinichert's data, as seen in Fig. 2.30b, revealed that it took place anywhere within two temperature ranges. Schwartz emphasized the "underlying second-order nature of the transition" and, without further justification, treated it as a second-order transition. Not surprisingly, a number of the observed facts were found to be just the opposite to what was expected (Sec. 3.4.7).

Measurements of the heat capacity of NH_4Cl were also carried out by Chihara and Nakamura [152]. The most attention was paid to the "λ-transition." The measurements were done only upon heating. Their objective was "to reveal the shape of the heat capacity anomaly in the vicinity of $T_λ$" with a greater precision. The instrumental precision was, indeed, very high, but three earlier established crucial facts were neglected: (a) different location of the λ-peak, if measured upon cooling, proven by Extermann and Weigle [148], (b) "strong sample dependence in the data," as Schwartz [103] pointed out, and (c) a two-phase state of the sample, demonstrated by Dinichert [110]. Considering this, the high technical precision was of little value. The "heat capacity" peak was subjected to a detailed theoretical analysis implicitly assuming the object to be homogeneous. The transition was interpreted as taking place at a fixed temperature point indicated as the position $T_λ = 245.502 \pm 0.004\ °K$ of the λ-peak (about 3 °C higher than actual T_0). A remark "the transition is very close to the first order" did not stop the authors from treating the phase transition as a critical phenomenon.

Essentially the same erroneous approach was taken in the subsequent set of three articles on calorimetric study of mixed crystals of NH_4Cl and NH_4Br by Callanan et al. [153]. This work is just one more example of a phenomenological approach in which no importance was attached to the fact that first-order phase transitions occur by rearrangement at interfaces rather than homogeneously. Almost 40 years after the Dinichert's work the two-phase state in the transition ranges was still not taken into account. The observed phase transitions were vaguely classified by "isothermal" and "λ-type." The "transition points $T_λ$" were determined as "the temperature at which "the heat capacity at the gradual transition reaches its maximum." The λ-peaks, erroneously taken for a heat capacity, were utilized to speculate on the character of molecular motions in the crystals.

3.4.5 Two additional ways to disprove "specific heat λ-anomaly"

The hysteresis of the λ-peaks was already strong evidence against their heat capacity origin, for heat capacity hysteresis is impossible. As shown in the previous section, it proved insufficient to even stir doubts in their heat capacity origin. Therefore, two quite unassailable additional experimental proofs are presented below (as well as in Appendices 2 and 3), one dealing with the position, the other with the sign of the peaks.

The idea of the experiment was to carry out calorimetric measurements of NH_4Cl along the temperature path A to G, as shown in Fig. 3.6a. The result is shown schematically in Fig. 3.6b. In the heating runs the "rounded" peak in NH_4Cl phase transitions appeared only in the A → B and E → G runs. There was no trace of a peak in the heating run C → D. This fact eliminates any argument in favor of a heat capacity nature of the peaks. In terms of the nucleation-growth concept the phenomenon has a trivial explanation: in order that the latent heat peak could appear again in a heating run, as it is in the E → G case, the specimen had to be beforehand transferred into the low-temperature phase.

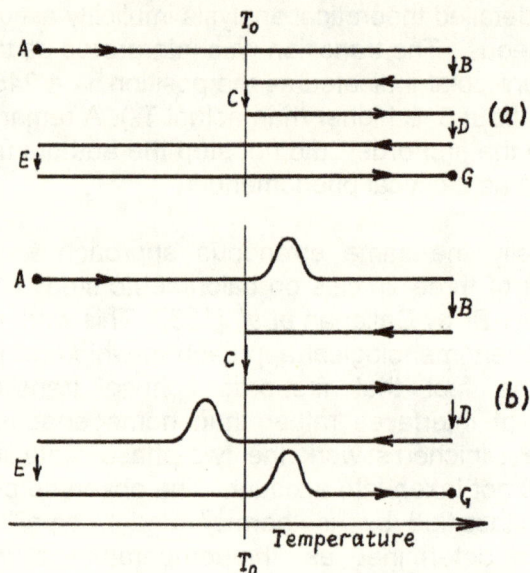

Fig. 3.6 The DSC experiment to disprove a heat capacity origin of the λ-peak in the NH_4Cl phase transition.
(a) The temperature pathway to follow in recording DSC signal.
(b) The result. A heat capacity peak had to appear in the C → D run, but it did not; the peak must not appear if it is a latent heat of the phase transition.
Note: In order to consider separately two phenomena - position of the peaks and their sign - the schematic deals here only with the former. Appendix 2 deals with the latter.

In this experiment a differential scanning calorimetry (DSC) was used owing to its capability, as opposed to adiabatic calorimetry, of recording upon cooling as easily as upon heating. In every respect all collected evidence matches the nucleation-growth phase transitions: the peak must not appear in the run from C to D and it did not; the peak must arise after the preceding descending run was extended beyond T_o to obtain the H → L peak, and it did; the peaks taken in a direct and the reverse runs must be separated on the temperature scale, and they were. Also, (a) the areas of all recorded peaks were equal within experimental error, (b) the latent heat of the phase transition estimated from the areas was Q_{lat} = 86±10 cal/mole, which practically coincides with 84±10 cal/mole reported in [151] as a "total" transition heat (see previous section for a comment on Ref. [151]), (c) the area under a peak was proportional to the specimen weight, (d) the areas in a direct and the reverse runs were equal, (e) only the peak area, not its amplitude, half-width, or exact location, was a stable characteristic of the transition.

Another, not yet mentioned, result of the expounded experimental work deals with the *sign* of the peaks: they were positive (endothermic) upon heating and negative (exothermic) upon cooling. This does not fit at all to a heat capacity origin of the peaks that are expected to be positive both in temperature ascending and descending runs. The issue is presented in all detail in Appendices 2 and 3. (The reason to consider this issue in appendices is not that it is less important, but because it stirred an instructive polemics with an anonymous opponent.) It was shown there that only differential scanning calorimetry is free of the instrumental limitations that prevented all previous workers from finding that the temperature-descending peaks were negative (exothermic).

3.4.6 Analyzing old literature data

Two experimental works published in 1942 in the same journal issue by Dinichert [110] and Extermann and Weigle [148] were already sufficient to establish the nature of the NH_4Cl phase transition and its "λ-anomaly." The experimental graphs from those articles, being positioned under one another in the same temperature scale (Fig. 3.7), reveal both why the λ-peak arises and why it is subjected to hysteresis. The only feature to be realized is how the graphs correlate: Fig. 3.7b is, in a proper scale, the first derivative of Fig. 3.7a. In other words, the curve 'b' is proportional to the rate with which the mass fraction m_r of the resultant phase increases at the expense of m_i. The release or absorption of a latent heat, inevitable in every phase transition, is proportional to this rate. Thus, it should have been possible to conclude in 1942 that the cause of the "C_p-λ-anomalies" is actually the latent heat of transition, as well as that hysteresis of the λ-peaks resulted from the hysteresis of m_r.

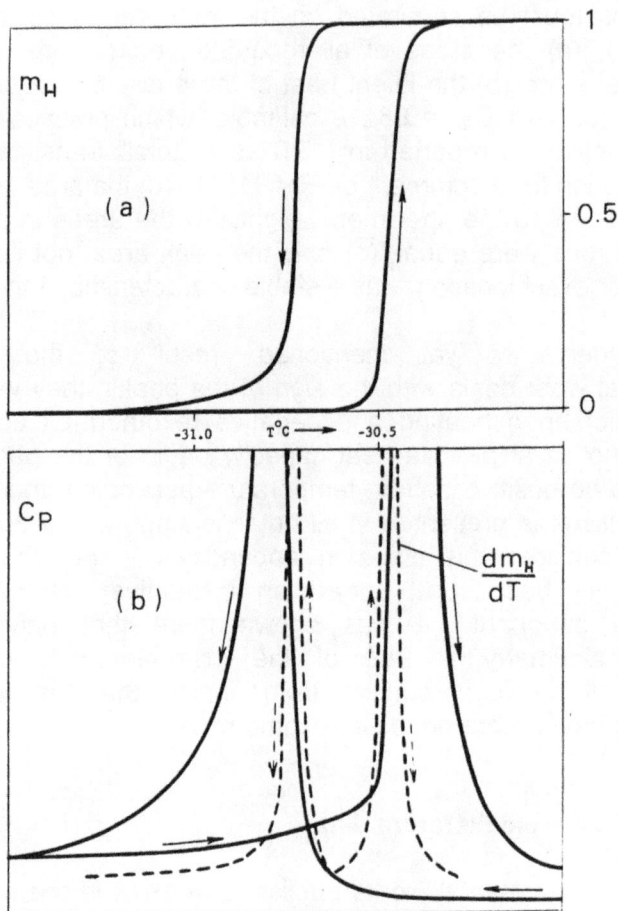

Fig. 3.7 Comparison of two experimental plots representing the measurements through the range of NH$_4$Cl phase transition, performed as long ago as 1942. Plot 'a' (already shown before in Fig. 2.30b), by Dinichert [110], represents the fraction m_H of H in the two-phase, L+H, range of transition; $m_L + m_H = 1$. The plot 'b' (solid line) by Extermann and Weigle [148] represents calorimetric λ-peaks. The two plots are shown in the same temperature scale to make it evident that shape and location of the peaks are very close to the first derivative (dashed curves) of m_H. The nature of the λ-peaks and their hysteresis becomes clear if one bears in mind that absorption / release of the latent heat is proportional to dm_H / dT.

3.4.7 Turning the observed inconsistencies into harmony

While interpreting experimental data in terms of second order phase transitions and critical phenomena, some authors noticed numerous inconsistencies. The calorimetric study by Schwartz [103] of the "specific-heat peaks" in NH_4Cl phase transition was an example. The author blankly called this phase transition "second-order," applied a theory based on the Ising model, calculated the "critical exponents," believed that a good C_p-peak must go up into infinity, etc. As to the observed facts, they contradicted that approach in every point.

The author himself listed the striking contradictions. "There is a strong sample dependence in the data that is not understood." "There is a latent heat involved in the transformation." "The heat capacity at the transition is finite." "...Distribution of transition temperatures in the samples." "Most striking are the differences in the values of the specific-heat peaks... This peak was 5-6 times smaller for samples 3 and 5 than it was for 2." "The rougher the treatment during thinning, the larger and narrower is the heat-capacity peak... However, we would then expect that a more perfect crystal would have a larger anomaly in the heat capacity. The trend we observe above goes in the opposite direction. The sample receiving the roughest thinning had the sharpest peak... We cannot try to explain this behavior since there were not sufficient data to determine what physical factors are controlling the critical behavior." The picture will be complete after we add hysteresis of the λ-peaks, clearly being present in the experimental plots.

After the preceding analysis there is hardly any need to explain all the above observations. These are the plain manifestations of a nucleation-growth phase transition, its *contact* mechanism in particular. For example, the rougher the preliminary mechanical treatment is, the more crystal defects there are to serve as heterogeneous nucleation sites, the narrower the temperature range of transition becomes. The result is a higher and narrower "heat capacity" peak, for the same amount of latent heat is now absorbed or set free over that narrower range.

3.4.8 On linear correlation between specific heat and coefficient of thermal expansion

A linear correlation $(C_p / α)$ = constant between the λ-peaks of $C_p(T)$ and $α(T)$ in the NH_4Cl phase transition has been noted by Pippard [154] who even suggested thermodynamic relations accounting for the phenomenon. However, no special theory was required. Pippard did not know that "$C_p(T)$" actually describes latent heat rather than specific heat. Both "$C_p(T)$" and $α(T)$ are delineated by the same parameter: gradual change of m_r in the range of transition. As can be seen from the equations describing the dilatometric α(T) (Sec. 3.3) and calorimetric Y(T) (Sec. 3.4.2) curves, they roughly describe, in different scales, the same function dm_r / dT, which is a λ-peak. The only difference is that they rest on different baselines (Figs. 3.1b and 3.7b).

3.4.9 Summary on the mechanism of phase transition in NH_4Cl

After all, what is the molecular mechanism of phase transition in NH_4Cl and, particularly, how does the "ordered" phase L transforms into the "disordered" phase H, and vice versa? Well, it is much simpler than it was assumed in the innumerous attempts to find answer to this question. It is neither cooperative, nor of a λ-type. This is a regular nucleation-and-growth phase transition, of which the molecular mechanism was described in Chapter 2. There are two crystal phases, one (L) being stable at $T < T_o$, the other (H) at $T > T_o$. A phase transition can occur only provided that T_o has been passed far enough into the "preference" area of the latter. This is due to the nucleation requirements which also imply that the nucleation spreads over a temperature range. More exactly, there is a separate range on each side of T_o, depending on the direction of transition. Contrary to the traditional interpretation, in the range of transition the specimen does not represent intermediate states in a thermodynamic sense: it is not homogeneous. The phases coexist without a dynamic equilibrium between them. Each has its own independent set of properties, including separate $C_p(T)$. The fact that the H phase is more "disordered" than L is merely another difference between them, bearing no relation to the mechanism of molecular rearrangement at the interfaces. The rearrangement proceeds according to the universal contact mechanism, irrespective of how much "disorder" or "order" is attributed to each phase (see Sec. 2.7). As to the λ-peaks, they have nothing to do with the heat capacity. Their area indicates latent heat of this phase transition to be about 85 cal /mole.

3.5 "λ-Anomaly" of dielectric constant

3.5.1 The discordant facts

"λ-Anomalies" were sometimes also found in measurements of a dielectric constant $\varepsilon(T)$ through crystal phase transitions. Their "critical phenomenon" nature was taken for granted [2]. As will be shown below, these λ-peaks bear no relation to critical phenomena and do not even

- 184 -

represent a dielectric constant. True $\varepsilon(T)$ curves do not exhibit λ-peaks (these peaks will be conditionally called "ε-peaks").

Fact 1. Widely recognized "λ-transitions" do not necessarily (and usually do not) exhibit ε-peaks, as is the case in the NH_4Cl "λ"-transition [155, 156] (Fig. 3.8a). This fact can hardly be naturally explained in terms of the orthodox views. It would seem, one should expect the ε-peaks to consistently appear, together with other λ-peaks, such as of $C_p(T)$, in every transition of a "λ-type." Otherwise one deals with a critical phenomenon regarding one property, but not another property.

Fact 2. The ε-peaks (Fig. 3.8b,c) are found mainly, if not exclusively, in ferroelectric phase transitions. Ferroelectrics used to be a major example of critical phenomena, but not at the present time (Sec. 4.2.1). The most prominent ε-peaks have been found in KH_2PO_4 (KDP) and $(NH_4)_2SO_4$ classified now as definitely first-order. Zheludev [157] commented on the ε-peak in KDP [159]: "The dependence [of ε] on temperature used to be an illustration of a second-order phase transition... In fact ... the phase transition in these crystals is of first-order." Thus, the ε-peaks are found in "non-critical" phase transitions. There is a specific reason for the appearance of ε-peaks in ferroelectrics (Sec. 3.5.2).

Fact 3. The ε-peaks are found to appear only in some particular crystallographic directions, but not in others (Fig. 3.8c). If these peaks manifest critical behavior, one has to assert that such phase transitions can be a critical phenomenon only in some selected direction(s), which would be a highly unrealistic idea. What kind of a phenomenon is the phase transition in the "non-critical" directions? Can a phase transition be "cooperative" in direction a and nucleation-growth in direction b?

3.5.2 Electric conductance during epitaxial phase transition

Since the ε-peaks are characteristic of first-order rather than second-order phase transitions, they obviously do not belong to critical phenomena. It remains now to account for the ε-peaks and related observations in terms of *contact* nucleation and growth, which it will be recalled, manifests itself in two modes. One involves non-oriented nucleation and growth, which in a common event of multiple nucleation results in the conversion of a single crystal into a polycrystal. There, only a short-range propagation of individual interfaces takes place - until the growing grains meet. The second mode (Sec. 2.8) involves nucleation

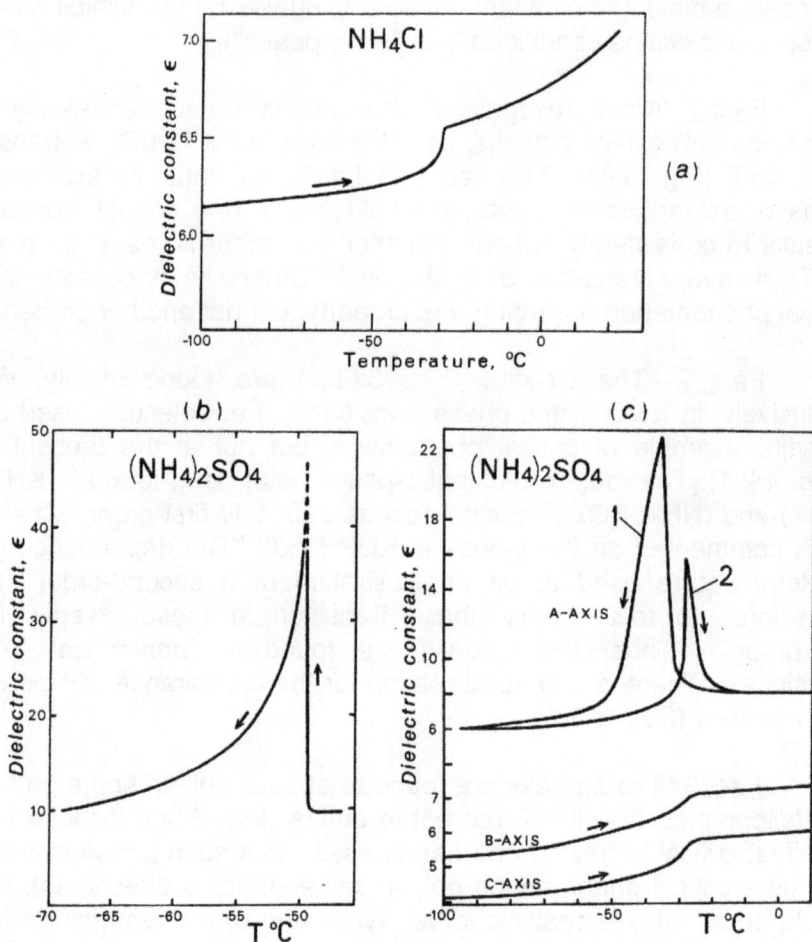

Fig. 3.8　Dielectric constant ε vs. temperature in two typical instances referred to as λ-transitions.

(a) (By Kamiyoshi [156]) There is no λ-peak of ε (T) in the classical NH₄Cl λ-transition.

(b) (By Hoshino et al. [159]) The ε-peak in (NH₄)₂SO₄ phase transition known to be ferroelectric and first-order. The peak was measured, upon cooling, along the orthorhombic c-axis which is the polar axis in the ferroelectric phase.

(c) (By Kamiyoshi and Miyamoto [160]) The data for (NH₄)₂SO₄ in three axial directions of the crystal. The ε (T) plots show a peak in only one of the directions. The plots in two other directions are S-curves (Table 3.1). The ε-peak was recorded both upon heating and cooling. The following three facts should be noted. (1) Hysteresis. (2) Asymmertric shape of the peaks. (3) Mutually reverse asymmetry of the peaks in the heating and cooling runs: the side of the peak being approached first has a steeper slope (cf. curves 1 and 2).

and growth in an epitaxial fashion, giving rise to the lamellar-domain structures that are characteristic, in particular, of ferroelectrics. Examination of several particular instances where ε-peaks were found allowed the conclusion to be made that *the ε-peaks appear only in epitaxial transitions and only in the directions in which the constituent layers/domains protrude through the-crystal*. Fig. 3.9 shows the orientation of a KDP crystal in the measuring capacitor, giving rise to the ε-peak.

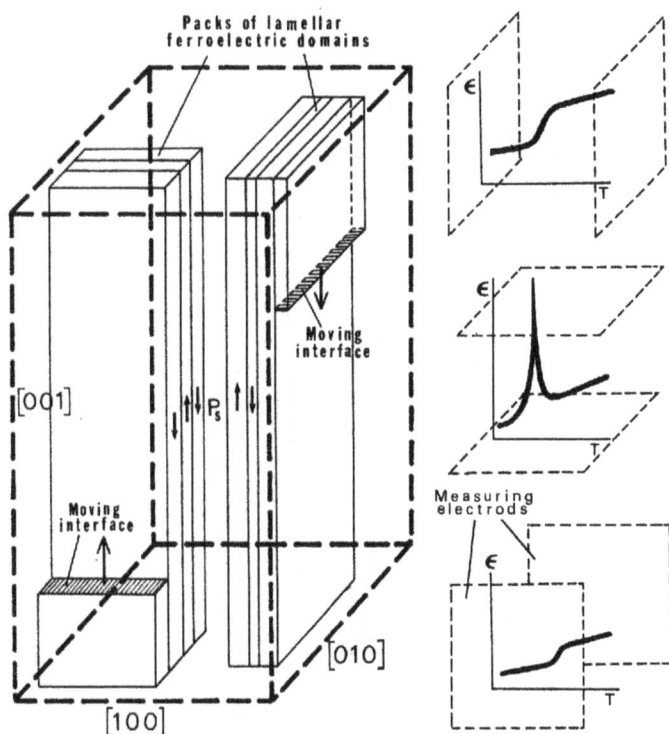

Fig. 3.9   A sketch illustrating the particular direction of measurements of a KDP single crystal in which ε(T) peak appears. The direction is parallel to the spontaneous polarization Ps. More importantly, this is the direction of interface movement from one electrode of measuring capacitor to the opposite one. The plots of ε(T) in two other axial directions are of the S-type. The scheme on the right sums up a character of the response in the three directions of measurements.

An epitaxial phase transition is controlled by nucleation in the tiny interlayer wedge-shaped cracks located predominantly at certain crystal faces. These are the faces composed of the layer / domain ends and adherent to the capacitor electrodes when the ε-peak is recorded (Fig. 3.10a). The interfaces, independently developed in each lamella from the nucleation sites, quickly run through the whole crystal from one electrode to the opposite one, completing the phase transition separately in each lamella. *This physical movement, fraught with the possibility of electric charge transfers between the electrodes in the temperature range of transition, is the probable cause of the ε-peaks.*

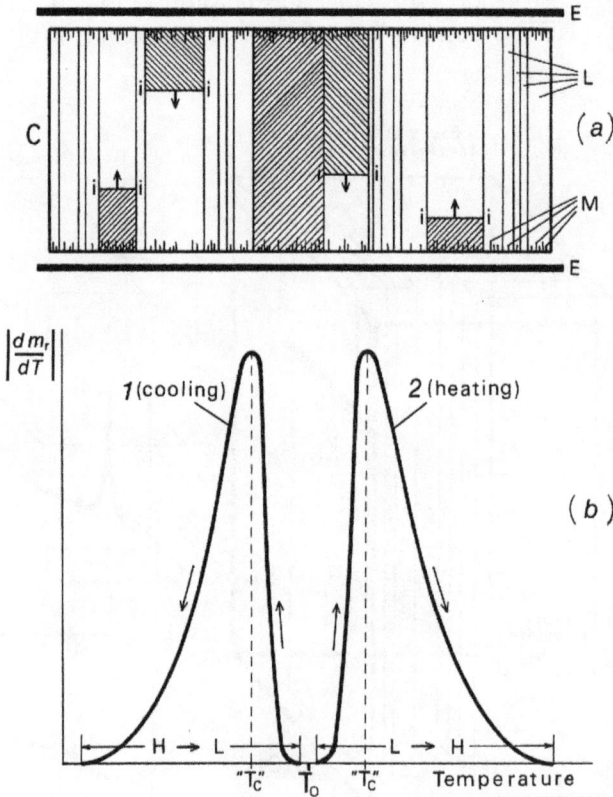

Fig. 3.10   The features of epitaxial phase transitions accounting for   1) electric conductance upon ε(T) measurement,   2) the asymmetric shape of ε(T) peaks, 3) dependence of that asymmetry on whether the phase transition took place upon heating or cooling.

(a) A crystal in the position to produce a peak in ε (T) measurements. Shown are: electrodes E of the measuring capacitor, crystal C, constituent lamellae L of the crystal, areas of the 'r' phase (shadowed), interfaces i -- i across the lamellae, direction of interface propagation (arrows), microcracks M serving as nucleation sites. The phase transition is controlled by the rate of nucleation in M and proceeds by quick interface propagation to the opposite electrode, separately in each lamella.

(b) The asymmetric shape of the peaks of dm_r / dT (m_r is mass fraction of 'r' phase) due to the simple fact that transition occurs in thicker lamellae first. This is true for both cooling and heating, thus giving rise to the reciprocal reverse asymmetry of the peak in a direct and the reverse runs. The slope is always steeper on the side of the peak that is approached first.

Assuming the ε-peaks to be caused by electric conductance, we are able to account for the facts listed in the previous section. The apparent ε-peaks

- do not appear in non-epitaxial phase transitions because in practice such transitions do not involve propagation of the interfaces - charge carriers - through the whole crystal from one measuring electrode to the opposite one,
- are found in ferroelectric phase transitions, because they are dielectrics of a layered epitaxial type,
- appear only in the direction in which the domains stretch from one measuring electrode to the opposite one, because the interface movement responsible for the electric charge transfer proceeds only in this particular direction,
- have not been found in NH_4Cl transition [156 (Fig. 3.8a), 161]; though it is of epitaxial type, it does not form layered structures stretched from one measuring electrode to the other, because nucleation and growth from different nucleation sites proceeds in three equivalent directions of the initial cubic lattice.

Our attention will now be turned to the function dm_r/dT in epitaxial phase transitions. Within the range of transition such a crystal represents a pack of lamellae, some having changed into the resultant phase, while others still had not. The electric conductance due to the interface motion between the two electrodes will be proportional to dm_r/dT (Fig. 3.10a). The lamellae in the pack have very different thickness. The thickest undergo phase transition first, that is, under relatively lower overheating or overcooling ΔT, for they contain the greater number and variety of nucleation sites (microcracks). This effect is responsible for the asymmetric shape of the dm_r/dT peaks, with a sharper rise and more gentle descent after the maximum is passed. Such asymmetry, together with its inversion when heating is substituted by cooling (Fig. 3.10b) lends support, as will be shown, to the above-described idea of the origin of ε-peaks.

3.5.3 ε-Peak in $(NH_4)_2SO_4$ phase transition

Since NH_4Cl reveals no ε-peak (Fig. 3.8a), we had to turn to another example. Rather arbitrarily, $(NH_4)_2SO_4$ was chosen, only because it was intensely studied [159, 160,162-170]. In spite of the studies, the nature of the ε-peak was not clarified. But the reported evidence turned out to fit well to the foregoing electrical conductance hypothesis:

● This is a ferroelectric phase transition, hence of the epitaxial type. The ε-peak was found only in one of three basic directions, which is along the layers, the plots in two other directions being S-curves (Fig. 3.8c).

● The phase transition is now classified as of the first-order; therefore it proceeds by interface propagation.

● A higher applied frequency in dielectric measurements, as well as slowing down the temperature change, resulted in a decrease of the ε-peak amplitude up to its elimination. These observations by themselves, even without an explanation of their cause, prove that the peak has nothing to do with the dielectric constant. As to the cause of these effects, it is possibly related to the time needed for the charge to be transferred by a moving interface before the electrodes change their polarity. The higher the frequency, the smaller the charge that can be caught by the departing interface that transfers it to the opposite electrode. In this respect the method of demagnetization of a magnet by slowly removing it from a high-frequency alternating magnetic field comes to mind. The phenomena under consideration are consistent with the idea on charge transfer by the interfaces; speculation on the fine details of this process is not our objective.

● The shape of the observed ε-peaks generated a great deal of discussion [159,165,166,169]. The peak was always asymmetric. But while some authors found the asymmetry as in Fig. 3.8b, others found the inverse asymmetry. One experimental difference in the measurements had been overlooked: the former peaks were taken upon cooling, while the latter upon heating. Their shape merely reflected the shape of dm_r/dT shown in Fig. 3.10b. The experimental result explicitly exhibiting the inversion of the asymmetry when heating replaces cooling [160] (Fig. 3.8c) was unnoticed.

● In accordance with the foregoing account, Ohshima and Nakamura [169] concluded that the ε-peaks were due to a "movement of domain walls."

Electric charge transfer by moving interfaces in layered crystals is a simple and plausible explanation of the "ε"-peaks. More importantly, their "non-critical" origin is beyond question.

3.6 " λ-Anomaly" of light scattering

3.6.1 Static source of "critical opalescence"

The problem of the narrow peaks of light and neutron scattering centered at the "critical temperature T_c" is called the *central peak problem*. In the light scattering alone, extensive literature was devoted to this phenomenon called also *critical opalescence*. A notorious "λ-anomaly" can be recognized in the central peak (Fig. 3.11). The significance of this phenomenon for the theory of critical phenomena was expressed in the 1979 article "Quasielastic Light Scattering Near Phase Transitions" by Lyons and Fleury [171]: "Our understanding of structural phase transitions has evolved in the last decade through several stages from the simple soft mode and mean field theories to the modern coupled-mode, renormalization group and dynamic scaling ideas. Crucial to this evolution has been an increasingly detailed interpretation of the experimental data, particularly those resulting from scattering experiments." A disconcerting circumstance, however, was that by 1979 it was already a well-proven fact that the central peaks were caused by scattering from *static* centers. At that stage Lyons and Fleury had to suggest that the scattering centers were rather *long-living clusters* than *entropy fluctuations*, but failed to show to what extent this fact discredits the dynamic theories.

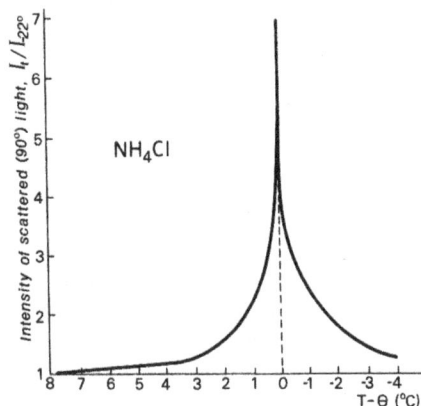

Fig. 3.11 The apparent "critical opalescence" peak of light scattered at 90° to the primary light beam by a NH$_4$Cl single crystal during its low-temperature ($T_0 = -30.6$ °C) phase transition (after [177]). The representation of the temperature scale as a distance from the position of the peak maximum obscures the fact that the peak is not located at T_0.

The 1981 review by the same authors [172] revealed a certain evolution in the representation of the subject: though the *critical phenomena* approach was left intact, a separate section entitled "Static and dynamic central peaks" appeared. There it was stated that the first instance of the central peak of light scattering, reported for quartz by Yakovlev at al. [173], was later proven [174] to be due to "scattering from static microdomains, rather than to true critical opalescence," and that "in most other cases confusion still persists...." Also mentioned was KDP where the central peak has been observed [175], but shown [176] to be "largely of static origin." Omitted from this account was the case of NH_4Cl (Fig. 3.11, [177]) usually considered a valuable asset in illustrating critical phenomena: the scattering centers there were also static [178].

Ginzburg at al. in their review [179] listed, besides, other examples of the central peaks caused by static scattering centers ($SrTiO_3$, $Pb_5Ge_3O_{11}$, $KH_3(SeO_3)_2$). It was explained that "extraction of the part due to thermal fluctuations from the total scattering intensity is generally a rather hard task." But evidence that such a part existed was absent: it was *postulated* by the authors.

3.6.2 Revealing experiments by Durvasula and Gammon

In 1975, one especially revealing study has been reported. Durvasula and Gammon [180] correctly pointed out that light scattering investigations, first, failed to measure temperature carefully and, second, paid little attention to the fact that the phase transitions clearly exhibit first-order behavior. In their combined examination of the phase transition in KH_2AsO_4 the light scattering experiments were complemented by dielectric measurements; simultaneously, the emergence of the second phase (ferroelectric domains) was visually monitored. Both heating and cooling runs were carried out. A hysteresis loop (Fig. 3.12) was detected; in every respect it was of the type shown in Fig. 2.30.

It was found that the sharp increase in light scattering starts at that same moment when the domains appear in the illuminated area of the specimen. This was a direct demonstration that the source of the light scattering responsible for the central peak was the interfaces emerging in the two-phase transition range. This finding was quite revealing and specific. However, the leading theorists in the field would continue using vague language discussing the "static origin" of the central peaks and speculating about the nature of the "long-living clusters" or "inhomogeneities." The nature of the "inhomogeneities" as a two-phase state had already been established, but Ginzburg *et al.* would suggest years later [179] that they are crystal defects.

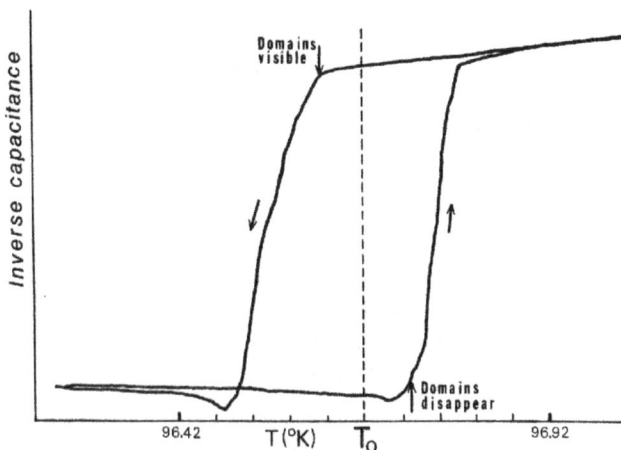

Fig. 3.12 (By Durvasula and Gammon [180]) Inverse capacitance *vs.* temperature for a KH_2ArO_4 crystal plate; also shown are the temperatures of appearance and disappearance of domains of the low-temperature ferroelectric phase. The curve is a typical hysteresis loop (*cf.* Fig. 2.30). It is the appearance and development of the second phase that constitutes the process of phase transition and produces a "central peak" of the scattered light. There is no "critical point." The position of T_0 was added to the original picture. (The authors' marks in the figure need a correction: "domains visible" must be substituted by "first domains appear," and "domains disappear" by "domains start to disappear").

3.6.3 Light scattering by nuclei and interfaces

The "central peaks" were initially regarded as a direct consequence of the "critical" nature of phase transitions. Later, when all evidence indicated a static cause of the central peaks, it was maintained that the static effect overshadows the scattering by critical fluctuations. Such assertions were not corroborated by any reliable evidence. They were aimed at saving the previously developed theories that assumed light scattering to be a result of "critically fluctuating" matter. These obsolete theories became the obstacle to correct interpretation of solid-state phase transitions.

The nature of the central peaks is plain. One merely deals with nucleation and growth over a certain temperature range. The following are sources of the light scattering:
• Scattering by the 'r' nuclei of the size comparable with the wavelength of the incident light. Every growing 'r' nucleus must pass this size. The scattering takes place because the nuclei are

inclusions of different density. The effect practically contributes when nucleation is multiple, as was the case in the light scattering measurements of quartz and NH_4Cl (Sec. 3.6.5).

- Scattering owing to refraction, reflection and birefringence at the multiple interfaces that separate two different crystal states in the grainy two-phase structure within the transition range. A ray of light in such a structure encounters a succession of the interfaces and will therefore be scattered over and over again.

- Scattering, owing to the same effects, by the crystal imperfections and distortions, both elastic and plastic, caused by the strains accompanying interface propagation. We classify these distortions and imperfections, including cracks, twinning and layer stacking faults, as secondary phenomena. Some extent of adverse effects is inevitable in every phase transition, but which kind and to which extent depend on many particular parameters, such as perfection, size and shape of the 'i' crystal, mechanical properties of the polymorphs, difference in their densities, OR, the number and distribution of growing 'r' crystals, and rate of heating / cooling.

These are the sources of the increased light scattering during crystal-crystal phase transitions. No special effort is required to explain why the scattered light manifests itself as a peak. The nuclei, interfaces and a good part of secondary phenomena appear in the beginning and disappear at the end of a two-phase temperature range of transition. If sometimes the scattered light do not fall to the initial level after passing its peak value, it is because not all secondary phenomena are reversible. Not infrequently a phase transition leaves the initial single crystal noticeably deteriorated, and any visible crystal defect is a source of light scattering. Domain and grain boundaries are another source of residue scattering.

All the above-listed contributors to the central peak are evident to one who devoted some time to visual observations of phase transitions in single crystals. A human eye is also a detector of scattered light. Anything within a crystal is only seen because it scatters or absorbs light. What our eyes detect upon phase transitions was described and shown in the photographs in Chapter 2. As the temperature of a single crystal changes, our eyes detect no increase in the light scattering up to the temperature when the first small 'r' crystal emerged. That is when any other detecting device would start recording an increase in the light scattering. In a SC → SC transition the reflected and refracted light will be detected only under certain discrete angles. Then the interfaces (natural faces of the 'r' crystal) will start reaching the external surfaces

(faces) of the crystal and the light scattering will go down. In case of multiple nucleation when OR is absent, the transition will be SC → PC. Here (a) every ray will be deflected many times, and (b) scattering by the final grain boundaries will remain after the transition is completed.

Critical fluctuations (of entropy, order parameter, refractive index, matter, or whatever) have not revealed themselves in the light scattering experiments - not because they are relatively weak, but because they are nonexistent. That is why "extraction of the part due to thermal fluctuations from the total scattering intensity" [179] will never be successful.

3.6.4   The correct solution has been proposed, but ignored

The correct explanation of the central peaks for two prominent cases was proposed by Bartis as early as 1973 in the articles "Critical Opalescence in Ammonium Chloride" and "The Transitional Opalescence of Quartz" [178,181]. Some assumptions leading to this explanation were not quite valid. Phase transition in NH_4Cl was mistakenly considered a three-step process, of which the 1st and 3rd are continuous, and the intermediate step is two-phase. The author was unaware of the real cause of the temperature range of two-phase coexistence, attributing it to internal stresses in the sample. However, the rational part, namely, the temperature range of two-phase coexistence, was sufficient to advance the following idea: "To understand the increased scattering of light we focus our attention on the interface between the two crystalline forms in the intermediate stage. Inasmuch as the two forms have substantially different properties, light incident on the interface is bound to experience some scattering."

Here is a resume of Bartis' reasoning. Interest in critical phenomena had led to the odd turn of events due to "the discovery of a discontinuous intermediate stage in a score of transitions previously believed to be second order." The intermediate stage [in fact, the range of transition due to a spectrum of nucleation lags - Y.M.] should produce a sharp rise in the scattered light that could easily be mistaken for critical opalescence. Ginzburg, and then Yakovlev et al., had made this mistake when they attributed the opalescence of quartz during its α - β transition to critical fluctuations. Even though Ginzburg's theory predicted $\sim10^4$ increase in the light scattering, as was actually observed, it would be proven later that the effect had nothing do with critical fluctuations. Two forms of quartz coexist over 1°. The transitional opalescence in quartz was caused by light scattering at the interfaces, with the light emerging at a 90° angle being proportional to their area.

The above ideas suggested a quick solution to the whole problem of central peaks. It was, however, inconsistent with the orthodox theory developed over previous years. The almost self-evident solution was ignored. It was neither discussed, nor taken into account even in subsequent publications directly devoted to the central peaks in quartz and NH_4Cl. The Bartis' articles were not included in the comprehensive list of about 700 references accompanying the 1981 review by Fleury and Lyons on the topic [172]. The cause of the central peaks had already been named, but the search for it was continuing. Lyons and Fleury stated [171] that further theoretical work is needed. Ginzburg et al. [179] modified the theory to include "static inhomogeneities and defects" as the primary contributors to the central peaks. The new theory involved critical fluctuations and did not mention nuclei and interfaces; only an awkward explanation was suggested as to why crystal defects are present at all temperatures, but come into play only within a narrow range near phase transition.

3.6.5 Light scattering in NH_4Cl and quartz phase transitions

The detailed optical microscopic study of the NH_4Cl phase transition by Pique, Dolino and Vallade (PDV) [161] unambiguously identified the cause of its central peak. Both polymorphs are of cubic, but not the same, symmetry. PDV observed nucleation and growth both in H → L and L → H transitions. The growing 'r' crystals were lamellae parallel to (*111*). The transitions were epitaxial with four symmetrically equivalent (*111*) directions to grow. PDV also found a rather low limit of crystal elasticity. These facts predetermine low hysteresis and quantity nucleation.

PDV stressed the importance of the heterophase region (which they also called "mixed state" and "coexistence state") in understanding the phase transition. Their observation about the central peak was: "During the coexistence state ... there is such an intense light scattering that...no light remains in the [primary] beam of the He-Ne laser." Other details showed that the primary beam was deflected by the (*111*) planes.

One point should be clarified. Considering the morphology of the phase transition, it is obvious that the incident light will scatter when the multiple $(1,1,1)$, $(-1,1,1)$, $(1,-1,1)$ and $(1,1,-1)$ lamellae of the 'r' phase appear in its way. It is to be noticed, however, that after the transition the original single crystal will not be a single crystal, but, rather, a conglomeration of the lamellae of the above four crystallographic orientations. Consequently, some boundaries will remain after the transition is completed. It would seem, one should expect a higher level of light scattering after the transition than prior to it. But it is not the case: due to a cubic symmetry the specimen will regain optical isotropy of the original single crystal - in accord with the actual observation.

A meticulous investigation of the α - β phase transitions in quartz has been carried out by Dolino [182]. Light scattering measurements were combined with optical microscopic observations. In the vicinity of the transition (~573 °C) a small temperature gradient was applied along the investigated quartz single crystal. A two-phase band moving from the α-end of the crystal to its β-end and in the reverse direction was observed. Dolino arrived at the conclusion that the light scattering centers were nuclei of the α-phase in the β-phase on cooling, and β-phase in α- phase on heating. The central peak resulted from light scattering by this heterophase band. Some secondary phenomena that could not be exactly identified were also observed (a possibility of microtwinning was suggested) and, perhaps, also contributed to the central peak. The remaining uncertainties, however, are of little importance in the present context. Nucleation and growth were responsible for the observed central peak.

Light scattering "central peaks" in crystal phase transitions have underwent their own transition - from best friends of critical fluctuations to their worst enemies. The reader is also referred to Appendix 5.

3.7 "λ-Anomaly" of neutron scattering

The central peaks of neutron scattering do not need to be discussed in detail. Their story is a carbon copy of what happened to the central peaks of light scattering.

- Original excitement over the discovery of a phenomenon seemed to offer an intimate insight into the critical dynamics of solid-state phase transitions.

The first observation of a neutron scattering central peak was reported in 1971 [183], 15 years after the first light scattering peak was reported [173]. To that time the "static nature" of the light scattering peaks had been established.

- Subsequent disappointment after finding that the neutron central peaks were scattering from static centers.

It was even tried to represent the phenomenon as a particular case of "critical dynamics" - as a response of the vibrational system that occurs at approximately zero frequency.

- Search for the nature of the static centers anywhere except the self-evident place: nuclei and interfaces in the two-phase range.

The same idea on "coupling" of all kinds of permanent crystal defects with imaginary fluctuations was put forward [184]. The Bartis' suggestion [118] that the transitional neutron scattering can be successfully explained in terms of a region of the phase coexistence was not considered.

- Dealing with the first-order phase transitions as if they were of second-order.

As usual, the treatment was marked by lack of understanding that first-order phase transitions occur by nucleation and structural rearrangement at the interfaces over a certain temperature range.

- Desperate experimental search for a "dynamic" component in the "largely static" central peaks.

The scope of experimentation was limited by the much higher cost of the neutron scattering experiments, as well as by problems with having sufficiently intense neutron sources. As a result, the experimental data were less reliable than in light scattering, thus leaving a wider field for speculations.

- Statements that more hard work and time is needed in order to completely understand the origin of neutron central peaks.

The source of both the neutron and light scattering "anomalies" is the same (refer to Sec. 3.6.3).

3.8 On solid-state phase transitions, their molecular mechanism and anomalies - final notes with some philosophical and psychological overtones

Is there anything left of the problem of phase transition anomalies after the analysis presented in this Chapter? Nothing. It is understandable how exciting it was for experimentalists to discover such anomalies as the λ-peaks, for they seemed to promise a breakthrough in a previously unexpected direction. These anomalies were no less exciting for theoretical physicists, who found an application for their talents, their knowledge of statistical mechanics and their belief in its general power and the dynamic nature of everything. But Nature had its own agenda, namely, to make its natural processes (a) universal, (b) simple and (c) the most energy-efficient. Being uncompromising in these principles, Nature produced better processes than the most brilliant human beings, even Nobel Prize winners, could invent.

Solid-state phase transition is such a process. It is more universal, simple and energy-efficient than critical-dynamic theories can offer. It is universal because it is just a particular manifestation of general crystal growth. It is also as simple as crystal growth. It is energy-efficient because it needs energy to relocate one molecule at a time, and not the myriads of molecules at a time as a cooperative process requires. An important lesson can and should be drawn from this. The whole effort was largely misdirected. Great amounts of time, hard work, resources and talent were wasted. Insufficient attention to facts, such as the disregard of nucleation and growth as a mechanism inherent in all solid-state phase transitions, was substituted by excessive theoretical creativity. Vast contradictions were tolerated. Correct solutions were ignored. Desperate efforts to defend failed theories prevailed over scientific objectivity. Trial and error is a normal way of a scientific advancement; it is only honorable for a scientist to recognize being incorrect. But the latter does not happen frequently and has not happened (yet?) in the present case. As a result, the solution to the problem of solid-state phase transitions was unnecessarily delayed for more than half a century.

As for the "λ-peaks," their nature is so unpretentious that they do not deserve to be called "anomalies." We demonstrated this by examination of the most well-known "anomalies": of thermal expansion, heat capacity, dielectric constant, light scattering, neutron scattering. The general cause of all these seeming anomalies is the inevitable spreading of a phase transition over a temperature range, narrow or wide, where the ratio changes between the two coexisting phases. In C_p measurements, for instance, this gives rise to a gradual absorption of the latent heat in the form of a peak mistaken for a genuine C_p. Table 3.1 offers the key to the "anomalies" of any other properties that have not been mentioned.

Misunderstanding the nature of solid-state phase transitions had important ramifications, in particular, preventing the development of a correct theory of ferromagnetism and ferroelectricity.

CHAPTER 4. FUNDAMENTALS OF FERROMAGNETISM AND FERROELECTRICITY*

4.1 Unsatisfactory state of the theory** and suggested new solution

This chapter is not about fundamentals of magnetism. The electron structure of atoms, spins, electron shells, spin-orbital interactions, etc., are beyond its scope. The topic of this chapter is *ferro*magnetism and ferroelectricity. The starting point is a crystal built from the atoms carrying magnetic moments in ferromagnetics; in case of ferroelectrics it is a crystal containing dipole molecules or dipoles in the form of positive and negative ions.

Judging from the textbooks on physics one may conclude that the theory of ferromagnetism is rather successful. In these books and other concise presentations of the theory every effort was made to portray it as basically valid and a great achievement, while contradictions, blank areas, and vast disagreements with experiment are either omitted as "details" or only vaguely mentioned. As a result, a new student gets wrong impression about the real status of the theory. But the more detailed the source is, the more these drawbacks are exposed. There are experts who recognized the existence of very essential shortcomings (see Appendix 5). Yet, no one has been doubtful about the basics of the theory. In particular, the existence of the molecular / exchange field remains unchallenged. However, *it is the basic assumptions of the theory that are mistaken*. There is a *crystal field*, rather than a specific "molecular field"; a ferromagnetic phase transition is not a "critical phenomenon"; a "Curie point" is simply a temperature range of nucleation of the nucleation-and-growth phase transition between orientation-disordered crystal (ODC) and crystal states; magnetic domain structure is more of a structural than a ferromagnetic origin; magnetization occurs by a structural rearrangement at the *contact* domain interfaces rather than by a gradual dipole rotations over the thick "Bloch walls," etc.

---

* Depending on the context, these terms can be used in a general sense, as in this title, that is, including "antiferro-","ferri-," and any other dipole order.

** This chapter is not a source of detail referencing on the general state of affairs in the science of ferromagnetism. The reader can turn, *e.g.*, to [65, 116, 121, 185].

As for ferroelectrics, they exhibit essentially the same properties as ferromagnetics, with the only difference that their dipoles are electric rather than magnetic. The similarity is so close that the term *ferroics* denoting both ferromagnetics and ferroelectrics will be used. However, a "classical" theory of ferroelectricity does not exist. As The New encyclopedia Britannica stated, "In general, ferroelectricity is not well understood..." [186]. No unified theory applied to all ferroelectrics, much less to both ferroics, has been developed. By contrast, the concept presented in this book is that of ferroics. Hereafter, as a rule, a particular ferromagnetic problem can be expressed in ferroelectric terms, and *vice versa*.

The theory of ferromagnetism went off course many decades ago when Weiss advanced his idea that in addition to the usual magnetic interaction between atomic magnetic moments there is a much stronger "molecular field" lining them up parallel to each other; at the Curie point this parallel alignment is destroyed by thermal agitation.

The existence of the "molecular field" is the principal point of the theory of ferromagnetism; its nonexistence is the principal point of the current presentation. The essence of the problem was already touched upon in Sec. 1.8. Owing to its importance, let us trace the logic behind the Weiss' theory in some more detail and pinpoint where it is flawed. It had to be explained how an assembly of parallel dipoles can be thermodynamically stable if their antiparallel order is preferable. This alone led to the idea that there is a certain additional field, besides the internal magnetic field, which holds the dipoles in parallel. If so, how strong this "molecular field" should be to make a ferromagnet stable up to the actual Curie points (769 °C in Fe)? Calculations showed that the molecular field has to be thousands times stronger than the magnetic dipole interaction alone. This was considered as undeniable evidence of the "molecular field." The previously unknown "molecular field" was able, it seemed, to account for both the uniform dipole orientation and the high-temperature location of the Curie points in ferromagnetics.*

Two opposing factors were considered by the Weiss' theory: the "molecular field" causing alignment of an ensemble of elementary magnets and thermal agitation destroying this alignment. The error was that the role of a ferromagnetic crystal was implicitly reduced only to

---

* The term "molecular field" specifically denotes the Weiss' "supermagnetic" field that is thousands times stronger than pure magnetic interaction provides. It is not to be confused with any other field within a molecular, atomic, ionic or dipole assembly.

providing a positional, but not orientational, order to its magnetic dipoles. A system of atomic magnetic dipoles was assumed to be a dipole system only. The objects of thermal agitation were the elementary magnets, and not the atoms carrying them. The *crystal field* was overlooked. There are powerful bonding forces combining molecules, ions, atoms, magnetic or not, into a crystal 3-D long-range order, both positional and orientational. It is the crystal field that imposes one or another dipole order in crystals containing dipolar molecules; it does this simply by holding the molecules in fixed orientations. As to ferromagnetics, there is no reason to assume that crystal structure allows the *atomic orientations* not to participate in its 3-D order. Atomic carriers of magnetic moments are not spherical symmetric entities and will be oriented in the crystal structure together with their magnetic moments in a certain orderly fashion. *The crystal field is the powerful field providing one or another positional and orientational dipole order in ferromagnetics and ferroelectrics.*

To be fair, Weiss can hardly be blamed for the mistake, considering that orientation-disordered crystals (ODC) had not yet been discovered at that time. At present, however, there are plenty of known phase transitions from a normal crystal into a higher-temperature ODC state (Sec. 2.7). One can now realize that a transition *ferromagnetic→paramagnetic* is just one of them. In other words, the dipole thermal rotation resulted from rotation of the atoms carrying these dipoles. To induce this rotation, the ordering power of the crystal field (atomic chemical bonding) has to be overcome by thermal agitation. To compare this with non-ferroic crystals, no "molecular field" is needed to hold their constituent particles in a 3-D orientational order at all temperatures below melting point (or transition to ODC state if there is one).

It has become a cliche that Heisenberg gave the quantum-mechanical account for Weiss' "molecular field." His theory assumed that the electron shells of neighboring atoms overlap and give rise to a collinear orientation of the magnetic moments. The main parameter in the quantum-mechanical formula was "exchange integral." Its positive sign led to ferromagnetism, and negative to antiferromagnetism. The value of the exchange integral was quantitatively compared with the Curie temperature. It is now evident that the theory actually failed on every account: (a) on the absolute value of the exchange integral, (b) on its sign, (c) on overlapping of the electronic shells, and (d) on collinear orientation of the magnetic moments. It has been found that (a) the value of the exchange integral for Ni is lower by about two orders of magnitude needed to account for the real Curie temperature, (b) it has the wrong sign, (c) the appropriate electron shells in the rare-earth metals, which are ferromagnetic, do not overlap, and (d) there is a great variety of non-collinear magnetic structures. Furthermore, even being unable to account for simpler ferromagnetism in Fe, Ni and Co, the 'exchange field' theory encountered the materials where magnetic moments were too far apart to make any direct exchange possible. It was expanded to those cases anyway, to become "superexchange".

The actual diversity of dipole patterns in crystals demanded new explanations from the theory which initially seemed straightforward. This time we mean the phenomenon of weak ferromagnets that were given the name "ferrimagnetics" and explained as the case of a "non-compensated antiferromagnetism." In a simple case, the dipoles in such a structure have alternatively opposite directions, but they do not completely compensate one another. Here the theorists had to turn to the crystal lattice, but they represented it as two sub-lattices inserted into one another, each one being ferromagnetic. No justification was given to representation of a *single* crystal in the form of *two independent* sub-crystals. To illustrate how artificial this method is, it will now be shown that this method and the Weiss-Heisenberg theory are mutually exclusive. Let us imagine a magnetic structure with N arbitrarily oriented dipoles in a unit cell. Applying the above method, this magnetic order can be represented as N ferromagnetic sub-lattices (Fig. 4.1), each one being the subject of the Weiss-Heisenberg theory. In other words, it is the "exchange field" that makes every sub-lattice ferromagnetic. Thus, while the exchange field theory was intended to explain why the dipoles maintain strict parallelism, the same theory plus the "sub-lattice method" can explain why they do not maintain any parallelism. The theorem is proven.

In the meantime, it was established experimentally that a collinear order of the atomic magnetic moments in ferro-, antiferro- and ferrimagnetics represents only some particular cases and that there is, in fact, a great variety of non-collinear magnetic structures as well. These are some types of magnetic structures in crystals: "simple ferromagnetic," "simple antiferromagnetic," "ferrimagnetic," "weakly ferromagnetic," "weakly non-collinear antiferromagnetic," "triangle," "simple helical," "ferromagnetic helical," and there are more. Only in the metallic heavy rare earths the following magnetic structures were listed [185]: "ferromagnet," "helix," "cone," "antiphase cone," "sinusoidally modulated," "square-wave modulated." A plausible reason for this great diversity is that *any dipole pattern is possible.* The diversity in the mutual positions and orientations of the magnetic moments can only be matched by the diversity in the world of crystal structures. The molecular / exchange field, assumed to be extremely powerful, was unable to provide a parallel alignment in innumerable magnetic structures. This is one more argument that crystal structure, rather than "molecular field," imposes and holds the orientations of its dipoles.

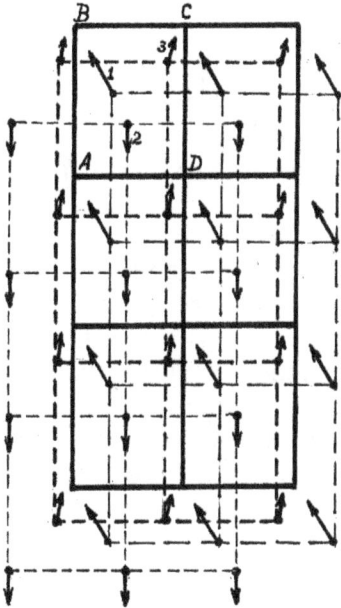

Fig. 4.1 (Schematic) A crystal structure with three dipoles (arrows 1, 2 and 3) of arbitrarily chosen orientations in its unit cell ABCD. This structure is represented as three "ferromagnetic sub-lattices" (dashed lines). The figure is to illustrate that while the Weiss-Heisenberg theory had been developed to justify a parallel dipole alignment, the same theory plus the "method of sub-lattices" justifies the opposite.

A brief note has to be made regarding the *sign* of the "exchange integral." According to the theory, a *positive* sign leads to a collinear ferromagnetism and a *negative* sign to a collinear antiferromagnetism. One may ask: do the exchange integrals in those diversified non-collinear structures have signs? Of course they have. In every particular case it is either plus or minus - but neither is consistent with the exchange field theory.

As said before, the Weiss-Heisenberg theory basically dealt with a system of elementary magnets rather than the crystal lattice particles carrying them. However, in ferromagnetic phenomena the role of a crystal was too evident. The concept "anisotropy energy" was added. It assumed that the magnetic dipoles *prefer* to be oriented in certain crystallographic directions. For this purpose, another property had to be attached to the exchange field: its orientational asymmetry relative to the crystal axes. A straightforward statement by Bozorth [65] that the direction of magnetization in the absence of magnetic field is determined by the crystal structure actually represented a certain recognition of the role of crystal field. But the point is how *strong* the relation between the crystal field and the dipole orientation is. The conventional theory assumes that the dipoles can have any orientation relative to the crystal axes under the action of a sufficiently strong magnetic field. Most importantly, it does not abandon the Heisenberg's exchange field. But evidence, if properly interpreted, always shows a permanent orientation of the magnetic moments inherent in the particular crystal. A belief that the magnetic moments can change their orientation, beyond a reasonable elastic deformation, without a corresponding structural rearrangement, is not based on any solid observations. In the framework of the present new concept the notion "anisotropy energy" is meaningless. *Crystal structure firmly determines, rather than influences, the orientation of its spins.*

The phenomenon of antiferromagnetism constituted a problem to the part of the theory (Landau and Lifshitz [146]) accounting for the origin of the domain structure in ferromagnetics. The breaking of a ferromagnetic crystal into domains was explained by minimization of the energy due to magnetization of the neighboring domains in the opposite directions. This account was too narrow, however. A similar explanation could not be applied to antiferromagnetics and antiferroelectrics, which also have domain structure. The origin of domain structure in ferroics will be clarified in Sec. 4.5.

The "classic" theory of ferromagnetism had already been basically developed when physics was confronted with the phenomenon of ferroelectrics. Application of the "ferromagnetic" ideas to ferroelectrics would later be an additional test for the theory of ferromagnetism, because the ferroelectric properties were a carbon copy of those observed in ferromagnetics. The result of the test turned out to be negative: the explanations involving a molecular / exchange field were inapplicable to ferroelectrics, since no such field was concluded to be present there. This situation opened two possibilities: (1) to find a different explanation of the nature and properties of ferroelectrics or (2) to recognize that both ferroics have to be explained in terms of general principles without "molecular fields." Attempts to develop a theory of ferroelectrics basically followed the former direction. We follow the latter direction. Many books were published on ferroelectricity [55,157,187-190]. The most realistic was that by Zheludev [157]. It contained a comprehensive review of experimental data and theoretical problems, but left most questions without correct answers. The extensive monograph by Lines and Glass [55], on the other hand, was based on the erroneous "soft-mode" concept (see Sec. 1.6). No consistent theory of ferroelectrics existed to date.

The conventional theory of ferromagnetism failed to see, in particular, that (A) magnetic phase transitions are essentially first-order (nucleation-growth) structural transitions, (B) critical Curie point is neither *critical*, nor a *point*, and rather a nucleation temperature range which, due to its nature, is unfit to play the crucial part in that theory. (C) a crystal field is the creator of all existing kinds of magnetic structures, (D)the crystal field can provide overall stability to a ferromagnet even when the magnetic interactions by itself elevate its free energy, (E) dipole reorientation always results from the structural rearrangement, (F) nucleation lags are the primary cause of magnetic hysteresis. In full measure the points A to F are valid as applied to ferroelectrics as well.

On the other hand, a better understanding of ferromagnetic and ferroelectric phenomena could not be achieved until the general mechanism of phase transitions in solids was clarified. The categories described in the previous chapters of this book, such as contact interfaces, edgewise mode of crystal growth in a solid state, epitaxial and non-epitaxial nucleation and growth, nucleation in microcavities, "pre-coded" information stored in certain lattice defects, temperature range of phase transitions, hysteresis loops of phase transitions, structural domains, etc., are vital in the interpretation of magnetic phase transitions, domain structures, domain boundaries, magnetization process, magnetic hysteresis loops, magnetostriction, Barkhausen effect, etc.

The time to abandon the theory of molecular / exchange field and its attributes is long overdue. This will make it possible to consistently clarify all details of the nature and properties of ferromagnetics and ferroelectrics. The present chapter provides the principles of their new interpretation.

4.2 Ferroelectric and ferromagnetic phase transitions

4.2.1 Ferroelectric phase transitions

Preparatory to reviewing ferroelectric phase transitions, different dielectric states should be defined. Some of them have been defined clearly, others have not.

Pyroelectrics. The dielectrics with a permanent spontaneous polarization $P_s \neq 0$ which is not reversible in applied electric fields. In a simple pyroelectric all dipoles are aligned in parallel.

Non-polar dielectrics. Those with a distribution of positive and negative electrical charges in the centrosymmetrical unit cell, so that there are no dipole moments in the cell.

Polar-neutral dielectrics. Those having dipole moments in the unit cell with their vector sum being zero.

Orientation-disordered dielectrics. Those in ODC state (Sec. 2.7), with their dipoles being in a thermal orientational disorder, so that $P_s = 0$. They are always a higher-temperature phase relative to the normal ordered crystal phase.

Paraelectrics. The dielectrics with $P_s = 0$, which change into a ferroelectric phase upon cooling.

The above definition of paraelectrics is not quite complete until ferroelectrics are defined. It remains also to define ferrielectrics and antiferroelectrics. This will be done in the following sections and answer the central question of what they really are. For now it suffices to accept that ferroelectrics are those with dipoles in their structure and the ability to be polarized and repolarized in electric fields.

Ferroelectric phase transitions are classified in literature into two types: "displacive" and "order-disorder" (Sec. 1.5.3). The former are believed to occur by a "displacement" of some ions relative to others in the initially centrosymmetrical unit cell (the term "distortion" is also used to describe this process). At this point the reader should already know, however, that solid-state phase transitions do not proceed by "displacement" or "distortion," but only by nucleation of new crystals in the body of the parental crystal and subsequent growth until the latter ultimately disappears.

It is typical for current literature to confuse cause and effect by stating that a ferroelectric transition is *accompanied* by change of the crystal structure. Actually, it is the structural rearrangement that gives rise to change of the physical properties. With only few exceptions, the transitions from a higher-temperature phase lead to a lower-symmetrical phase. In the particular case of a 'non-polar *paraelectric* → *ferroelectric*' transition this involves a loss of the center of symmetry, resulting in $P_s \neq 0$.

Similarly, the dominant role of structural rearrangements in "order-disorder" ferroelectric phase transitions was not recognized; the thermally disordered state of the dipoles in the paraelectric phase was not related to the well-known orientation-disordered crystal state. The phase transition in question is simply a structural transition from the ODC mesomorphic state, in which the molecules (in general, the constituent particles) are engaged in a hindered thermal rotation, to the crystal state: only the substance in this case happened to contain polar, rather than electrically neutral molecules. As was demonstrated in Sec. 2.7, the "ordering" and "disordering" proceed by nucleation and growth.

Together with the above-considered classification, ferroelectric phase transitions were also classified by first / second order. No correlation existed between the two classifications: some "displacive" transitions were designated as of the first order, others as of the second order. The same occurred to the "order-disorder" transitions. The question of why there was such inconsistence was not discussed. Why, for example, ferroelectric transition in $BaTiO_3$ is denoted to be of the first order, and in $SrTiO_3$ of the second order? Our answer is: those of the second order were classified incorrectly.

It is presently recognized that the overwhelming majority of ferroelectric phase transitions are of the first order (which does not mean, however, that they are treated as such). The remaining "second-order" transitions will be reclassified. The following examples are indicative. $BaTiO_3$ was used by Landau and Lifshitz [36] to exemplify a second-order phase transition, but was later reclassified into the first-order. The same happened to KH_2PO_4 (KDP), although "for years this crystal had been considered as a typical representative of the ferroelectrics undergoing second-order phase transition" [157]. One more example is TGS (tri-glycine sulfate). The transition in this crystal was believed to be the most typical second-order ferroelectric phase transition. As usual, the second-order has been assigned on the basis of "smooth" changes of some crystal parameters and physical properties upon the phase transition. However, the following factors can explain why sharp changes were not recorded: (a) phase transition occurred over a two-phase temperature range owing to a non-simultaneous multiple nucleation, (b) transition was of the epitaxial kind and therefore exhibited narrow hysteresis, (c) emerging ferroelectric phase consisted of lamellar domains of two alternating orientations. As soon as small unipolar (monodomain) TGS samples were used, a first-order phase transition was found [191]. There is another proof of the first-order transition: jumps in the electric susceptibility at the "Curie point" upon measurements along the three crystallographic axes X, Y, Z were recorded, and the temperature hysteresis ~ 0.2 °C was also detected [157].

Thus, the three "most reliable" second-order ferroelectric phase transitions turned out to be first-order phase transitions and, consequently, to proceed by nucleation and growth. The problem with the second- and higher-order phase transitions was that they were assumed to exist without unambiguous evidence supporting that. We predict that the remaining "second-order" phase transitions, and those which will be named so in future, ferroelectric or otherwise, will be eventually reclassified into first-order, and *the first / second order classification will become only an unfortunate episode in the history of science*.

4.2.2 Ferromagnetic phase transitions

As stated by Vonsovskii [121], the theory of second-order phase transitions provided an "impetus" to studies of magnetic phase transitions. Interest in this topic is explained by a hope to shed light on the nature of ferromagnetism. In addition, in view of the incessantly shrinking availability of second-order phase transitions, ferromagnetic transitions are believed to be the most reliable case justifying their existence [8]. In 1965 Belov wrote in his monograph "Magnetic Transitions" [64] that ferromagnetic and antiferromagnetic transitions are "concrete examples" of second-order phase transitions. His work was devoted to the investigation of spontaneous magnetization and other properties in the vicinity of the Curie points. The problem was, however, how to extract these "points" from the experimental data which were always "smeared out" and had "tails" on the temperature scale, even in single crystals. Unfortunately for this and other authors, they were actually dealing with all the effects that accompany first-order nucleation-growth phase transitions described in the previous chapters, namely, the temperature ranges of phase transitions and related pseudo-anomalies.

Just a few years later, when the comprehensive review book on magnetism by Vonsovskii [121] was published, it was already widely recognized that many ferromagnetic phase transitions were of the first order. In the book about 25 such phase transitions were listed. They were interpreted in the usual narrow-formal manner as those exhibiting "abrupt" changes and / or hysteresis of the magnetization and other properties. Some of these first-order ferromagnetic transitions Vonsovskii described as "apparent," explaining that structural transition occurs before the "true" temperature of transition from the ferromagnetic to paramagnetic state can be reached. (Disassociation of magnetic transition and structural rearrangement is a major misconception.) But existence of genuine first-order ferromagnetic transitions was also recognized. The puzzling fact of their existence led to the numerous theoretical and experimental studies that were surveyed in that book. The conventional theory was in a new predicament: the Curie point was not a point any more, and was rather a range of points and, even worse, was a subject to temperature hysteresis. As it happens in such cases, in order to save the failing theory attempts were made to complicate it even more. The ideas to make the exchange field dependent on the lattice deformation, interatomic parameters, energy of magnetic anisotropy, etc., were tested. *It was not realized that a first-order phase transition meant nucleation and growth, and not a critical phenomenon* (Sec. 1.2). The theorists could not resist a temptation to apply their only available theoretical tool - statistical mechanics - to even first-order phase transitions.

Another peculiarity of these works was that magnetization was assumed to be the cause of ferromagnetic phase transitions, while changes in the crystal parameters, density, heat capacity, etc.- the accompanying effects. The idea that change in the state of magnetization is secondary to change in the crystal structure has not been advanced. The problem of the first-order ferromagnetic phase transitions has not been resolved. The thermodynamic theory that treats them as those of the second order has lost its grounds; it cannot be justifiably applied even to such basic ferromagnets as Fe, Ni and Co. They were not included in the Vonsovskii's table of 25, but their ferromagnetic transitions are also of the first order.

A "discontinuity" of the Mösbauer effect in Fe, incompatible with a second-order phase transition, was first reported in 1962 by Preston *et al.* [192], and later in more detail by Preston [193] who found it being sharper than 0.3 °C. The statement by Preston that this "might be interpreted as evidence for a first-order transition" would probably be more conclusive if he knew that a first-order transition occurs over a temperature range rather than at a point. In case of Ni, such titles as "Mössbauer Study of Magnetic First-Order Transition in Nickel" [194], or "Structural Phase Transition in Nickel at the Curie Temperature" [195] speak for themselves. Neither of the mentioned works related first-order transitions to rearrangements at interfaces.

The solution to the problem of first-order ferromagnetic transitions resides in the fact, already stated, that *all* phase transitions in solids are of the nucleation-and-growth type. In every particular case this can be proven if subjected to scrutiny (see also Sec. 2.6.8). Its implication is fundamental for the currently existing theory of ferromagnetism, making it clear that this theory is based on premises not rooted in reality.

The foregoing can be regarded as a historic excurse. The internet search for "first order magnetic transition" and "first order ferromagnetic transition," taken in 2007, produced 4,000,000 hits – but nobody, it seems, tries to theoretically explain them now.

4.2.3 Ferromagnetic non-second-order non-phase transition in iron?

The phase behavior of Fe, as presented in literature, is marked by inconsistencies. Fig. 4.2 summarizes available data on its temperatures of phase transitions and phase designations. The transition between paramagnetic and ferromagnetic phases occurs at ~769 °C. This temperature is treated as a Curie point, and the transition as of the second order. Above it, there are two more phase transitions, both of a paramagnetic to paramagnetic type. A paramagnetic state is an orientation-disordered crystal (ODC) state, the one where the constituent particles are engaged in a hindered thermal rotation (Sec. 2.7), in this case - rotation of the spheric-asymmetrical atoms as a whole. These two ODC-ODC phase transitions occur at ~910° and ~1400°. A well-detected hysteresis was observed at the ~910° transition, which means that it is of the first order and proceeds by nucleation and growth.

Below 769° there is an aberrant situation: both the paramagnetic and ferromagnetic phases are designated as the same crystal α-phase. It is body-centered cubic (bcc) with the unit cell parameter $a = 2.86645$ Å (20 °C). Only conditionally the "paramagnetic part" of the α-phase is sometimes called the β-phase. This brings up two questions: Is the β-phase a *phase*? Can the "α to α" transition be a phase transition of the *second order*?

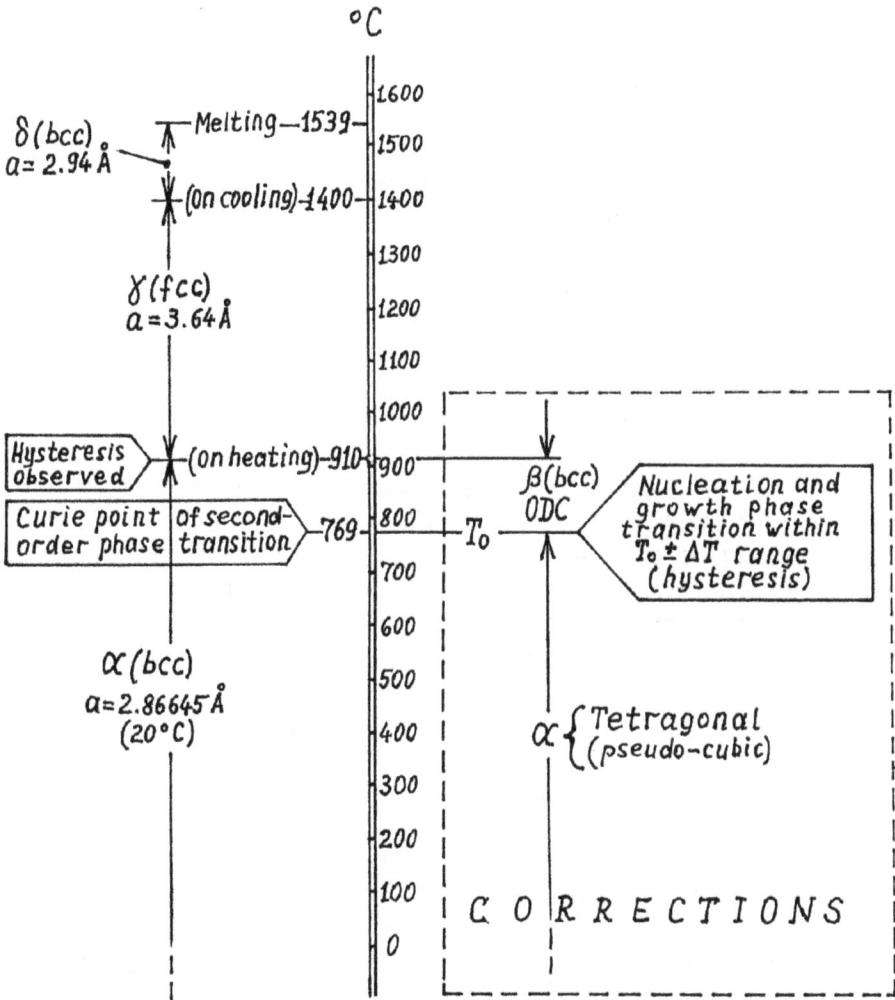

Fig. 4.2   Phase transitions in Fe as they are presently believed to be. On the right - corrections eliminating existing contradictions. The difference is in the interpretation of the α- and β-crystal structures and their ferromagnetic transition at ~769 °C. "ODC" is the orientation-disordered crystal state of the paramagnetic α-phase. The temperatures indicated as measured only on heating or cooling contain some systematic error.

To help answer these questions, the notions of a solid-state phase and solid-state phase transition are to be addressed.   One definition [39] was already discussed in Sec. 1.4: "A *phase*, in the solid state, is characterized by its structure. A solid state *phase transition* is therefore a transition involving a change of structure, which can be specified in geometrical terms." According to this definition, the ferromagnetic transition in Fe is *not a phase transition*. Nevertheless, the conventional theory regards it a (second-order) *phase transition*. What is the cause of the contradiction?

Now we turn to another contradiction. Everyone seemed to agree that "transition from the ferromagnetic to the paramagnetic state is a phase transition of a second kind. At the Curie point, where the spontaneous magnetization disappears..., the symmetry of a ferromagnet changes sharply" [121]. It was Landau who defined second-order phase transitions as a sudden change in the crystal symmetry [34]. But, allegedly, there is no symmetry change in the ferromagnetic transition α (bcc) → α (bcc) in Fe. Again, what is the cause of the contradiction?

These inconsistencies have existed in a dormant state for many decades. No attempt to solve them has been made. The point is that they could not be eliminated in terms of the conventional theory. The solution involves two corrections (see "corrections" in Fig. 4.2): (1) the β-phase is a distinct crystal phase and (2) the ferromagnetic phase transition is not of the second order, but occurs by nucleation and growth.

The following is a description of the polymorphic transitions, *paramagnetic→ferromagnetic* in Fe, free of the previously described contradictions. The β-phase is paramagnetic and has a rather typical for ODC genuine body-centered cubic lattice. Its transition to the ferromagnetic α-phase is a structural rearrangement ODC → CRYSTAL. It proceeds by multiple nucleation and growth of the ferromagnetic domains. Because nucleation temperatures in different nucleation sites are not the same, there is some temperature range of transition. This explains the failure of all attempts to find a "true" Curie point. If the transition is reversed, some hysteresis can be revealed. There is a very small, but finite, density change, even if it was not detected yet. The resultant ferromagnetic phase is pseudo-cubic, but it is actually tetragonal (this subject will be dealt with in Sec. 4.7). The lattice parameters of the α- and β-phases differ, but are very close. Two reasons make both the range of transition and hysteresis relatively narrow: the phase transition is of a CRYSTAL → ODC type (Sec. 2.7) and the nucleation is epitaxial (Sec. 2.8). The cause of a relatively large hysteresis in the β → γ transition is a significant difference between the structures involved, which makes epitaxial nucleation impossible.

The confusion about the "α to α" ferromagnetic transition resulted from theoretical prejudice and from the insufficient sensitivity of the X-ray technique.

4.3 Spontaneous magnetization and spontaneous polarization: why are they spontaneous?

This is a fundamental question. Why are spontaneously magnetized and spontaneously polarized states thermodynamically stable? It is definitely the case in ferromagnetics, antiferromagnetics, ferroelectrics, antiferroelectrics and pyroelectrics. But this is not all: any of the above can be not only collinear, but also non-collinear of many different kinds. The answer to that question has already been given in Sec. 1.8 and will be briefly outlined again at the end of this section as the alternative to what the conventional theory could offer. The present state of the subject is as follows.

Ferromagnetics. It is a molecular / exchange field that is supposed to maintain a strict parallelism of the magnetic dipoles. But as indicated in Sec. 4.1, the theory has failed quantitatively, produced an erroneous sign of the exchange integral, could not account for the existence and diversity of non-collinear magnetic structures and could not be legitimately applied to the whole group of the rare earth ferromagnetics.   Even assuming conditionally that the theory qualitatively accounts for the spontaneous magnetization in the collinear ferromagnetics, the fact is that in most cases they are not collinear. As a result, crystal structure considerations had to be additionally invoked. Thus, the crystal field was implicitly recognized to be a force as powerful as the exchange field.

Antiferromagnetics. If the exchange integral in the Heisenberg theory happened to be negative, this should produce a collinear spontaneous "antimagnetization."   But the correct sign cannot be inferred from the theory and therefore has to be assigned.   Also, everything being said about non-collinear magnetic orders in ferromagnetics is valid here too, i.e., the theory has the additional problem of covering those cases. In reality, the 'exchange field' is not required to explain the thermodynamic stability of these systems: the dipole attraction by itself lowers the crystal free energy.

Ferroelectrics. Zheludev [157] considered it very important to account "at least qualitatively" for the thermodynamic stability of

- 215 -

spontaneously polarized structures. Application of the Weiss theory to electric dipole systems has led to the conclusion that no molecular field existed there. Yet, spontaneous polarization of ferroelectrics was a fact, the parallel order of the dipoles increasing their free energy. Zheludev had to admit that "...the theory of spontaneous polarization is not existing yet...." Besides, all the ferroelectric structures with non-collinear dipole arrangements were also waiting for justification of their stable existence.

Antiferroelectrics. There is not even a negative "exchange integral" in order to somehow justify ferroelectric spontaneous "antipolarization." It is also to be remembered that there are many non-collinear antiferroelectrics too. Meanwhile, the antiparallel dipole order without any additional theory will provide more stability to the crystal structure due to purely electrostatic dipole interaction.

Finally, there are pyroelectrics, collinear and non-collinear, whose stable existence is the living solution to the problems in question. The solution is simple and applies to all ferroics and antiferroics, both collinear and non-collinear. The reader may turn to Sec. 1.8 where it was put forward. It may be summed up as follows. Any particular dipole pattern in a ferroic is just an element of the three-dimensional crystal order. It is the crystal field of the atomic / molecular chemical bonding that imposes both positional and orientational dipole order. This is why such a great diversity in the dipole patterns is observed. Since the contribution of magnetic or electrostatic interaction to the crystal free energy is relatively small, in some cases the crystal structure becomes thermodynamically preferable in spite of the destabilizing action of the magnetic or electrostatic interaction. All crystals with $M_s \neq 0$ and $P_s \neq 0$ fall into this category. As for antiferroics, their dipole interaction makes the crystal free energy even lower.

By eliminating such basic attributes of the orthodox theory of ferromagnetism as "exchange field" and "anisotropy energy," the concept presented here is the foundation of a new understanding of the general nature of both ferroics, reducing their differences to some particular details of secondary significance.

4.4 "Curie point," a misnomer

A Curie point is the cornerstone of the classical theory of ferromagnetism and has been transferred to ferroelectrics in a similar capacity. It was introduced into the theory as a fixed critical temperature point at which a ferromagnetic order of the magnetic dipoles, established by the molecular field, is overcome by their thermal agitation, or *vice versa*. It has been concluded that the strength of the interaction between magnetic dipoles has to be several thousand times of that of their purely magnetic interaction to hold the ferromagnetic dipole order up to the high temperatures where the "Curie points" were located. That is how the "molecular field" was born.

The "Curie point," however, was always besieged with problems. The theoretical physicists were unable to describe the specific heat and other physical properties near the Curie point, comprehend the nature of the sudden transition, etc. Feynman in his "Feynman lectures on physics" [196] stressed the importance of solving these problems for the physical science, considering them very interesting, exciting, and worthy to work on (see Appendix 4). Moreover, the Curie point was always at odds with the experimental facts. They revealed that it is not a "point," its position is not quite reproducible, it exhibits hysteresis, and even its approximate location is in many cases far away from where the theory predicted it to be. The validity of the notion itself had to be questioned, but this did not happen: without the "critical" Curie point the molecular / exchange field theory would collapse. But some complaints were serious:

- "The Curie point is not always defined in accordance with the Weiss theory but in other more empirical ways..."[65].
- "Many important questions ... remain unsettled or in dispute. These include ... the actual temperature behavior of the spontaneous magnetization near the Curie point, the causes of the 'smearing out' of the magnetic transition...the existence of 'residual' spontaneous magnetization above the Curie temperature, and the nature of temperature dependence of [different] properties near the Curie point. It even remains unsettled what we should take to be the Curie temperature, and how to determine it" [64].
- "The Neel temperature T_N [the "Curie point" in antiferromagnetics] often vary considerably between samples, and in some cases there is large thermal hysteresis" [197].
- "It is difficult to establish some *definite* temperature at which the *magnetic transition 'ferromagnet↔paramagnet'* takes place, since this transition becomes 'blurred'..., it is only possible to specify some *transition temperature range"* [121, p.466].

The problem of Curie points and related phenomena is not just solved, it is entirely eliminated by the simple fact that the "point" itself is a misnomer. While struggling with the uncertainties of the Curie point, the above-quoted authors precisely described the temperature region of nucleation-growth phase transitions. The crucial circumstance that was overlooked by the Weiss theory and not corrected afterwards was that the magnetic dipoles were not suspended in vacuum: they were a specific property of magnetic atoms, and the latter were combined into a crystal structure. The appearance or disappearance of spontaneous magnetization could not be achieved without a corresponding structural rearrangement CRYSTAL↔ODC, the magnetic changes being a consequence of this restructuring. It was a mistake to assume that only the dipoles, and not their atomic carriers, acquire or loose their orientational order at the temperature of *paramagnetic ↔ ferromagnetic* phase transition. It is worth repeating that both the positions and orientations of the atoms in a ferromagnetic phase are firmly determined and fixed by the system of chemical bonding in the crystal.

Solid-state phase transitions proceed by nucleation and growth. The nucleation was analyzed in Sec. 2.5. Its features easily account for the "poor behavior" of "Curie points." A Curie point, as defined by the Weiss theory, does not exist in nature. Moreover, the very idea, permeating all the literature, that a phase transition in solid state occurs (or has to occur under ideal conditions) at a fixed temperature *point* is in error. Any phase transition can start only after passing the temperature T_0 at which the two phases have equal free energies. The actual lags are not fixed either; the two lags - above and below T_0 - constitute hysteresis. In case of multiple nucleation (which is most common) the transition occurs over a two-phase temperature range which can be wide or narrow, but always exists. This range is also not quite reproducible and located above T_0 upon heating and below T_0 upon cooling (hysteresis). Hence the hopeless problem to locate the "point," the above-mentioned "smearings," "residual magnetization," as well as other apparent anomalies observed around "Curie points." They were accounted for in all detail in Chapter 3. The reader is also advised to look through Secs. 2.5 and 2.6.

4.5 Origin of domain structures

Properties of ferroics are determined to a large measure by the fact that they have domain structure, *i.e.*, they consist of tiny spontaneously magnetized or polarized blocks. The origin of the ferroic domain structures is therefore another fundamental issue. In this case, too, the existing theory has little to say.

The only account for the division of a ferromagnet into a system of spontaneously magnetized domains of mutually opposite magnetization was proposed by Landau and Lifshitz (L & L) in 1935 [46] and was incorporated into the theory as valid and even self-evident. The domain structure is considered to be a specific *product* of ferromagnetic interactions. The standard sketch showing the division of a spontaneously magnetized body into a number of parallel domain strips, which can be found in every pertinent book and review, is to illustrate the idea by L & L that the demagnetizing influence of the surface of the body is the only reason for the domain structure to occur. Magnetic poles of a large single crystal produce a large external magnetic field, and division of the crystal into domains lowers the energy of this field; the process continues until the energy required to form domain boundaries becomes greater than the reduction of the magnetic energy.

The L & L theory is not as impeccable as it is presented in the literature:

(1) The theory was based on the erroneous assumptions of *exchange field* and *anisotropy energy* (Secs. 1.8, 4.1, 4.3).
(2) Explanation of *origin* of the domain structure had to have two elements: the thermodynamic reason and the actual mechanism of domain formation. The theory in question dealt only with the former, being silent about the latter and the fact that the initially homogeneous system becomes heterogeneous. It would be reasonable to suggest that this transformation occurs by nucleation and growth of the individual domains of two (or more) magnetic orientations in the body of the paramagnetic phase. Actually, nucleation and growth of ferromagnetic domains is not a suggestion, but experimental fact. This means, however, that a ferromagnetic phase transition is *in principle* not of the second order. This conclusion, in turn, undermines the notion of a "Curie point," a cornerstone of the Weiss' molecular field theory (Sec. 4.4). One may also recall that the theory of second-order phase transitions (which must proceed homogeneously) was developed by Landau, one of the authors of the above-mentioned domain theory.
(3) While the L & L theory predicts antiparallel magnetization on the opposite sides of a domain boundary (so called "180°-walls"), in most cases it is not so. Thus, in Fe there are more "90°-walls" observed in the domain patterns than "180°-walls."

(4) If the L & L theory were correct, the domain structure would not be observed in antiferromagnetics, as was asserted by Hubert [198]: "While experimental observation of antiferromagnetic [domain] structures is common, the reason for their occurrence is still rather elusive. As opposed to the case of ferromagnetic domains, they cannot lower the energy of the leakage field..." Such a statement is rare in the relevant literature: the issue of origin of antiferromagnetic domains is not mentioned even in the comprehensive books on ferromagnetism.

The explanation of ferroelectric domains was only slightly better. Zheludev [157] could not find any significant difference between pyroelectrics and ferroelectrics* except for the fact that the former do not have domain structure, while the latter do, and concluded that "the true cause of difference in the behavior of linear pyroelectrics and ferroelectrics is, probably, more complicated and is rooted in the very nature of spontaneous polarization." After this incorrect assumption, this author suggested "a very general outline" of the thermodynamic cause of domain formation as a competition between the energy of the electrostatic leakage field and that of domain boundary formation - by analogy with ferromagnetics, but without a "molecular field" and "anisotropy energy." Probably, the theoretical development of this idea was not successful enough to go beyond its general outline. At least, this type of reasoning did not seem attractive to the authors of another book on ferroelectricity [187] who stated: "We have no reason for a domain structure to occur in ferroelectrics."   Moreover, the above generally outlined idea could not also be applied to antiferroelectrics for the same reason as it could not be applied to antiferromagnetics.

In the same monograph, however, Zheludev made a step toward the correct solution. He suggested that the domain structures may result from the equally probable polarization along the equivalent directions in different parts of the paraelectric crystal. This occurs upon passing the Curie temperature. Separately, a role of nucleation in this process was mentioned. Zheludev related this possibility only to the phase transitions of the first order, since "in ferroelectric phase transitions of the second order nuclei do not form." This actually made the fore-mentioned fruitful idea useless, for ferroelectric phase transitions were consistently described in the book as a "displacement," "distortion," "deformation" or "ordering," rather than nucleation and growth. Lack of proper information about solid-state phase transitions

---

* Russian equivalent "segnetoelectrics" was used in the original.

prevented this author from correct explanation of domain structure in ferroelectrics.

Next step in our survey is to take into account the essential fact, mentioned without comments by Kittel [197] that domains form not only in ferromagnetics, but also in antiferromagnetics, ferroelectrics, antiferroelectrics, ferroelastics, superconductors, and sometimes in metals. We can add: and sometimes in ionic crystals, organic crystals, minerals, even liquid crystals - in fact, in any type of crystals irrespective of their electric or magnetic properties. Obviously, naming a general cause of domain formation - wherever they are observed - should deserve more credibility than explaining the domain structure in ferromagnetics by ferromagnetism, in antiferromagnetics by... (antiferromagnetism ?), in organic crystals... (by what?)... and so on. This general cause was already revealed in Sec. 2.8.6. *Any domain structures come into existence by means of multiple epitaxial nucleation in two or more equally probable crystal orientations - either upon phase transition or liquid crystallization.* If the crystal is ferromagnetic or ferroelectric, the internal magnetic or electrostatic field can modify the emerging domain pattern toward minimizing the magnetic or electrostatic energy. This secondary effect, although important (Sec. 4.9) was previously mistaken for the primary cause of the domain formation in ferromagnetics. Taking this into account, it may be stated that L & L actually dealt with that secondary, but real, phenomenon.

A clarification has to be made concerning the conditions of the epitaxial (or "coherent") nucleation responsible for the formation of domain structures. Formation of laminar domains in layered structures (Sec. 2.8.6), especially characteristic for ferroelectrics, was a good illustrative example, but it is not the only case when domains form. There is a certain correlation between two conditions for the nucleation to be strictly epitaxial: (1) degree of cleavage and (2) proximity of structural parameters of the phases. The closer the parameters, the less pronounced the cleavage can be in order for epitaxial nucleation to occur. The parameters of α and β-phases in Fe are so close to each other that they were erroneously believed to be a single phase (Sec. 4.7). As a result, there the nucleation is always epitaxial even in the absence of significant cleavage: any appropriate crystal defect offers epitaxial formation of a domain nucleus in one of the six equivalent crystal-structural (and the corresponding dipole) orientations.

4.6 Two basic components that make a ferroic

According to conventional views, magnetic ordering affects crystal structure and all physical properties of this matter: thermal, mechanical, electrical, etc. Wile it is true that all physical properties of a crystal are interrelated, in reality all of them, including the magnetic state, are primarily a function of its structure. The essence of any solid-state phase transition is the changing of crystal structure, and the resultant changes in magnetic and all other physical properties are secondary to that restructuring.

There are two basic components that make a ferroic. The first component is having dipoles in the crystal structure. Let us consider a dielectric with polar molecules which has an orientation-disordered high-temperature phase (ODC, Sec. 2.7). In this mesomorphic state the molecules retain only their three-dimensional positional order. Their dipoles are oriented in all directions due to molecular thermal rotation, so that, on average, the resultant $P = 0$. Upon cooling, this phase changes to the normal crystal phase where molecular orientations are fixed as well. The actual crystal structure, including dipole positions and orientations, is determined almost entirely by the minimum energy of chemical bonding rather than dipole interaction. Therefore, any dipole pattern imposed by chemical bonding is possible: parallel, antiparallel, non-collinear, helical, etc. The second component is that the phase transition must be epitaxial, in which case the resultant phase will have a domain structure capable of undergoing structural rearrangements at domain interfaces in external electric fields. In other words, it will be ferroelectric. The foregoing is true for both ferroics. The general principle is

DIPOLE CARRIERS + DOMAIN STRUCTURE = FERROIC

Table 4.1 illustrates how four independent factors indicated below can result in piroelectrics, permanent magnets, ferroelectrics, ferromagnetics, ferrielectrics, ferrimagnetics, antiferroelectrocs, antiferromagnetics, any non-collinear ferroics, or simply non-ferroic domain structures. These factors are:
- What is the initial phase?
- Does it contain dipoles?
- Does phase transition occur and is it epitaxial?
- What dipole order is imposed by the crystal structure?

Table 4.1 Formation of ferroics

	Initial phase	Phase transition	Dipole order	RESULT
1	ODC	EPITAXIAL	No Dipoles	NON-FERROIC DOMAIN STRUCTURE
2	Liquid	THERE IS NO PHASE TRANSITION	↑↑	PERMANENT MAGNET / PYROELECTRIC
3	ODC	EPITAXIAL	↑↑	FERROMAGNETIC / FERROELECTRIC
4	ODC	EPITAXIAL	↑↓	ANTIFERROMAGN. / ANTIFERROELECTRIC
5	ODC	EPITAXIAL	↑↓	FERRIMAGNETIC / FERRIELECTRIC
6	ODC	EPITAXIAL	↑↓ ↑↓	NON-COLLINEAR, HELICAL, etc., FERROICS

Thus, one does not need a molecular or exchange field to account for the above ferroics and their diversity, or, for example, suggest [157] that there is a difference in the nature of spontaneous polarization of piroelectrics and ferroelectrics. The crystal structure of pyroelectrics is simply not layered. No domain structure appears during their liquid crystallization, nor could it appear during phase transition, for there is no phase transition. The difference in question does not exist. In case of any doubt, here is how to make a pyroelectric from a ferroelectric. A defect-free single-domain ferroelectric crystal should be prepared by growing it from a solution at the temperature where it is stable. In the absence of the appropriate defects for nucleation, this crystal will actually be a pyroelectric: it would not be possible to induce its repolarization by an electric field.

4.7 Understanding the cause of magnetostriction

Understanding the cause of magnetostriction is an important element in the clarification of fundamentals of ferromagnetism. The phenomenon is defined as change in the dimensions and / or volume of a ferromagnetic body caused by magnetization. The change in the length in the direction of the applied magnetic field is typically between 0.0001 and 0.01% and can be either elongation or contraction. A change in the transverse direction also occurs and has the opposite sign. The change in volume is smaller than the change in length by orders of magnitude; therefore magnetostriction is basically a change in the shape. The bulk of performed experimental studies related to polycrystal samples of Fe, Ni, Co and different alloys, but single crystals were also investigated. The theoretical work for the most part had empirical or half-empirical character and shed no light on the cause of the phenomenon.

Magnetostriction was commonly considered to be an elastic deformation of the crystal lattice by magnetic and exchange forces. The standard theory does not have a sound explanation for this phenomenon, as the following summary exemplifies: "The phenomenon arises from the dependence of crystalline anisotropy energy on the state of strain of the lattice; it may be energetically favorable for the crystal to deform slightly if doing so will lower the anisotropy energy more than the elastic energy is raised" [199]. The cause of magnetostriction, as will be shown, is much simpler than that and easier to understand. Fe crystals are chosen for illustration.

Misunderstanding the cause of magnetostriction can be traced to the absence of a correct general picture of solid-state phase transitions. According to Vonsovskii [121], when a ferromagnet is cooled down below its Curie point, spontaneous deformations, associated with emergence of spontaneous magnetization, appear in it; magnetostrictive deformations (strains) occur in each domain. In this description the phase transition is implied to be cooperative second-order, which it is not. The *paramagnetic* → *ferromagnetic* phase transition occurs differently and produces a different result. Instead of the Curie point, there is a temperature range of multiple epitaxial nucleation of ferromagnetic domains in the paramagnetic crystal matrix. In Fe, the matrix is truly cubic (bcc), while the ferromagnetic domains have a tetragonal unit cell with the parameters very close to those of the cubic matrix. The ferromagnetic domains grow in the matrix until they meet one another and form 90° and 180° contact interfaces. The phase transition is not a deformation of the paramagnetic unit cells into spontaneously magnetized unit cells. This phase transition is no different from other solid-state phase transitions analyzed in Chapter 2: it is a crystal growth. The resultant phase simply has a different crystal structure and should not be viewed a strained or deformed parental phase. As in any solid-state phase transition, some internal strains may develop, but they are not an inherent feature of the ferromagnetic domains.

We shall take a close look at the crystal lattice of α-Fe. The fact that its unit cell is not of a cubic symmetry was not discovered experimentally, being deeply buried in the polycrystalline data and a polydomain structure of the crystal grains. The domains had close pseudo-cubic symmetry. Only very high precision measurements of the lattice parameters, preferably in a single domain, could reveal the actual structure not being cubic. Every domain is spontaneously magnetized. Its unit cell and magnetization directions are shown in Fig. 4.3. Its spontaneous magnetization is an integrated part of the crystal forces participating in the establishment of the final equilibrium interatomic distances. The structure is in its natural equilibrium state: it is not strained, deformed or distorted.

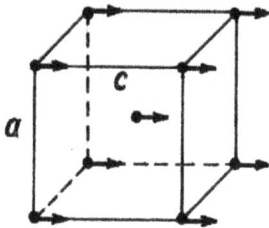

Fig. 4.3 A unit cell of ferromagnetic α -Fe crystal and its spontaneous magnetization. It is tetragonal with $c > a = b$. This difference is the source of the magnetostriction.

While every source stated that α-Fe lattice has a cubic symmetry, a uniform alignment of its atomic magnetic moments reveals that this cannot be the case. In terms of the magnetic dipole interaction there must be some difference in the unit cell parameters; $c \neq a = b$. All atomic dipoles have the same direction along the c-axis. With this dipole arrangement, the magnetic interaction should cause contraction along the c axis and expansion along the a and b axes, giving rise to a tetragonal unit cell with $c < a = b$. This consideration has only formal significance, namely, to state that α-Fe crystal lattice cannot have a strictly cubic symmetry in terms of the geometry of the atomic centers. The magnetostriction measurements of polydomain single crystals have shown that there is an elongation, and not a contraction, in the direction of spontaneous magnetization, i.e., $c > a = b$. The disparity proves that another cause, rather than the magnetic interaction, makes α-Fe unit cell tetragonal and holds the dipoles in their particular parallel order.

Some researchers, looking for the cause of magnetostriction, estimated the lattice "distortion" caused by the magnetic interaction alone and concluded that it was too small by an order of magnitude. On the other hand, calculations by Van Vleck [200] showed that the spin-orbit coupling could be responsible for the effect. His results may explain why the *equilibrium* (and not strained, deformed or distorted) crystal lattice of α-Fe has a tetragonal rather than a cubic symmetry, as well as why the period c is longer, and not shorter than a = b.

As opposed to the conventional interpretation, magnetization is not just a change of dipole orientation in the same crystal lattice. When the term "magnetization reversal" is used in the literature, it means only that: rotation of the dipoles in the crystal lattice. Similarly, a "domain rotation" actually means rotation of the dipoles in the domain crystal structure. This misinterpretation of magnetization has been a major barrier to the identification of the cause of magnetostriction. *The magnetization process is a change of the dipole directions by atom-by-atom rebuilding the lattice itself.* It is agreed [65] that a "magnetization reversal" by 180° in a Fe crystal does not produce magnetostriction, but a change of the direction by 90° does. This means that the α-Fe unit cell, being oriented either parallel or antiparallel to the magnetic field, has the same length which exceeds its dimension in the transverse direction - whether the magnetic field is applied or not (Fig. 4.4). Again, the conclusion is that c > a.

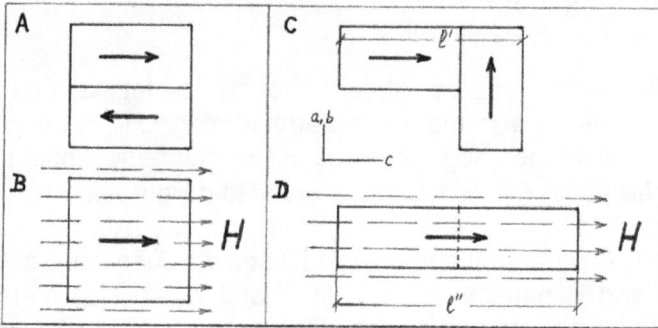

Fig. 4.4 Magnetostriction in a crystal consisting initially of the antiparallel domains (A,B) and another crystal consisting initially of the domains making 90° with each other (C,D). In the former case the magnetostriction is zero; the only effect of the applied magnetic field H is to move the interface downward until it disappears. In the latter case, rebuilding the domain on the right into the one with its c to be parallel to H gives rise to the magetostriction (ℓ'- ℓ) / ℓ' = 0.5 (c-a) / c. (Note: the line between the domains in C is not the domain interface: ref. to Sec. 4.9).

In a polydomain Fe crystal, the c-axes, initially distributed equally in six equivalent directions, acquire preference in the direction closest to that of the applied magnetic field. This is the cause of the changing shape of the sample. Quantitatively, the difference between axes c and a was not large enough to be directly detected. But it was sufficient to be measured as a cumulative effect called *magnetostriction*. Inasmuch as this phenomenon is a change in the shape of the same quantity of the matter, the *length* magnetostriction is accompanied by a *transverse* magnetostriction of the opposite sign, leaving very little to the *volume* magnetostriction. The latter is, evidently, a partial accommodation of the structural changes by the internal strains.

Now one can get a perception of the difference between the c and a parameters of the α-Fe unit cell in Ångstöm units. The reported maximum length magnetostriction (fractional elongation) for a polydomain single crystal is $2 \cdot 10^{-5}$. Only about 2/3 of all domains contribute to it (Fig. 4.5). Also taking into account that not all domains acquire optimum orientations in the applied field H and that some elongation will be absorbed by strains, it is reasonable to suggest that the fractional difference $(c - a) / c$ should be about 4×10^{-5}. In terms of this difference, the reported value of the α-Fe unit cell $a = 2.86645$ Å (see Fig. 4.2) is, evidently, the pseudo-cubic lattice parameter representing the midpoint between $a = 2.8664$ Å and $c = 2.8665$ Å. Indeed, $(c - a) / c = 0.0001 / 2.8665 = 3.5 \times 10^{-5}$. This tiny difference between a and c could easily escape X-ray detection.

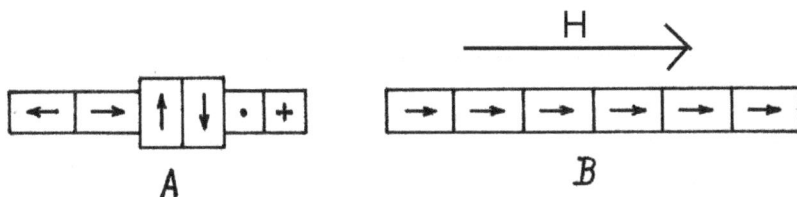

Fig. 4.5 Only about 2/3 of all the domains in the initially unmagnetized polydomain Fe crystal (A) participate in the magnetostriction resulted from application of the magnetic field H. The domains already parallel or antiparallel to H do not contribute to it (B).

The Fe simple crystal structure and the simple pattern of its spontaneous magnetization were very helpful in the foregoing analysis and illustration of the cause of magnetostriction. This section provides a clue for an examination of magnetostriction in other ferromagnetics. One might notice, for example, that Ni ferromagnetic phase is also believed to be truly cubic. We predict with confidence that it is not so.

The counterpart of magnetostriction in ferroelectrics had to be "electrostriction," but this is not the case. The notion "electrostriction" was defined differently - not specific to ferroelectrics. As opposed to the definition of magnetostriction, this name includes the tiny elastic deformation of any dielectrics when they are placed in an electric field E. Because the (+) and (-) charges of the electric dipoles are, or can be, spatially separated in the crystal unit cell of a dielectric, application of E either causes its polarization, or affects the value of its dipole moments if they were already present. This, in turn, produces some elastic deformation of the crystal. The analogous effect – induced magnetization and the corresponding elastic deformation – is essentially absent in ferromagnetics where the poles of the elementary magnetic moments cannot be split apart.

Even though the theory does not have a name for the ferroelectric counterpart of magnetostriction, it is real. Let us name it, just for the purpose of the present discussion, "ferroelectrostriction" (FES). The essence of what was said about magnetostriction holds true for FES. The case of $BaTiO_3$, for instance, is rather similar to that of Fe. The cubic paraelectric phase of $BaTiO_3$ becomes tetragonal below 120 °C after its (nucleation-growth) phase transition into the ferroelectric phase. The only difference is that the c-axis exceeds the $a = b$ by 1%, *i.e.* much greater than in Fe. Formation of the domain structure by epitaxial nucleation in the six fixed orientations within the cubic paraelectric matrix occurs in the same manner. Domain rearrangements during polarization in electric field are accompanied by FES.

FES is a change in the shape of a polydomain ferroelectric body as a result of crystal rearrangements of the domains in the process of polarization under the action of an applied electric field. In the combined effect of induced polarization and FES, the former is much smaller.

4.8 Domain interfaces. Tearing down the Bloch wall

A "Bloch wall" (Fig. 4.6a) was a derivative of two erroneous premises, namely, the molecular / exchange field and anisotropy energy. This type of a boundary between ferromagnetic domains was not discovered experimentally, neither did any credible circumstantial evidence exist. It was invented to help the Weiss-Heisenberg theory overcome a predicament: prohibitively high energy of the domain boundaries. In terms of the theory the exchange energy between neighboring magnetic dipoles on the opposite sides of a 180° domain border (*i.e.*, when the dipoles are antiparallel) is at a maximum. A gradual change of the dipole directions over a number of lattice periods would reduce this energy, but their deviation from the crystallographic "direction of easy magnetization" (Sec. 4.12) would have the opposite effect. The "Bloch wall" was the result of a theoretical optimization of "exchange field" and "anisotropy energy" effects. This did not mean that the "Bloch wall" was a low energy boundary: it was only minimal under the above particular assumptions. If they are dropped, the energy of the boundaries becomes dramatically lower*, as is the case with ferroelectric domain boundaries. Some authors tried to find visual evidence of a gradual change of spin orientation across the wall, but the results neither proved, nor disproved the point. Calculations have set the "wall" thickness from 50 to 200 nm in Fe. Kittel [197] came up with the number of about 300 lattice spacings (~86 nm) required for the magnetic moments to turn to the antiparallel position.** There have been no experimental data validating this number.

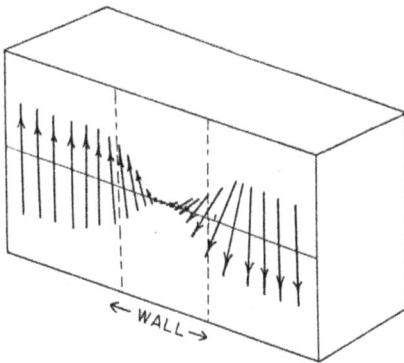

Fig. 4.6 (a) (left) Typical schematic of the "Bloch wall" between neighboring domains. "180° - wall" is shown, but there are also "walls" of different angles of dipole rotation, e.g., "90° walls." Rotation of the dipoles in the crystal lattice is implicit. The theory of "Bloch walls" failed to notice, much less take in account, a different orientation of the crystal lattice on both sides of the boundary.
(b) (below) Inconsistency between estimates of the "Bloch wall" thickness and the dimensions of ferromagnetic domains by the current theory.

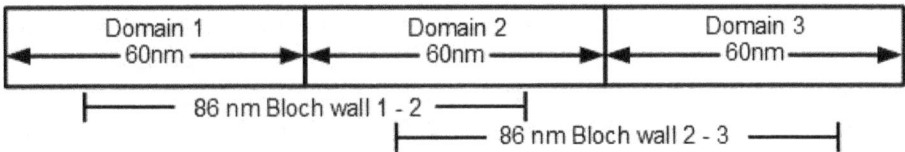

The "Bloch wall" and its schematic picture is widely represented in the literature. Any doubts in its reality can hardly be found in spite of the contradictions in its theory. It will be revealing to compare the

---

* We consider this statement self-evident, leaving to theorists to come up with actual numbers.

** According to Hubert [198], the "Bloch walls" in soft magnetic materials (to which Fe crystals belong) are "several thousands" crystal spacings wide. Understanding "several" as a minimum of three, this gives rise to the "walls" in Fe at least ten times wider than stated by Kittel.

~86 nm width of the "Bloch wall" with the size of domains in Fe inferred also from the standard theory. Bozorth's estimation [65], in agreement with other authors, was ~10 nm for the maximum size of a Fe crystal remaining as a single domain. This means that a larger crystal, for example, 12 nm will, or at least can, consist of two domains. But Bozorth warned us not to take the estimate "too literally." Therefore, we assume that not 12, but a 120 nm particle can consist of two, and for better illustration, a 180 nm particle of three domains. A comparison of the result with the 86 nm "Bloch walls" between them is presented in Fig. 4.6b. As can be seen, the resulting picture is nonsensical. Even changing one of the values under comparison by another order of magnitude (*e.g.*, increasing the three-domain particle to 1800 nm) would not produce a satisfactory result, for even then as much as 14% of every domain will be occupied by the "walls." The theory silently coexists with this vast inconsistency.

Since both the "exchange field" and "anisotropy energy" are attributes of the erroneous theory, the "Bloch wall" – their derivative – becomes meaningless. A stationary domain boundary is a *structural twin interface* of zero thickness. The dipoles on the opposite sides of a domain interface in ferroics are either antiparallel, or their projections on the interface are antiparallel. The observed domain patterns (to be discussed in Sec. 4.9) are entirely consistent with the crystal twin nature of the domain interfaces. The antiparallel dipole order at the twin interface provides an additional advantage by reducing its energy due to the magnetic attraction. The energy of the twin interface, evidently, constitutes only a minuscule fraction of that associated with the "Bloch wall." The twin interfaces propagate according to the general *contact mechanism* (Sec. 2.4). A confirmation of this mechanism comes from the experimental observations of edgewise motion of tiny steps along domain interfaces (Sec. 4.10). The concept of "Bloch wall" cannot explain why the boundary motion proceeds in such a manner, and not by its continuous movement as a whole. While the theoretical literature on Bloch walls is voluminous, the edgewise mode of its motion was neither predicted, nor even mentioned. But the reader of the previous chapters already knows that the edgewise mode of interface motion is universal for all crystal rearrangements. Relocations of atoms at a domain interface occur one by one at the steps on the interface surface. Every relocation carries a reorientation of one magnetic moment. This mechanism is the same whatever the angle happened to be between the magnetic moments of the domains. Not so with the "Bloch wall." As usual with artificial constructions, its every application needed a modification. For example, for the angles 180°, 109.47° and 70.53° at the domain boundaries in Ni, one has to assume three different types of the "walls." What's more, the "Bloch wall" was initially designed specifically for ferromagnetics, and not to represent domain boundaries in antiferromagnetics, any ferroelectrics, or any other domain structures listed in Sec. 4.5. The theory of "domain walls" became so cumbersome that an entire book was needed to provide a general overview of it [198]. Wrong premises made this theory useless.

Investigations of domain boundaries in ferroelectrics initially followed the path taken in ferromagnetics. They ended in impasse after it was concluded that no exchange field or its analog exist in ferroelectrics.* The solution that was finally found turned out to be not only correct, but also would be the solution for all ferroics and antiferroics. The conclusion was: the "wall" in ferroelectrics has *zero thickness*. For instance, it was shown by calculation of the 180°-boundaries in $BaTiO_3$ that the crystal lattice remains undistorted, the dipoles on the both sides of the boundary remain antiparallel, and the boundary does not have any thickness. Zheludev [157] summarized the results as follows. In the case of ferroelectrics there is no analog to the magnetic exchange energy; the presence of electron and ion polarization is reduced only to electrostatic interaction between the polarized ions; the difference between the interaction energies of parallel and antiparallel dipole arrays in dielectrics is rather small; therefore, a domain boundary is narrow and cannot be treated as a continuum.

Paraphrasing the foregoing Zheludev's summary as applied to ferromagnetic domain boundaries, one would come up with the following correct statement. In ferromagnetics there is no magnetic exchange energy; the presence of magnetic dipoles is reduced to only their magnetic interaction; the difference between the interaction energies of parallel and antiparallel arrays of magnetic dipoles in ferromagnetics is rather small; therefore, the domain boundary is narrow and cannot be treated as a continuum.

---

* While erroneous theories become sometimes instantly accepted, as was in the "molecular" and "exchange" fields cases, it is not necessarily so with correct solutions. Thus, Hubert in his book on domain walls [198] continued to assert existence of some "correlation energy" playing part of the "exchange field" in the nonmagnetic systems. The nature of the "correlation energy," he admitted, was not quite clear.

4.9 Domain equilibrium in ferroic structures

A number of elaborate techniques have been developed and applied by different researchers to reveal the domain patterns on the surface of ferromagnetic samples. But it will be the patterns themselves to which we now direct our attention. The most interesting and informative, not surprisingly, were those observed on rather perfect plate-like and "whisker"-like little crystals prepared by chemical reduction, rather than phase transition from the paramagnetic phase. Fig. 4.7a is a diagram representing main features of one of the fine pictures received by de Blois [201]. The sample was a small thin-plate Ni crystal. It will be demonstrated that there are certain "rules" incorporated in the framework of this kind of domain patterns. Application of a weak magnetic field shifts the domain boundaries in such a coordinated way that those rules remain unbroken.

The vectors of spontaneous magnetization M_s of the domains in Ni (fcc) structure have eight different [*111*] directions and thus do not lie in the principal plane (100) shown in Fig. 4.7a. The basic rules shaping domain patterns will be illustrated with a simpler instance, a cubic Fe crystal where M_s can be parallel to any of the six equivalent [*100*] directions.

<u>On free energy F of ferromagnetic continuum.</u> Ferromagnetic continuum is at its F_{min} when it is a uniform crystal medium without domains and their interfaces. But ferromagnetic materials almost exclusively resulted from a paramagnetic to ferromagnetic phase transition. Let us consider paramagnetic phase as a uniform continuum, disregarding that it is always polycrystalline in practice. Ferromagnetic phase appears as a result of multiple nucleation in the crystal defects of the truly cubic paramagnetic matrix. Because the nucleation is epitaxial, the domain nuclei emerge in six equivalent [*100*] directions. There are many parameters affecting the final domain pattern, such as the concentration and distribution of the nucleation sites in the matrix, the rate of cooling, and the actual nucleation temperatures encoded in those sites. If the sites are far apart, the emerging domains form independently. In such a case, vectors of M_s appear with equal probability in any of the six directions. The closer they are to one another, the greater the probability for M_s to be antiparallel. Nucleation in different sites is not simultaneous: the domains that emerged earlier grow first and shrink the area where other potential nucleation sites can be located. After the *paramagnetic* → *ferromagnetic* phase transition is completed, additional time is needed for the domain system to be

finalized. (When this time is long enough, it is called "magnetic aging"). While both non-ferromagnetic and ferromagnetic domain structures have the same crystal-structural origin (Sec. 4.5), the latter have an additional feature: the internal magnetic field modifies initial domain structure to minimize the energy of magnetic interaction. This process leads to the overall magnetization M = 0 in the absence of external field. As described above, this process starts when the original domain structure is initiated by epitaxial nucleation, and continues as adjustment of the domain boundaries after the transition is completed.

Eventually, the system is in its relative potential well, in other words, in a metastable state. This can be demonstrated by application of weak magnetic field H in different directions: every time when H returns to zero, the domain system moves toward its initial state. The subsequent slow adjustments in the domain pattern will only deepen the metastable potential well. (In case of polycrystals, a part of the "magnetic aging" should be assigned to motion of the grain boundaries.) A proof that the system is not at its lowest energy level comes from the fact that its actual free energy would be lower if all domain boundaries were of a 180°-type. The latter is valid both in terms of the Landau and Lifshitz domain theory [146] and the very different approach being presented now. Clearly, antiparallel dipole attraction is strongest. But there are always more 90° than 180°-boundaries. Indeed, let A and B be two independent (separated by a sufficient distance) neighboring nuclei in the paramagnetic matrix. From the six equally probable B orientations, one is parallel to A and leads to the merging A and B into a single domain. From the remaining five orientations one leads to the formation of 180°-boundary and four to 90°-boundaries. Thus, on average, the boundaries of the latter type form more frequently.

On minimum energy of domain interfaces. A ferromagnetic phase transition is completed at the moment when all growing domains meet one another and the remnants of the paramagnetic matrix disappear. One can hardly expect the initial lines of their meeting – initial domain boundaries – to be as highly structured and logically organized as those in Fig. 4.7a,b. Rearrangement toward minimizing the free energy of the domain system begins. It proceeds in three directions:
  (a) to provide the overall magnetization M = 0;
  (b) to provide best structural match at domain interfaces;
  (c) to provide maximum magnetic attraction at the domain interfaces.

Of these three factors, the structural matching at the domain boundaries (domain interfaces) is not taken into account by current theory, which is understandable since the boundaries are believed to be about 300 lattice periods thick. Nevertheless, structural matching at domain interfaces offers a better interpretation of interface geometry in the patterns presented in Fig. 4.7a,b. These patterns should be treated as those of a "final," "ideal" type, showing the direction in which the post-phase-transition domain readjustments proceed, even though in most cases such final states are hardly achievable.

Fig. 4.7b is not an actually observed pattern, it is rather a composite drawing incorporating actual features of the clearest domain patterns in Fe-type ferromagnetic crystals with M_s parallel to [100] directions. The arrows showing M_s directions lie in the (100) plane which is parallel to the drawing plane. The pattern is organized according to the following rules. (Reminder: the terms "180°-boundary" and "90°-boundary" denote a 180° or 90° angle between the dipoles on the opposite sides of the domain boundary, not the direction of the boundary.)

- All 180°-boundaries are (100)-type planes in the Fe pseudo-cubic crystal structure and are parallel to M_s. More precisely, they are (001) planes in the Fe true lattice, which is tetragonal (Sec. 4.7). These boundaries have the lowest possible energy owing to both maximum magnetic attraction and perfect structural matching. The boundary is a crystallographic twin interface of zero thickness.
- All 90°-boundaries are (110)-type planes in the pseudo-cubic lattice, but they are only approximately (110) planes in the true tetragonal lattice. Nevertheless, there is perfect structural matching, as Fig. 4.7c explains. These boundaries are twin crystal interfaces of zero thickness as well. As to the magnetic attraction, it is only 0.7 of that at 180°-boundaries.
- Projections of the M_s vectors onto each border from its two sides are opposite in sign. Thus, the domain marked with an asterisk (*) in Fig. 4.7b has one 180°-boundary and five 90°-boundaries with its neighbors. Whichever boundary is chosen, the projections in question are opposite in sign. This rule is somewhat "tricky," for it also has to be applied to any of the six neighbors toward their neighbors... and so on.

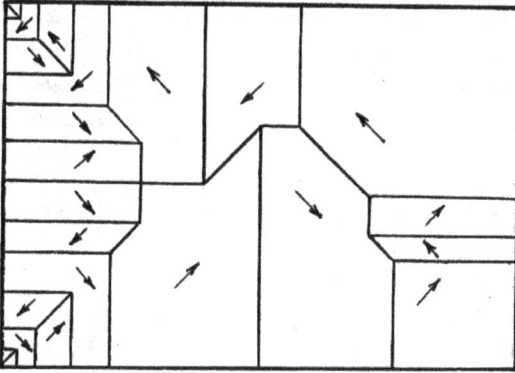

Fig. 4.7 (a) Main features of one particular domain pattern on the surface of thin-plated Ni crystal. The arrows indicating **Ms** directions are in fact only projections of actual [111] **Ms** directions on the (100) plane shown in the drawing. As a result of intricate interdependence of the domain boundaries with one another, change in the position of any one requires coordinated movements of many others. (The drawing is based on work by de Blois [198]).

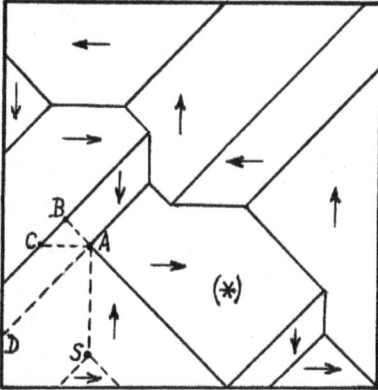

Fig. 4.7 (b) A composite schematic incorporating main features of domain patterns in ferromagnetic cubic crystals with **Ms** along 6 equivalent [100] directions (4 in the shown (100) plane). Only AS, and not AB, AC or AD, is the correct boundary.

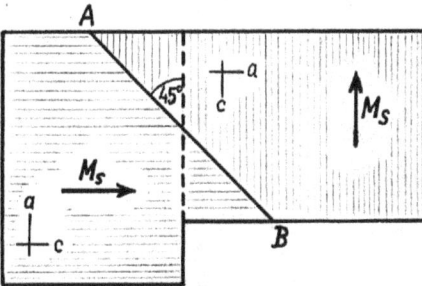

Fig. 4.7 (c) "90°-Boundary" (AB) in a tetragonal crystal lattice with **Ms** parallel to c-axis. It makes 45° with axes a and c in both neighboring domains, thus providing a twin structural matching, even though it is not exactly a rational crystallographic direction.

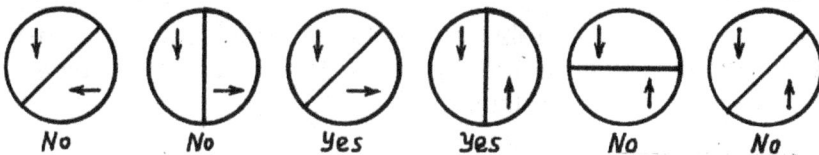

Fig. 4.7 (d) The "allowed" (Yes) and "not allowed" (No) directions of domain interfaces in domain patterns of the type shown in 'b'.

- 235 -

To illustrate the "puzzle" (which Nature solves, as it does everything: naturally), suppose that the boundaries shown with dashed lines did not appear on the photograph for some experimental reason and one had to determine the direction of the missing boundary from point A. One will find that out of the 4 possibilities (AB, AC, AD, AS) only AS complies with the rules. Fig. 4.7d shows the "allowed" and "not allowed" directions of domain interfaces in the "ideal" domain patterns under consideration.

One more feature of ferromagnetic domain structures is the "domains of closure" (see the triangle domains in Fig. 4.7b). It is understood that they close the magnetic flux and thus lower the free energy of a magnet. Missing in this interpretation is the crystallographic part of it, and especially the role of nucleation. These domains do not always appear. The conditions for their formation are (a) the possibility of epitaxial nucleation in the transverse direction and (b) the availability of nucleation sites. In Fe, for example, the two conditions are satisfied, including the availability of nucleation sites on the surfaces in the form of microscopic chinks. The internal magnetic field determines which one of the 6 available domain orientations, otherwise equally probable, will close the magnetic flux. In ferroelectrics, however, the domains of closure can rarely be found. This is because paraelectric crystals are usually layered crystals with only one set of the crystallographic planes (h,k,l) offering epitaxial nucleation of the ferroelectric domains; the formation of domains in the transverse direction in such cases is impossible.

4.10 Barkhausen effect as manifestation of crystal growth

Discovered in 1919 by Barkhausen, this effect still cannot be consistently incorporated into the conventional theory of ferromagnetism. The basic facts are as follows. A magnetization (or remagnetization) in a static or changing magnetic field H, seemed to be smooth and continuous, is actually a stepped process of short advances and stops if it is recorded with a sufficiently sensitive instrument (Fig. 4.8). Some steps are relatively large, but every such step consists of numerous smaller steps. It is the latter small steps which remained without any plausible explanation.

Let us turn first to the two most comprehensive review books on ferromagnetism. Bozorth [65] presented a rather detailed description of experimental studies, such as measurements of sizes of the

Barkhausen "discontinuities." The large ones could be associated with remagnetization of individual domains or their clusters, but no explanation for the tiny steps was offered. Vonsovskii [121] devoted to the Barkhausen effect only a few lines in his voluminous work, asserting that the volume of the "magnetization-reversal" regions of the individual "jumps" agree with theoretical estimates of average domain sizes. However, in the next sentence this author invalidated this claim by stating that, "as a rule, individual jumps correspond to reversals just of parts of domains".

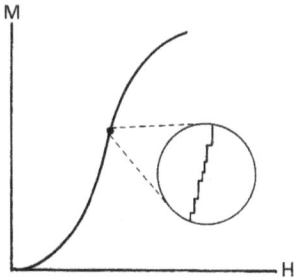

Fig. 4.8   Barkhausen effect.

Shur [202] qualitatively explained the Barkhausen effect as stops of smoothly moving domain boundary when it meets crystal irregularities of different kinds, such as cracks, strains, etc.; the obstacles are overcome after the magnetic field increased. This explanation, presented as fact, was actually an opinion not based on real observations. The Barkhausen effect is observed in a static magnetic field as well. More importantly, at no moment do domain boundaries move smoothly as a whole. This circumstance made Kittel [197] note that motion of domain walls in ferroelectrics "is not simple": they do not move as a whole perpendicular to itself, but resulted from the repeated nucleation of steps along the parent wall, adding that "this is not unlike the usual situation with ferromagnetic domain walls."

The current theory was unable to explain origin of the Barkhausen effect. The Weiss-Heisenberg theory did not assume it; the domain theory could try to account only for the largest magnetization jumps; the theory of the "Bloch wall" domain boundary did not even try to explain how crystal defects could change a smooth translational motion of this boundary, which is about 300 lattice periods thick, to the intermittent "stop-run" mode.

In more recent times the Barkhausen effect was attracted much more attention, mostly due to its connection with the magnetic hysteresis. Most of work – both experimental and mathematical - was devoted to its phenomenological description and did not shed light on its nature.

The Barkhausen effect, being foreign to the conventional theory, follows directly from the concept advanced in the present book, namely, that magnetization proceeds by a structural rearrangement. The failed conventional formula :

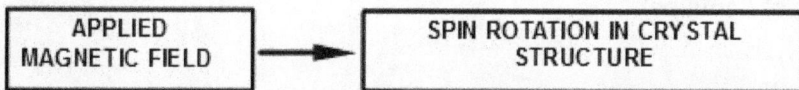

APPLIED MAGNETIC FIELD	→	SPIN ROTATION IN CRYSTAL STRUCTURE

must be replaced by :

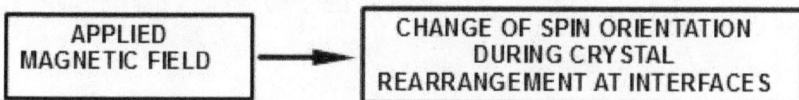

APPLIED MAGNETIC FIELD	→	CHANGE OF SPIN ORIENTATION DURING CRYSTAL REARRANGEMENT AT INTERFACES

In general, the application of a magnetic field to a ferromagnetic body increases its free energy, thus inducing its crystal rearrangement. The latter gives rise to a change of the crystal structure or its reorientation and, in doing so, to the reorientation of the constituent spin carriers. The structural rearrangement takes place at the contact domain interfaces. The layer-by-layer edgewise mechanism of crystal→crystal rearrangements was described in Sec. 2.3. This process is crystal growth. Quick rearrangement of every layer at the domain boundary produces a Barkhausen magnetic "jump." The rearrangement of every successive layer is delayed by nucleation. The layers can be as thin as one lattice space, or they can be conglomerates of numerous elementary layers that produce larger steps ("avalanches") on the magnetization curve. This process accounts for the Barkhausen effect. A quick restructuring of a whole domain will, evidently, produce a relatively large step, but it will inevitably consist of a great quantity of small ones.

The motion of domain interfaces, after it is understood, is simple. The real nature of the Barkhausen effect, so foreign to the traditional theory, is clarified in terms of our crystal growth concept. (For more on Barkhausen effect, see Addendum H).

4.11 Nucleation in single-domain particles

Small ferromagnetic particles have been intensively studied both experimentally and theoretically. The materials containing or composed of them were found to exhibit high coercive force H_c*. Naturally, the cause of the phenomenon needed to be explained. But the theoretical implication had a more general significance. It has been suggested, and substantiated in experiments, that sufficiently small particles (their size will be discussed later) consist of single domains. The theory had to come with the mechanism of remagnetization of single-domain particles in a magnetic field. The resolution was: it occurs by *rotation* of the magnetization vector M_s**. Magnetization by "rotation" has become an additional component of the conventional theory.

The term "rotation" is used in literature in three different ways that need to be clarified. (1) "Rotation of magnetization vector" implies rotation of the dipoles in a domain crystal lattice without changing the crystal or its orientation. (2) "Rotation of domains" does not mean the rotation of domains: it again denotes rotation of dipoles in the domain crystal structure. (3) "Rotation of spins" over the region of a "Bloch wall" dividing two neighboring domains. In brief, magnetization was always reduced to rotation of magnetic moments *in* the domain crystal structure. The conventional theory assumes that it is accomplished either by propagation of domain boundaries, or, as in case of small single-domain particles, by pure dipole rotation not involving nucleation and growth. Magnetization by propagation of domain boundaries was well experimentally established and recognized as the dominant mechanism; magnetization by pure rotation was a product of imagination based on invalid premises and flawed logic.

This point is fundamental: is there one general mechanism of magnetization, or there are two? There must have been compelling reasons to come with the second magnetization mechanism. We find these reasons groundless. The "pure rotation" idea was introduced as *the only alternative* for single-domain particles to be remagnetized. Bozorth [65] explained it as follows: in sufficiently small particles there is no boundary formation, so that magnetization by boundary displacement cannot occur, "therefore, any change of magnetization must occur by domain rotation." Let us analyze this logic. (A) Formation of a boundary

---

* "Coercive force" H_c is the magnetic field of opposite polarity required to demagnetize a thoroughly magnetized sample.

** Subscript "s" stands for "spontaneous" and / or "saturation".

has a greater destabilizing effect in smaller crystals; so, one can imagine a crystal small enough is which, indeed, the boundary would be prohibited. This is true, however, if magnetic field H is not applied. Sufficiently strong **H**, directed away from the spontaneous magnetization **Ms**, will make boundary formation feasible. (B) There is, however, a strong argument against formation of a "Bloch wall" under any conditions: it is so wide that in many cases can accommodate whole single-domain particle within. But domain boundaries are not "Bloch walls": they are *contact interfaces of zero thickness* (Secs. 4.8 and 4.9).

Most estimates of the maximum size at which particles remain single-domain came up with the number ~ 100 Å in Fe, but ranging up to several times larger in different materials. There has been no explanation of the fact that there is always a *range* of single-domain particles, including much larger ones. In terms of the "rotation" concept it would be reasonable to expect approximately the same H_c for all particle sizes in the single-domain range, but this was not the case. It was found that the smaller ones give rise to a sharply higher coercive force H_c. Moreover, a progressive sharp increase in H_c begins long before the particles become single-domain, in one particular study, for instance, from ~ 250 000 Å, and continued over the single-domain range.

The theoretical literature on the issue operated with such parameters as exchange energy and anisotropy energy. The estimates of the maximum size of single-domain particles were too low and arbitrarily extended up to 10000 Å. Whether the theory was successful in accounting for H_c can be seen from the following note by Vonsovskii: "...The observed maximum of H_c lies considerably below that predicted by theory" [121].

The theorists were mistaken in believing that "rotation" was the only way the single-domain particles can be remagnetized. The problem has a simple solution: it is again nucleation. The width of the emerging interface is not an obstacle: it is zero. Nucleation can account for all actually observed phenomena related to the size of particles, their shape, magnetic irreversibility, and the H_c values. The problem is reduced to nucleation lags only, for propagation of domain interfaces in such tiny crystals is completed almost the moment it starts. Nucleation in single-domain particles can be epitaxial, as is prevalent in formation of multidomain crystals, or non-epitaxial. In the former case, **Ms** cannot be completely realized in a system of arbitrarily oriented particles. Magnetization of every individual particle will be crystallographically related to its initial orientation and, in general, be closer, but not parallel, to **H**. In the non-epitaxial (non-coherent) nucleation, which is common in temperature polymorphic transitions (see Chapter 2), orientation of the nuclei will be directed by the magnetic field. A sample of many particles can be magnetized almost to full saturation. Formation of a non-epitaxial nucleus requires, evidently, a higher activation energy, *i.e.*, application of a stronger magnetic field.

In terms of the nucleation concept, the observed increase in the coercive force H_c as particles size becomes smaller is the natural consequence of diminishing probability of finding a crystal defect for heterogeneous nucleation in the smaller volume. The probability also depends on the strength of the field H: if the given H did not initiate nucleation, a stronger field has to be applied until it matches the nucleation value encoded in the "smaller" defects.

According to our basic concept, the dipole orientations are firmly established by the crystal. While a dipole rotation under the action of field **H** may occur within the limits of elasticity, these limits are negligible as magnetization is concerned; the increased crystal free energy will inevitably be released by a nucleation-growth restructuring. To sum up, the following diagram shows the logical chain from the initial erroneous premise to the unfounded idea on magnetization by "rotation."

REMAGNETIZATION BY ROTATION?
(Standard theory and comments)

THEORY OF MOLECULAR/EXCHANGE FIELD

Comment: molecular/exchange field does not exist. There is only crystal field with magnetic dipole interaction as its part.

$\downarrow$

"BLOCH WALL" STRUCTURE OF DOMAIN BOUNDARIES

Comment: domain boundaries are low energy contact interface of zero thickness

$\downarrow$

FORMATION OF "BLOCH WALL" IN SMALL PARTICLES IS IMPOSSIBLE

Comment: Formation of zero thickness domain interfaces in crystals of any small size is possible

$\downarrow$

REMAGNETIZATION BY "ROTATION" IS THE ONLY OPTION

Comment: remagnetization occurs by nucleation-growth

When the size of the particles becomes smaller than 50 -100 Å, the H_c goes down. These particles are on their way from crystal to amorphous state and their structure is not strictly three-dimensional any more. The crystal field is weaker and not quite uniform. This effect is analogous to the fact that X-ray reflections from such very small particles lose their sharpness. The nucleation concept allows us to suggest that in order to further increase the H_c of small particles one must increase their crystal perfection, all other things being equal.

4.12   Magnetization stages. Meaning of "easy" and "hard" directions

In order to reveal the basic processes of magnetization it was expedient to turn to single crystals, and among them to those of Fe as exhibiting the simplest crystal and magnetic structure. Magnetization of Fe crystals was subjected to a very detailed experimental investigation by Honda and Kaya (H & K) as long ago as 1926 [203]. It was an outstanding work, unsurpassed to date as to the quality and richness of the results. Magnetization curves for single crystals of Fe, as well as for Ni and Co from the subsequent works by Kaya [204, 205], can be found in any contemporary presentation of ferromagnetism.   They are accompanied by statements that there are "easy" and "hard" directions of magnetization in the crystals. The meaning of these terms will be clarified in this section. One of the curves from the H & K work is shown in Fig. 4.9.

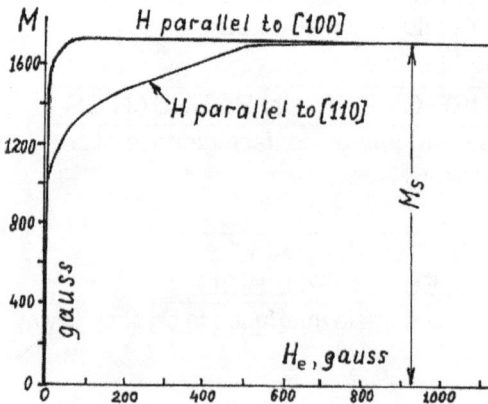

Fig. 4.9   Magnetization curves $M(H_e)$ of iron polydomain crystal in the "easy" [100] and "hard" [110] directions of the pseudo-cubic lattice (Honda & Kaya [203]). H_e is effective field. Saturation magnetization $M_s \cong 1720$ gauss.

The H & K work deserved more detailed attention. No sound interpretation of their experimental data could be done by the authors themselves, as at that time they had not yet accepted the Weiss' hypothesis on the domain structure of ferromagnetics. Below is our analysis of some selected results of the H & K work in order to provide a graphic illustration of the magnetization processes. The alternative to some common views will be presented.

The title of the H & K article "On the magnetization of single crystals of iron" was not quite correct. Crystal matter can be subdivided into polycrystals, single crystals, and polydomain crystals, and H & K dealt with the latter. A polydomain crystal is not a true single crystal, but a conglomerate of single crystals (domains) attached to one another in such a manner that the neighboring domains are crystallographic twins. There are six equally probable domain (and dipole) orientations in a polydomain Fe crystal (Fig. 4.10) with the total magnetization $M = 0$ in the absence of magnetic field $\mathbf{H}$:

$$M = M_{S1} + M_{S2} + M_{S3} + M_{S4} + M_{S5} + M_{S6} = 0,$$

where $\mathbf{M}_{S1}$ to $\mathbf{M}_{S6}$ are summary magnetizations of the corresponding domain groups $\mathbf{M}_1$ to $\mathbf{M}_6$ in the directions of their spontaneous magnetization. Magnetic field $\mathbf{H}$ causes growth of the domains whose magnetization directions are closer to $\mathbf{H}$ at the expense of others, resulting in a certain total magnetization $M \neq 0$.

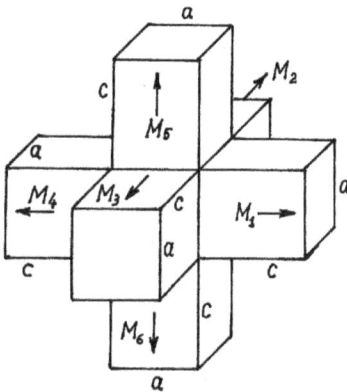

Fig. 4.10 Six mutual directions of spontaneous magnetization $\mathbf{M}_s$ that make 90° and 180° to one another in a polydomain iron crystal. While the unit cell dimensions c and $a = b$ are very close (Sec. 4.7), it is axis c which is the $\mathbf{M}_s$ direction. (The figure is not intended to illustrate domain boundaries)

We shall consider the H & K experiments with a sample cut out along (100) planes of the pseudo-cubic crystal. Its magnetization M was measured as a function of the strength of field H applied under different angles θ in the (100) plane, as shown in Fig. 4.11. At $\angle(\mathbf{H},\mathbf{M}_1) = \theta$, the other domain groups make with $\mathbf{H}$ angles 90° - θ, 90° + θ, 180° - θ, and

90°. At $\theta = 0°$ the **H** direction is [*100*], and at $\theta = 45°$ it is [*110*]. The domains with the closest **Ms** direction to **H** are those of the lowest energy (**M₁** in Fig. 4.11). In general, the resultant magnetization **M** had neither a rational crystallographic direction, nor the **H** direction, and H & K paid as much attention to this problem as to the absolute M value. The magnetization was measured both along **H** (parallel component M_p) and in the transverse direction in the (100) plane (normal component M_n).

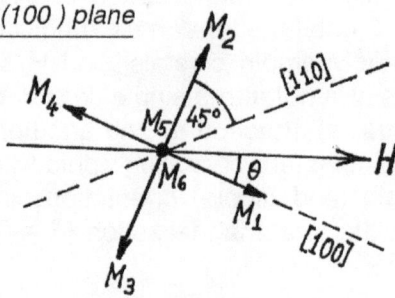

Fig. 4.11 The orientation of Fe polydomain crystal in magnetic field **H** in one of the H & K experiments under discussion in the text. Magnetization along **H** and at a right angle to it was measured *vs.* H strength under different angles θ. (Spontaneous magnetization of **M₅** and **M₆** domain groups is normal to the (*100*) plane and opposite to each other.

It was seen from the H & K data that magnetization advanced by stages, depending on the strength level of the effective magnetic field H_e. These levels are: weak (up to ~2 gauss), medium (up to ~100 gauss) and strong (over ~100 gauss). In the weak magnetic fields more than one half of the saturation magnetization M_S is attained (M_S corresponds to all the magnetic moments in the sample being parallel). This magnetization resulted from displacements of the existing domain interfaces without disappearance of any domains or formation of new ones. It is reversible* in the sense that after returning to H = 0 the internal field basically reinstates the initial domain pattern with M = 0.** According to the H & K data, at this level the **M** and **H** directions coincide ($M_n = 0$); the feature is characteristic of a cubic crystal only. One deals with the "weak" magnetization level when a piece of soft iron is touched by a permanent magnet: the piece becomes magnetized, but the magnetization disappears after the permanent magnet is removed.

We now turn to the magnetization process at the medium level of the magnetizing field. In general, $M_n \neq 0$ at this stage. As will be now demonstrated, its existence resulted from the fact that the corresponding *structural regrouping occurs predominantly between the*

---

* This is not the reversibility defined by thermodynamics which requires the system in the reverse run to pass exactly the same sequence of states.

** Complete reinstatement is impossible owing to hysteresis, which is unavoidable (Sec. 4.13).

fixed domain orientations shown in Figs. 4.10 and 4.11. At that, some domain groups may entirely disappear. Let us assume that H causes complete structural rearrangement of **M2, M3, M4, M5** and **M6** groups into **M1** whose magnetization is the closest to the **H** direction. This is realized by growth of the **M1** domains through all other, as well as by epitaxial nucleation of new domains of the **M1** orientation. (For brevity, this process will be termed "change... into"). As seen from Fig. 4.9, when $\theta = 0°$ the sample at 70-100 gauss is magnetized to Ms, which means that **M** is parallel to **H**. Evidently, M = Ms also at small θ, as shown in Fig. 4.12. Then $M_n = M \cdot SIN\ \theta$ and $M_p = M \cdot COS\ \theta$. Now we can account for the peculiar shape found by H & K of the $M_n(\theta)$ and $M_p(\theta)$ curves . Their principal features are shown in Fig. 4.13.

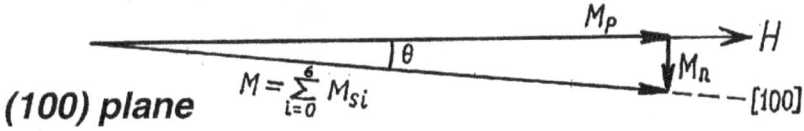

Fig. 4.12 The parallel M_p and normal M_n magnetization components in a Fe polydomain crystal when H makes a small angle θ with [*100*] in (*100*) plane. All 6 domain groups are aligned along [*100*]. Under this disposition, $M_p = M \cos \theta$ and $M_n = M \sin \theta$.

Normal component M_n. The curve $M_n(\theta)$ in Fig. 4.13 resulted from two opposite trends. On one hand, it is a sinusoid going from zero at $\theta = 0°$ to its maximum at $\theta = 90°$. On the other hand, at 45° it must be zero: when approaching $\theta = 45°$ (see Fig. 4.11), the alternative option - changing **M3, M4, M5** and **M6** into **M2** - becomes available until the magnetization is equally divided between **M1** and **M2** directions. At this point $M_n(\theta = 45°) = 0$; the internal field will equalize any difference in the magnetization in **M1** and **M2** directions caused by kinetics. It is easy to see that $M_n(\theta)$ between 45° and 90° has the opposite sign.

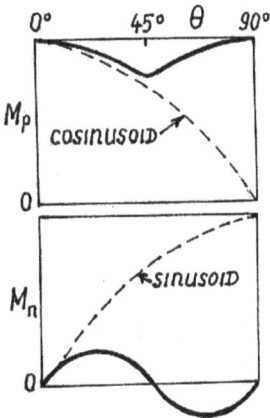

Fig. 4.13 The $M_n(\theta)$ and $M_p(\theta)$ curves reflecting the shape and features of those recorded by H & K under the geometrical arrangement shown in Fig. 4.11.

Parallel component M_p. M_p is maximum when **H** coincides with any of 6 directions of spontaneous magnetization. Thus, at $\theta = 0°$ it is maximum: all the domain groups change into **M₁**. From there M_p initially follows a cosine curve toward zero at $\theta = 90°$, but at θ approaching 45° the alternative orientation **M₂** becomes more and more available for the remaining four domain groups to change into (see Fig. 4.11). At $\theta = 90°$ magnetic field **H** coincides with the **M₂** direction and is maximum again. Hence the $M_p(\theta)$ curve with maxima at 0° and 90° and a minimum at 45°. *This successful description of the character of the $M_n(\theta)$ and $M_p(\theta)$ curves certifies that the application of a "moderate" magnetic field H to a polydomain crystal indeed gives rise to regrouping the domains between the pre-existing fixed orientations into an orientation, or orientations, closest to the H direction.* The next step is to determine whether this process corresponds quantitatively to the observed deviations φ of the vector **M** from the **H** direction. The curves in Fig. 4.13 suggest that φ should not be lower than 22.5° (the midpoint between $\theta = 0°$ and 45°, the two angles at which M_n and φ drop to zero). But the H & K tabulated data show that it was only 11.5° (found at $\theta = 28°$). The difference indicates an incomplete change of the remaining five domain groups into **M₁**. There is another variant of the dipole rearrangement in which $\varphi_{max} < 22.5°$ (Fig. 4.14). There φ_{max} is achieved at certain H strength, sufficient to change only **M₃**... **M₆** into **M₁**, but not **M₂**, since the latter makes the next lowest angle with **H**. Under these circumstances, as Fig. 4.14 shows, $\varphi_{max} = 16.7°$. While it is closer to the observed 11.5°, the remaining difference certainly exceeds the experimental error. Also, the $\varphi_{max} = 11.5°$ is found at H = 240 gauss, which exceeds the "medium" H strength. All accounts suggest that another magnetization process, peculiar to the "high" H level, was already partially involved at the "medium" level.

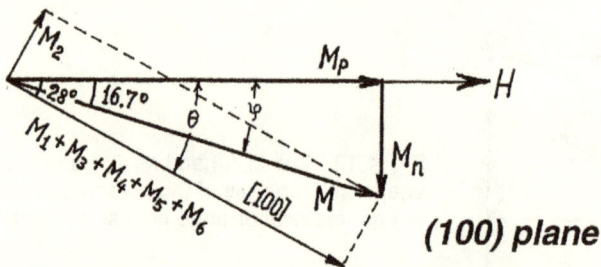

Fig. 4.14   The parallel M_p and normal M_n magnetization components in a Fe polydomain crystal when the domain groups **M₃**, **M₄**, **M₅** and **M₆** have changed into **M1**, but **M2** has not. The angle $\varphi = 16.7°$ is greater than the observed $\varphi_{max} = 11,5°$. The fact that $M_p = M \cos\varphi$ and $M_n = M \sin\varphi$ does not change the foregoing interpretation of the main features of the $M_p(\theta)$ and $M_n(\theta)$ curves in Fig. 4.13.

As seen from the foregoing, the domain restructuring in an applied field H of medium strength proceeds between the preexisting fixed magnetization directions, rather than into the direction of **H**. But it is also evident from the H & K data that φ in stronger fields becomes smaller until the magnetization vector **M** assumes a direction close to **H**. Let us follow the magnetization curve in Fig. 4.9 when **H** is parallel to [110], *i.e.* $\theta = 45°$. The magnetization M by restructuring between the six fixed orientations cannot exceed $1720 \cdot COS\ 45° = 1216$ gauss. In Fig. 4.9 it is about the point where M changes its path to grow more slowly. But the fact that M(H) continues to grow means that φ continues to become smaller. Consequently, a different magnetization process comes into action. When $M \approx M_s$, the directions of **M** and **H** have to be close. Some H & K data allow us to believe that strict **M** and **H** parallelism was not achieved there. Rather, in average, about 7° deviation of the M_i vectors from the H direction remained. Still, the fact of approaching **H** direction by **M** leaves no room for doubt. The question that remains is how. The standard answer was: by a *rotation* of the domains (meaning a rotation of the magnetic dipoles *in* the domains). This mechanism was discussed and rejected in Sec. 4.11. Magnetization of a polydomain crystal almost to full saturation M_s can be easily explained by *non-epitaxial nucleation* of new domains under the orienting action of a strong magnetic field. Their crystal lattice acquires a favorable orientation relative to the **H** direction, while the magnetic dipoles preserve their directions in the lattice. There is no need to introduce an additional magnetization mechanism (rotation of the dipoles in the lattice) that no one experimentally found.

The magnetization curves by H & K have been used as evidence of magnetization by "rotation" and the existence of "easy" and "hard" directions. Although that experimental study was very detailed, it was lacking data that could help choose between "rotation" and nucleation-growth rearrangement. This is because every reading was preceded by demagnetization. In order to choose between "rotation" and "restructuring" the following procedure is needed. First, a magnetization curve M(H) up to very strong H should be measured in a "hard" direction without demagnetization after readings. Then, without demagnetization, the measurements have to be taken a second time. If in the second run the same direction turns into "easy" ("easier" will suffice), a rearrangement of the crystal structure is proven.

In terms of the nucleation-growth magnetization process, the "easy" and "hard" directions of magnetization should be interpreted only in a practical macroscopic sense: a much stronger magnetic field has to be applied in "hard" directions to magnetize a polydomain crystal to saturation. Microscopically, however, a direction of spontaneous magnetization in the crystal is neither "easy," nor "hard." It is its permanent inherent characteristic that does not change upon remagnetization. In Fe, for instance, the alignment of its magnetic dipoles is always along the c-axis of its tetragonal unit cell (Fig. 4.3). The only way to reorient dipoles is to reorient the structure. A "hard" direction is such not because it is "hard" to rotate dipoles in the crystal structure, but because a non-epitaxial nucleation requires much stronger magnetic fields as compared to the epitaxial nucleation.

Magnetization of a polydomain crystal as a function of an applied magnetic field proceeds in three overlapping stages. They can be named as follows: simple shift of the existing domain interfaces, structural rearrangement of the domain groups between the pre-existing domain orientations, and non-epitaxial nucleation and growth. Polarization of ferroelectrics occurs in the same manner. The "rotation" is a feature of the ferromagnetic theory only. In ferroelectrics the possibility of such a phenomenon is not even discussed. In this respect, as in most others, we bring both ferroics to the same denominator.

4.13 Origin of magnetic hysteresis and formation of hysteresis loops. Magnetization as a counterpart of solid-state phase transition

4.13.1 General remarks

The cause of magnetic hysteresis has not been found by the current theory. Sixteen authors [cond-mat/0611542] asked "What causes magnetic hysteresis?" and stated that magnetic hysteresis is fundamental to magnetic storage technologies and a cornerstone to the present information age. They found that all the "beautiful theories of magnetic hysteresis based on random microscopic disorder" failed to explain their data. Their answer: "New advances in our fundamental understanding of magnetic hysteresis are needed." Neither their article, nor other extensive work on the topic, such as 2160 pages of the three-volume book "The Science of Hysteresis" [Academic Press, 2006] were able to find the cause of the phenomenon. Further attempts based on the current theory are destined to fail.

Ferromagnetic hysteresis loops of remagnetization in external alternating magnetic fields (Fig. 4.15) and the corresponding ferroelectric loops in electric fields are the most prominent feature of ferroics. In some respects the interpretation of these loops by some authors was correct. In particular, it was understood that (a) motion of domain boundaries is the main mechanism of remagnetization, and (b) hysteresis is the lags in that motion as well as the lags in nucleation of new domains. However, the cause of these lags has not been identified. It was not realized that hysteresis and its loops are unavoidable. These failures were rooted in the interpretation of the lags as those of a *magnetic rearrangement*, rather than a *rearrangement of the crystal structure*. Besides, it is erroneously accepted that magnetization can also occur by "rotation," without motion of the domain boundaries. There are other aspects requiring clarification or correction. It should be noted that an understanding of the hysteresis and its loops could not be complete without taking into account the specifics of nucleation in solid state presented in Sec. 2.5, as well as other information on the rearrangements at crystal interfaces given in the previous chapters.

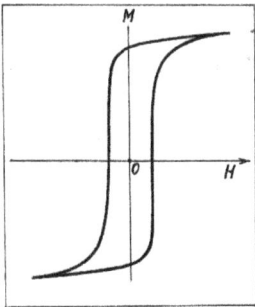

Fig. 4.15 Typical ferromagnetic hysteresis loop. M - magnetization, H - applied magnetic field.

One relevant circumstance remained unnoticed regarding the ferroic hysteresis loops: the same type of hysteresis loops have been observed in solid-state phase transitions when a physical property was measured *vs.* temperature T over the transition range. These "non-ferroic" hysteresis loops were analyzed in the previous chapters and shown in Figs. 2.30, 3.5, 3.7 and 3.12. Their origin and formation described in Sec. 2.6 is the key to the ferroic hysteresis loops. Three major characteristics of the non-ferroic loops are to be reiterated:
- they actually reflect the mass ratio of the two coexisting phases *vs.* T, regardless of the physical properties that are measured;
- nucleation lags and specific features of the nucleation (Sec. 2.5) are responsible for their formation and shape;
- since all phase transitions in solids proceed by nucleation and growth, the hysteresis loop, even if it is very narrow, is an inalienable feature of any solid-state phase transition.

The similarity between the ferroic and non-ferroic hysteresis loops is not accidental: they represent the same process of phase transition by nucleation and growth under the action of a variable. In the polymorphic transitions, which we dealt with in the previous chapters, the variable was temperature, while in ferroics it is the applied field. It should be remembered, though, that the former is a scalar, while the latter is a vector, but this difference is rather of a practical than principle significance. Changing temperature transfers a non-ferroic matter across the line in the phase diagram that separates its areas of stability and instability. It is the unstable state of the matter that is the driving force of the phase transition, *i.e.* nucleation and growth of new phase. Similarly, the application of a magnetic field to an unmagnetized ferromagnet makes it unstable. In case of a polydomain Fe crystal, for example, the applied magnetic field of any strength and direction turns at least 83% of the domains into an unstable phase, for the **H** direction cannot coincide with the magnetization of more than one of the six domain groups of the crystal. Magnetization of the crystal by nucleation and restructuring at the domain interfaces, considered in Sec. 4.12, is a solid-state transition from a higher-energy phase to a lower-energy phase.

When analyzing ferromagnetic hysteresis loops one has to take into account whether
- the sample is a single crystal, polydomain crystal or polycrystal,
- the magnetic field **H** is applied in the "easy" or any other direction,
- the H strength is sufficient to magnetize the sample to saturation,
- the loop is quasi-stationary, or recorded in a high frequency alternating field.

The actual shape of the hysteresis loops varies depending on these conditions, but, like their phase transition counterparts, they can always be entirely accounted for in terms of the structural categories of nucleation and growth.

4.13.2 Rectangular hysteresis loops

We start from the simplest case - rectangular loops (Fig. 4.16) - and consider the Sixtus and Tonks (S & T) experiments in which such loops were produced. They were described in a number of sources (*e.g.*, [65,121]), but because of the fundamental significance of these experiments we have to focus our attention on certain details in order to introduce corrections to their common interpretation. Why is it important? Any shortcomings in the interpretation of the production and formation of these most refined and revealing hysteresis loops are incorporated into the interpretation of all other - more complicated - ferromagnetic loops.

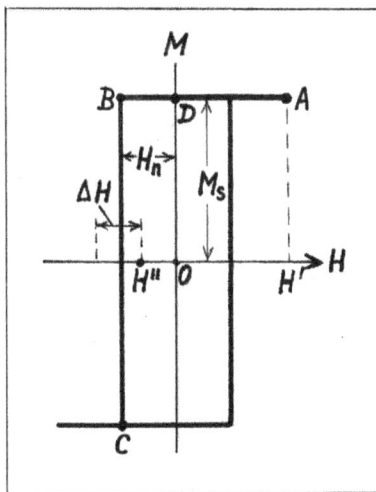

Fig. 4.16   Rectangular hysteresis loop of a ferromagnetic.

A schematic of the S & T experimental arrangement is shown in Fig. 4.17. A sample W in the form of thin ferromagnetic wire with positive magnetostriction was stretched along the axis of main solenoid S and subjected to elastic tension. The uniform magnetic field H' of the main solenoid magnetized the wire to its saturation M_s in the H' direction. Before turning to the results of the S & T experiment, it has to be explained how M_s was achieved. The wire in its initial state, prior to the applied tension, was polycrystalline. It consisted of numerous arbitrarily oriented grains, each one being a polydomain crystal. Among them there were grains in which the M_s of some domains was aligned along the wire axis. When a magnetic field was applied in the same direction, these domains became energetically preferable over all others. Crystal restructuring was initiated.   It involved two stages: growth of such domains that tended to spread over the grain, and growth of these grains that tended to comprise the whole sample.   The process was hindered by the accompanying change in shape (which is magnetostriction, Sec. 4.7) and could not be completed. The positive magnetostriction of the material would give rise to an elongation of these grains in the H' (and wire) direction, but the elongation was restricted by the surrounding grains. The application of a certain optimum tension corresponding to the magnetostriction released this mechanical restriction. As a result, the relatively few grains with their magnetization parallel to the magnetic field grew through the cross section and along the wire, consuming all other grains, thus transforming the polycrystalline sample into a quasi-single crystal magnetized to the saturation M_s in the direction of the magnetic field and wire axis. Such a process can be called "magnetic recrystallization": a recrystallization stimulated by magnetic field.*

A sample prepared in the above-described manner exhibited a rectangular hysteresis loop (Fig. 4.16). Reducing H' to zero and even applying negative field $H'' < H_n$, where H_n is the field necessary to start remagnetization, does not affect the M value. The line A-B is horizontal. An additional negative field ΔH, so that $|H'' + \Delta H| > H_n$, created by the short secondary coil C near one end of the wire (Fig. 4.17), triggered the formation of a nucleus with 180°-reversed magnetization, followed by the fast propagation of a domain interface over the whole sample. The line BC is vertical. Speed of the interface propagation in this and other experiments was measured by different authors. It varied depending on the sample and on the strength of the magnetic field. The maximum speed was well below that of sound in the material. The magnetic field H_n needed to create the nucleus is called "starting field" ("nucleation field" would be a better name); a somewhat weaker field, called "critical field," was sufficient to keep the interface moving.

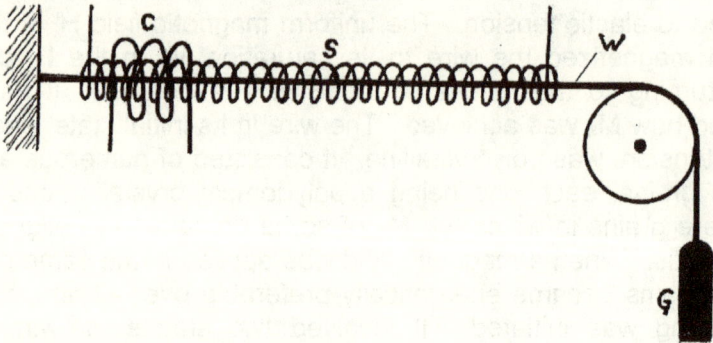

Fig. 4.17 Basic arrangement in Sixtus-Tonks experiments. W - Ferromagnetic wire (sample); S - main solenoid (source of uniform magnetic field); G - weight (provides elastic tension); C - secondary coil (creates local magnetic field of the opposite sign).

---

* Simple considerations suggest that the phenomenon of *magnetic recrystallization* must exist and play a part in the magnetization processes. It seems, it was not noticed previously. Not only it has not been taken into account in the interpretation of the S & T experiments, it is not mentioned even in the most comprehensive books on ferromagnetism.

Shortcomings in the previous interpretation of the S & T experiments [65] begin from the physical state of the sample. It was said that the domains in it are easily oriented by tension, so that in the absence of a magnetic field they lie parallel to the line of tension, half in each of the two opposite directions. In fact, the elastic tension of a polycrystalline wire can do little in orienting the grains and the domains of which they consist. Its function was to eliminate effects of magnetostriction and thus let *magnetic recrystallization* complete, as described above. The formation of the rectangular hysteresis loop was also not quite correctly presented. The reason why the material remains saturated when a strong positive magnetic field is reduced to zero and even becomes negative (line A-B in Fig. 4.16) was explained by a "great stability" of the domains directed parallel and antiparallel to the field. The cause of that stability, especially of the domains magnetized opposite to the field, was not specified. The nucleus was considered only as that of antiparallel magnetization, formed by a "180°-reversal." Propagation of the boundary from the nucleus over the whole sample was explained by the "magnetic influence" of the adjacent region at the boundary.

Now we will look at the same subject in terms of structural nucleation-growth concept. Remagnetization is not just a wave of magnetization reversal: it is brought about by the structural phase transition. Change in the M_s direction occurs by the structural rearrangement at the contact interface. One may argue that the crystal structures on two sides of the domain boundary are the same and, by definition, are not different phases. This argument is valid when the variable affecting the crystal free energy is a scalar, such as temperature or pressure, but magnetic field is a vector. The free energies of two structurally identical domains differently oriented in the magnetic field are not the same, which is the driving force of structural rearrangement at the domain boundaries. Unspecified local "magnetic influence" at the domain boundaries has nothing to do with their motion.

The next point to clarify is the reason why M_s, achieved at the strongest positive magnetic field H', remains unchanged after H' is reduced to zero and even farther into the negative side (horizontal line A-D-B in Fig. 4.16). The crystal structure is thermodynamically stable over the region A-D, indeed, for its magnetization remains parallel to H' and there is no driving force to cause change. In the region D-B, on the contrary, the sample is in the most unstable state, since the direction of its magnetization is opposite to **H**. It remains quasi-stable simply because no structural change can occur without nucleation, *i.e.*, until the negative field is sufficiently strong to increase the instability to the point when a structural nucleus of the opposite magnetization appears. This "starting field" H_n plays precisely the same part as the overheating / overcooling in temperature phase transitions. Because the H_n value is "pre-coded" in a structural defect, it is not exactly reproduced in different samples. This behavior is no different from that in any temperature solid-state phase transition: a thermodynamically unstable phase remains quasi-stable until conditions for the formation of a nucleus are satisfied.

Since the boundary motion was regarded in literature a "wave of magnetization reversal," it was a problem to explain why speed of this wave is so low. Eddy currents and spin-relaxation phenomena were suggested as the limiting factors. It has to be noted, however, that the boundary movements can be controlled with the magnetic field in the same manner as an interface in temperature phase transitions can be controlled with temperature. Structural phase transitions have also answers to the questions of why an excessive magnetic field ("critical field") is required to move the domain interface, why the latter is lower than a "starting field," and what the factors limiting the speed of interface propagation really are.

The molecular mechanism of structural rearrangement at the domain boundaries is the same nucleation and edgewise growth as in the structural phase transitions described in Sec. 2.4.3. The edgewise mode, by which domain boundaries propagate, was experimentally observed both in ferromagnetics and ferroelectrics. It is the general manner of any crystal growth and propagation of any solid-state interfaces (Secs. 2.3 and 2.4). It is intimately related to the cause of hysteresis, both in ferroics and temperature phase transitions. While more detailed facts and analyses can be found in the previous chapters, here is a summary as applied to ferroics.

Motion of a domain interface requires the formation of 2-D nuclei on it. The nucleation is heterogeneous, requiring the presence of a certain type of crystal defects. Each 2-D nucleation act forms an island which grows laterally, by "brick by brick" relocation (where a "brick" is atom or molecule), to the extremities of the surface. Then another suitable defect has to be found nearby to form next 2-D nucleus, and so on. This cannot occur without nucleation barriers. This is true all the more for the 3-D nucleation of new domains. If there is a sufficient supply of proper crystal defects, the magnetization (or polarization) can be completed in the sample at the fixed value of the applied field (as in the case of a rectangular hysteresis loop), provided that other crystal imperfections or internal strains due to the magnetostriction do not prevent it. Otherwise, after the building process of one or several thin layers is completed, the field strength has to be increased to initiate nucleation in the nearby potential sites. This is a simplified picture, for there are cumbersome effects described in Sec. 2.9.

As can be seen, remagnetization in a magnetic field is reduced to the crystal rearrangement of the same nature as in temperature polymorphic transitions. The process, in turn, is a variation of crystal growth. It involves 3-D and 2-D nucleation. It is the nucleation barriers, encoded in the structure of the potential nucleation sites, that make any such a process intrinsically non-equilibrium. These *nucleation barriers are the primary cause of all hysteresis loops: ferroelectric, ferromagnetic, or other crystal properties.* The 3-D nucleation of new domains involves different type of crystal defects and greater lags as compared to the 2-D nucleation in interface motion. As to the speed of the interface motion, no one should expect that crystal growth can proceed as fast as the propagation of a magnetic wave. Besides the primary cause of the hysteresis indicated above, there can be other obstacles to interface movements, such as different kinds of lattice imperfections, foreign inclusions and grain boundaries. But whatever the obstacle is, renewal of the interface motion requires repeated nucleation.

Polydomain "single" crystals (PDC) also exhibit rectangular hysteresis loops, but usually less perfect than those produced in the S & T experiments. At first glance, this may seem surprising, but there is a simple reason. Magnetization of PDC, as opposed to magnetization of wire samples in the S & T experiments, is not free from the effects of magnetostriction (ref. to Sec. 4.7). This observation points at magnetostriction as a factor in the formation of hysteresis loops of polycrystalline materials considered in the next section.

4.13.3 "Typical" hysteresis loops

There are variations in the actual shape of ferroic hysteresis loops, a typical one being shown in Fig. 4.18. This section deals with the factors affecting their shape. The rectangular hysteresis loop (Fig. 4.16) considered in the previous section provides a valuable starting position. Only nucleation and growth are involved in its formation. The conditions for the rectangular loop to form are: a monodomain crystal, the elimination of magnetostriction adverse effect, a sufficiently strong magnetic field applied parallel to the direction of spontaneous magnetization, quasi-stationary recording. The overall cause for the "typical" loop to not be rectangular is, evidently, that at least some of these conditions are not satisfied. When the "typical" hysteresis loops are reproduced in the textbooks and review articles, the above-listed conditions are not sufficiently specified, if at all. As a rule, the "typical" loops are related to polycrystals, the fact being given little or no attention, much less properly taken into account.

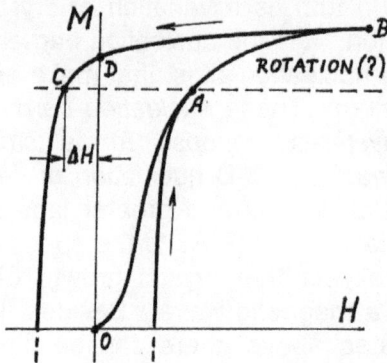

Fig. 4.18 A typical remagnetization hysteresis loop under consideration (its lower symmetric part is omitted).

Not infrequently the illustrative hysteresis loops look like the one in Fig. 4.19. In reality they have such a shape when they are not quasi-stationary. While the nature of quasi-stationary hysteresis loops has not been sufficiently clarified, non-quasi-stationary recording (*e.g.*, in alternating fields instead of point-by-point) adds two uncertainties. First, if the applied field H changes too fast, the domain interfaces do not have enough time to reach their quasi-stable positions corresponding to the H amplitude. The shape of such a loop depends on the frequency of the alternating field. Second, the relaxation time to partially dissipate the internal strains caused by the magnetostriction is short and a function of

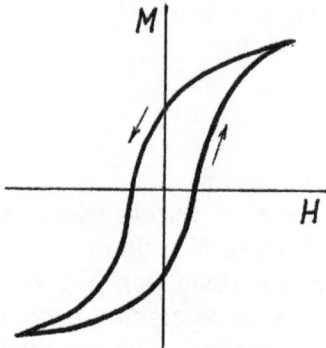

Fig. 4.19 The type of ferromagnetic hysteresis loop usually presented, with some variations, in the textbooks to illustrate the phenomenon; it is not best suited for that purpose.

the frequency too; these strains restrict magnetization during an ascending H phase and have a regressive effect during its descending phase. These effects change the shape of the hysteresis loops in the direction seen from the comparison of Figs. 4.15 and 4.19.

From this point on, only quasi-stationary loops will be discussed. Choosing the book "Ferromagnetism" by Bozorth [65] to represent the conventional version, a major issue in the following will be whether magnetization can occur by a "domain rotation." This magnetization process is asserted to take place in the part of a loop marked "rotation(?)" in Fig. 4.18. According to Bozorth, at point A the magnetization stage owing to motion of the domain boundaries is completed; the magnetic moments of all domains in the sample became uniformly aligned (magnetically saturated) in the "easy" direction of the crystal. Farther magnetization in the **H** direction (from A to B) proceeds by a "reversible rotation" of the magnetic moments from the "easy" direction into the direction of the applied magnetic field **H**.

This interpretation is inadequate:
(1) The polycrystalline material was treated as if it was a polydomain "single" crystal. There is a vast number of "easy" directions in the polycrystalline sample, determined by the orientation of every particular grain.
(2) A magnetic field alone, as we have seen (Sec. 4.13.2), was unable to magnetically saturate even a thin polycrystalline wire without the properly chosen elastic deformation being applied. There is no chance that the sample at point A can be magnetically saturated in a single "easy" direction.
(3) Yet, let us conditionally accept that at point A all magnetic moments are uniformly aligned in one "easy" direction and prove that the suggested "rotation" leads to contradictions.
   - Contradiction #1. The alleged rotation of M_s from the "easy" direction by the magnetic field can only be elastic, because the crystal forces will try to return M_s to the "easy" direction. In other words, magnetization by "rotation" has to be reversible, and the standard theory states that it is. However, the actual process is not reversible, as evident from the fact that the curve B → C does not follow B → A.
   - Contradiction #2. The "rotation" process leads to a conclusion that the physical states of the sample at points A and D must be identical. Indeed, (a) the domain boundaries stay still when the magnetization by "rotation" follows the curve A→B→D, therefore the crystal lattice experiences no change, and (b) at H = 0 the

- 257 -

direction of M_s is determined by the crystal structure, *i.e.*, has the same "easy" direction as at point A. But this is inconsistent with the fact that $M_D > M_A$.

(4) No evidence of any kind has ever been presented to validate the "rotation" process under any circumstances.

Introduction of the "rotation" magnetization was nothing more than a postulate based on the belief that no other way exists to account for the remagnetization of small particles (Sec. 4.11) and for the upper part of magnetization curves of polydomain "single" crystals (Sec. 4.12). This postulate can be ruled out, since there are two realistic processes to account for the apparent "rotation": *non-epitaxial nucleation* and *magnetic recrystallization*.

As in the case of rectangular loops, only nucleation and growth are the processes involved in the formation of all hysteresis loops. The loop in Fig. 4.18 is not rectangular because (apart from the possibility of not being sufficiently quasi-stationary) the initial physical state of the sample and the conditions for propagation of domain interfaces in it are different. The physical state of the sample at point A is not so uniform as described by Bozorth. The sample arrives at point A from point 0 still polycrystalline. Its every grain is a polydomain system with a few fixed "easy" directions of spontaneous magnetization. Within the grains, the domain restructuring between the "easy" directions into the one closest to **H** is in an advanced stage. In the case of Fe-type crystal structure, the resultant magnetization vectors **M'**, **M,"** **M'''**... of the grains are clustered about the **H** direction, most of them make angles less than 45° with it, thus providing about 0.8 saturation magnetization - in agreement with the real data.

When H strength increases farther, several processes may occur on the way from A to B: some final domain rearrangements within the grains, magnetic recrystallization by growth of the grains whose magnetization is close to the **H** direction, and non-epitaxial nucleation-growth of the domains with the magnetization along or close to **H**. All these processes will appear as a rotation of the total magnetization of the sample toward **H**. All of them will also create internal strains owing to magnetostriction. While at point B the sample is considered completely saturated in the direction of the applied field H, it is not quite so, even though numerically it may seem to be within the limits of experimental error. It is still polycrystalline, with the enlarged grains magnetized along **H**, and small grains not so precisely oriented. The system is permeated by internal strains prevented from relaxation by the magnetic field. A comparison with the S & T experiments supports this picture. Our sample at point B is not a monodomain crystal: the internal strains due to magnetostriction prevent a complete conversion. Only the mutual action of the magnetic field and properly applied deformation could produce it. If it were a monodomain, the reverse scan B → D would be a horizontal line, which is not.

The magnetization curve B → D shows that a major portion of the structural rearrangements that occurred on the way from A to B is retained, but there is some regression causing the observed slope. It should be noted that the very existence of rectangular loops proves that there is no *induced magnetization, i.e.,* essential dependence of a spontaneous magnetization M_s *per se* on the strength of magnetic field. Thus, $M_s(H)$ does not contribute to this slope. The cause of the slope is relaxation of the accumulated internal strains as H falls toward zero. Whichever of the above-listed three processes, or their combination, caused the magnetization A → B, it was accompanied by accumulation of internal lattice strains opposing this magnetization. A subsequent decrease in the H strength allows strains to relax by means of structural readjustments at the expense of the magnetization. By annealing the sample under the conditions marked by point B the strains can be eliminated. In this way the curve B → D → C can be flattened, even made horizontal.

It would be incorrect to maintain that any deviation of magnetic dipoles from their equilibrium orientations in the crystal is impossible, as it would be incorrect to deny a possibility of any other elastic deformation. But this process has narrow limits and hardly contributes into the slope of the curves A → B and B → C. Even if it does, it has to be strictly reversible according to the thermodynamic definition of the notion. It follows that the "rotation," if any, has nothing to do with the hysteresis.

Some decline in the level of magnetization over B → D → C is not a remagnetization yet. The latter begins only after H changes its sign to the opposite and exceeds a certain threshold value $-\Delta H$ to initiate nucleation of the oppositely oriented domains in the suitable crystal defects. But the sample is still polycrystalline and, in contrast to a rectangular loop where a single nucleation act caused a propagation of the domain interface over the whole sample, this time one nucleation act affects only one crystal grain. The "starting fields" H_n are different in different grains. It may be that the following redistribution of magnetization within a grain generates internal strains and thus facilitates nucleation in the neighboring grains, but the fact is that the process does not go far without increases in the H strength. Still, this is the most effective magnetization phase (I dM/dH I = max), ending at the point equivalent to point A.

A common misconception should be dispelled regarding the role of crystal defects in a magnetization process and the formation of hysteresis loops. The defects were always considered only as an obstacle to the motion of domain boundaries. In fact, their role is twofold. In a defect-free crystal neither motion, nor even formation of the boundary is possible (Sec. 2.5). We can imagine a ferromagnet exhibiting a very high coercive force because its crystal structure is "too perfect." Indeed, the shortage of adequate defects for nucleation in very small ferromagnetic particles requires very strong fields for their remagnetization (Sec. 4.11). On the other hand, different kinds of crystal defects that are not suitable to serve as heterogeneous nucleation sites may (and usually do) hamper propagation of domain boundaries.

There are two principles, or conditions, that must be kept in mind with regard to the formation of the hysteresis loops: (1) if a sample has been thoroughly magnetized to saturation by field +H, its remagnetization cannot start before H becomes negative and exceeds a certain threshold value $-\Delta H_{th}$, and (2) the hysteresis is inevitable not only in practice, but also theoretically. Both these conditions were not met, for example, in the case presented in Fig. 4.20.

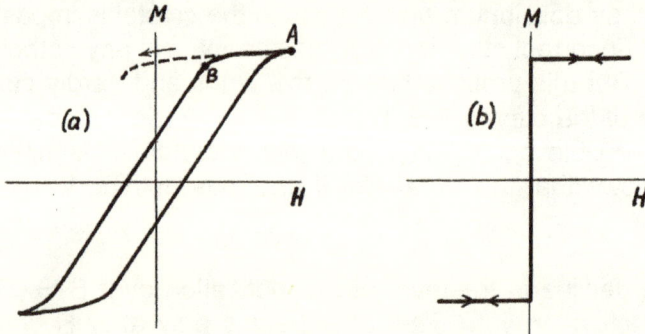

Fig. 4.20a,b   The phenomenon of ferromagnetic hysteresis loops as presented in [206]. Both pictures are incorrect.
(a) The ferromagnetic hysteresis loop (solid line) in which overall remagnetization from point A down begins at point B, when the magnetic field is still positive and strong. (The dashed line is added to show how the real process would go).
(b) The "hysteresis loop" for "maximum magnetic purity in the crystal." In fact, ferromagnetic hysteresis cannot be avoided under any idealized conditions. As opposed to this claim, the higher the crystal perfection, the wider the hysteresis.

4.13.4 Specificity of ferroelectric hysteresis loops

Almost the entire description of ferromagnetic hysteresis loops is directly applicable to the ferroelectric hysteresis loops of repolarization in electric fields E. The differences are reduced to the following two points.

(1) Polarization by a "rotation" is out of consideration. It is recognized that *induced polarization* of dielectrics occurs only by the deformation of the electron shells and displacements of the ions, while the *orientational* polarization is not possible. It is self-evident that the orientation of the electric dipoles in polar dielectrics is an element of their crystal structure. The terms "easy direction" and "hard direction" are not in use. Reorientation of the dipoles in a ferroelectric crystal can occur only by rearrangements of crystal structure at the domain boundaries. As we have argued, the orientation of the magnetic moments in ferromagnetics is fixed by the crystal structure as well, even though it was not so directly evident. In this respect the two ferroics are similar. The difference is in the following. What was perceived to be "rotation" in ferromagnetics actually occurs by crystal rearrangements involving the non-epitaxial nucleation. As a rule, the latter is not possible in ferroelectrics due to their pronounced layered structure (Sec. 2.8.6). Ferroelectrics, at least most of them, can be polarized to saturation P_s only in a direction determined by the crystal structure, and not in the arbitrarily chosen direction of the applied field E. It is conceivable to achieve P_s in polycrystalline ferroelectrics in the E direction through "electrostatic recrystallization" - growth of the grains that happened to be polarized in the E direction, but this is a slow process.

(2) While spontaneous magnetization M_s by itself does not appreciably depend on H, a spontaneous polarization P_s depends on E to some extent. As opposed to magnetic dipole moments, which are a property of atomic spins, the two electric charges of a ferroelectric dipole are spatially separated in the crystal unit cell. All ferroelectrics are also piezoelectrics. Their polarization, induced by the field E (reverse piezoeffect), is superimposed on that caused by the domain structural rearrangements and is noticeably present on the hysteresis loops. Specifically, the saturation polarization $P_s(E)$ continues to grow even after all the dipoles are parallel. This induced polarization is strictly reversible and has therefore nothing to do with hysteresis.

4.13.5 Double hysteresis loops

As stated in Sec. 4.13.1, hysteresis loops of polymorphic transitions, and those of remagnetization / repolarization, are inherently related. The ferroelectric double hysteresis loops (DHL) represent a peculiar intermediate case, a link between the two (Fig. 4.21). In polymorphic transitions two crystal phases have different crystal structure; they are changed into each other by changing temperature or pressure. In remagnetization and repolarization the participating "phases" have identical crystal lattice, differing only by their orientation in the applied field serving as the driving force. In the formation of a DHL the applied electric field E is the driving force of a *polymorphic transition* between antiferroelectric (AE) and ferroelectric (FE) structures.

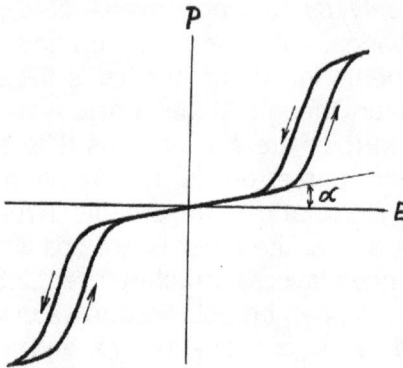

Fig. 4.21 Ferroelectric double hysteresis loop.

The DHLs in PbZrO$_3$ crystals [207] were observed in strong alternating electric fields of 20 to 60 kV/cm at the temperatures 5° to 30 °C below the temperature 230 °C of the phase transition from paraelectric (ODC) to antiferroelectric phase. As usual with ferroelectrics, the crystals were layered, and the resultant antiferroelectric crystals had a lamellar-domain structure of the type described in Sec. 2.8.6. The prerequisite for the AE → FE transition is the ability of the FE phase and inability of the AE phase to acquire the structural orientation with all the dipoles lined up in one direction under the action of applied electric field E. The following four components of energy basically determine which free energy is lower: chemical bonding in the AE phase U_{AE}, AE dipole attraction $U_{\downarrow\uparrow}$, chemical bonding in the FE phase U_{FE}, and interaction of the FE dipoles with E and between themselves $U_{\uparrow\uparrow E}$. The necessary condition of the AE → FE phase transition is

$$U_{FE} + U_{\uparrow\uparrow E} < U_{AE} + U_{\downarrow\uparrow}.$$

A glance at the DHL in Fig. 4.21 shows that it is not quasi-stationary (Sec. 4.13.3) and therefore it is inappropriate for a description of basic processes of its formation. For this purpose a corrected version (Fig. 4.22) will be used. Increasing E from zero initially affects only the length of the dipoles in the AE phase, inducing some polarization without changing the crystal structure (line O → S → B). This effect is strictly reversible. Nothing special occurs at point S corresponding to the field E_o at which the AE and FE free energies are equal; some additional increase, ΔE_n, is required to form initial FE nuclei (point B). Nucleation of the FE domains is strictly epitaxial and, as usual, occurs over a transition range, in this case – a range of E rather than T. Under the action of field E all the emerging domains have the same crystallographic (and dipole) orientation. Over the S-shaped curve B → C the crystal remains two-phase (Sec. 3.1). At point C it is completely ferroelectric with all the domains lined up in one direction ($P = P_s$). In the reverse scan the FE → AE transition will not start until the first AE nucleus forms at point N. At point D the sample is AE again. When alternating field E extends into the negative area, the second petal of the DHL forms; the two petals can be translated into each other by the center of symmetry at point 0.

The formation of two symmetric loops rather than one resulted from the fact that the field E, the dipoles in the participating crystal phases, and the polarization P of the sample are vectors. A picture in the P-E coordinates reflects both the relative quantities of the phases and the integral polarity P of the sample, as well as its sign. But if one is interested only in the relative quantities (mass) of the AE and FE phases, the DHL turns into a singular hysteresis loop which looks exactly as those of polymorphic transitions presented in the previous chapters, with E replacing the temperature as the driving variable (Fig. 4.23).

In the formation of DHL in $PbZrO_3$ the starting phase was assumed to be AE. But in terms of a broader view on DHL formation, it could be paraelectric (PE) ODC as well. It is feasible to have the DHL in a ferroelectric at slightly *higher* temperatures of a FE → PE phase transition. Application of strong alternating field E to the PE phase would make the FE phase preferable every time when |E| is strong and the PE phase preferable every time when E decreases to zero. Also, there is no reason not to have a ferromagnetic DHL. It will differ from a ferroelectric DHL in Fig. 4.21 only by $\alpha = 0$.

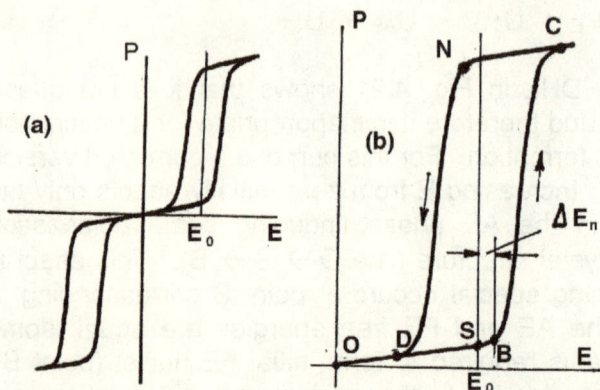

(a)

(b)

ΔE_n

Fig. 4.22 Major features of double hysteresis loop if it were a quasi-stationary.

(a) The whole double loop.

(b) Enlarged picture of its one petal.

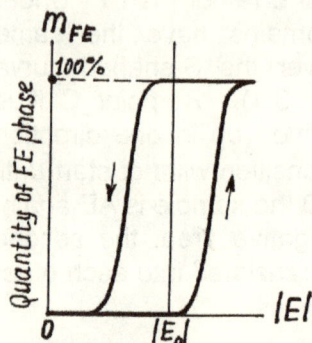

m_{FE}

100%

Quantity of FE phase

$|E|$

0 $|E_0|$

Fig. 4.23 Degeneration of a double hysteresis loop into a single hysteresis loop if information of interest were limited by the hysteresis of the polymorphic *ferroelectric* ↔ *antiferroelectric* phase transition only.

4.14 Crystal rearrangement of antiferromagnet by magnetic field

In terms of conventional science, this phenomenon must not exist. As Lavrov *at al.* [LKA,*] summarized, "the common perception [is that] magnetic field affects the orientation of spins, but has little impact on the crystal structure; one would least expect any structural change to be induced in antiferromagnet where spins are antiparallel and give no net moment." Nevertheless, LKA not only observed such an unexpected phenomenon, but reported it to occur "through the generation and motion of crystallographic twin boundaries." The reason for the above common perception is rooted in the belief that spins in an antiferromagnet are strongly bound together by the Heisenberg's exchange field and, therefore, the external field H cannot interact separately with the parallel and antiparallvel components of the spin system.

---

*A.N. Lavrov, Seiki Komiya and Yoichi Ando. "Magnetic shape-memory effects in $La_{2-x}Sr_xCuO_4$ crystals" (http://arxiv.org/abs/cond-mat/0208013)

However, according to the fundamentals of ferromagnetism presented in this Chapter, no exchange field exists. Magnetic interaction of the spins is relatively weak.The orientation of a spin is determined by the orientation of its atomic/molecular carrier, imposed by the chemical bonding of the crystal structure. Reorientation of a spin system is achieved by the lattice rearrangement at the interfaces. Transition *antiferromagnet* → *ferromagnet*, (A→F), under the action of temperature or pressure, a common phenomenon, is realized when the total free energy of the F crystal becomes lower in spite of some disadvantage of a parallel spin alignment.

In the case under consideration, we certainly deal with an A→F phase transition. The question is whether it can take place under the action of an applied magnetic field H. The answer is yes. It can occur for the same reason: free energy of the F phase can be lowered by having all spins aligned parallel to H as compared to only half of them in the A phase. This proceeds by re-orientation of every second spin carrier during particle-by-particle structural rearrangement at the interfaces.

4.15 Summary of the new principles

New fundamentals of ferromagnetism and ferroelectricity were put forward in this chapter. The main properties and manifestations of ferroics were accounted for in a consistent and straightforward way. The new approach was brought about by unearthing the true nature of solid-state phase transitions, described in the previous chapters. Like all solid-state phase transitions, ferromagnetic and ferroelectric transitions are those between crystallographically different phases.

The paramagnetic / paraelectric phase is orientation-disordered (ODC), where the *particles* carrying dipoles are engaged in a thermal rotation. Ferroic phase transitions are not a "critical phenomenon," and their temperature is not a "critical (Curie) point." They occur by formation of nuclei of new phase and their growth - as all solid-state phase transitions do. This undermines the basic assumption of the current theory of ferromagnetism on the existence of an extremely strong "molecular field" that establishes a parallel dipole alignment and maintains it up to the actually observed high "Curie points".

The new approach made it possible to eliminate vast discrepancies of the Weiss / Heisenberg theory, to solve the problem of "λ-anomalies," to explain domain structure, to unite ferromagnetism and ferroelectricity into one coherent picture, and more. The new answer to the fundamental question, namely, what makes magnetic and electric dipoles parallel (when they are), is simple, if not trivial. Any type of dipole alignment, including parallel, is established by the same crystal forces (chemical bonding) that provide the 3-D crystal order. Because the contribution of the dipole interaction into the total free energy of a ferroic crystal is very small, a ferromagnetic dipole alignment is imposed by the crystal forces if the total free energy of that particular crystal structure is minimal.

The following summary of the new principles is formulated for ferromagnetism, but can easily be adapted for ferroelectricity as well.

- No Weiss' molecular field (Heisenberg's exchange field) responsible for a parallel or antiparallel alignment of atomic magnetic moments in ferromagnetics exists. The interaction between the elementary dipoles is purely magnetic. As opposed to the Weiss / Heisenberg theory, a parallel alignment in itself elevates crystal free energy, and antiparallel lowers it.

- A particular dipole alignment ("magnetic structure") is determined by the requirements of crystal packing. The magnetic structure is an element of that 3-D packing, contributing a small positive or negative addition to the total crystal free energy. Ferromagnetism takes place in those cases when minimum free energy of the crystal packing requires atomic orientations with the magnetic moments not mutually compensated. Even though the magnetic interaction has some destabilizing effect, it is too weak to make any alternative structure preferable.

- A paramagnetic state is orientation-disordered state of a magnetic crystal. Thermal rotation of the magnetic moments in this specific state results from thermal rotation of the atoms that carry them.

- Ferromagnetic phase transitions are a nucleation-and-growth rather than a "critical" phenomenon.

- Ferromagnetic phase transitions are usually characterized by epitaxial (oriented) nucleation, small (but finite) range of transition, and small (but finite) temperature hysteresis. The existence of the range of transition and hysteresis is responsible for the effects erroneously believed to be "phase transition anomalies".

- Ferromagnetic domain structure results from two factors. It originates by multiple nucleation of the ferromagnetic phase in several different equivalent structural orientations within the paramagnetic matrix. Growth of these nuclei and subsequent "magnetic aging" proceed toward minimizing the magnetic energy.

- The neighboring domains are structural twins, which gives rise to alternating their magnetization directions. The domain boundary is a twin interface of zero thickness, rather than a thick "Bloch wall." While the presence of the boundary involves some energy due to violation of the 3-D crystal order, the antiparallel alignment of the magnetic moments (or their components parallel to the boundary) on opposite sides of the boundary contributes to the system stability.

- Magnetostriction upon magnetization of ferromagnetic Fe results from the difference between $a = b$ and c parameters of its pseudo-cubic structure.

- Magnetization in an applied magnetic field H occurs only by structural rearrangement. The latter proceeds, depending on H strength, by (a) motion of the existing domain boundaries, (b) epitaxial nucleation and growth of new domains in the (allowed by the paramagnetic matrix) domain orientation closest to the H direction, or (c) non-epitaxial nucleation-growth of new domains in the H (or close to H) orientation. No essential magnetization by dipole rotation in the same structure without its rearrangement is possible: crystal growth is the way magnetization occurs.

- Remagnetization hysteresis loops are a direct reflection of the structural rearrangements involved. The cause of the hysteresis is the nucleation lags. A magnetic hysteresis loop consists of two sigmoid-like curves. Their shape is due to the balance between the increase in the number of nucleation sites with temperature and the decrease in the amount of the original phase.

- All structural rearrangements during ferromagnetic phase transitions and magnetization in magnetic fields occur according to edgewise layer-by-layer molecular mechanism of crystal growth. The Barkhausen effect is its direct manifestation.

4.16 A few notes on superconducting phase transitions (SPTs)

The content of this book sheds no direct light on the physics of the superconducting state. However, the conclusion on the existence of a single general mechanism of phase transitions in solid state – nucleation and growth – directly relates to the investigation of SPTs. In particular, it eliminates two widespread misconceptions.

One misconception is the assumption that SPT is a critical phenomenon and occurs at a critical temperature T_c. Initially all SPT were believed to be of the second order. However, an internet search with GOOGLE taken in 2007 for "superconducting first order phase transitions" produced over 1.000.000 hits. There are plenty of SPTs that were recognized as first order. Yet, this did not change the unanimous assumption that all SPT occur at their critical points T_c. While SPTs are superficially assigned ether first or second order in literature, *all* of them proceed by nucleation and growth as described in Chapter 2. One implication of this fact is that a SPT necessarily occurs over some temperature range where the non-superconducting and superconducting phases coexist. In particular, this explains the phenomenon of heterogeneous coexistence of ferromagnetism and superconductivity.

Another misconception is the belief that finding the mechanism of how a non-superconducting phase turns superconducting can reveal, or help to reveal, the nature of the latter. But the mechanism is the nucleation and growth described in Chapter 2. It is the same whatever the physical property (in this case – electric conductivity) changes. The conclusion is as follows: while knowledge of the SPT mechanism is practically beneficial, a superconducting state should be investigated *in itself*. This does not mean that the comparison of the initial non-superconducting and the resultant superconducting states would necessarily be useless.

The particulars of nucleation, presented in Sec. 2.5, suggest the following procedure of *bringing a superconducting crystal to room temperature T_r*. A superconducting crystal of very high quality should be grown at the low temperature where it is stable. It will remain superconducting at T_r in the absence of the two conditions which would return it to the non-superconducting phase: availability of a crystal defect suitable for initiation of the reverse phase transition and availability of a sufficient concentration of microcavities for the sustained interface motion. Whether this idea can ever be practical is an open question.

APPENDIX 1: 300 DIFFERENT KINDS OF SOLID-STATE PHASE TRANSITIONS FOUND IN LITERATURE

(either "transition," or "transformation" should be added to every term)

ABSORBING
ALLOTROPIC
ANDERSON
ANDERSON-MOTT
ANTIFERRO-ORDER
ANTIFERRO-DISTORTIVE
ANTIFERROELASTIC
ANTIFERROELECTRIC
ANTIFERROMAGNETIC
ANTIFERROQUADRUPOLAR
ANTIPOLAR
ANTI-THERMODYNAMIC
AUBRY

BAND JAHN TELLER
BAND-CROSSING
BAROELASTIC MARTENSITIC
BEREZINSKII-KOSTERLITZ-THOULESS
BOND-TYPE
BROKEN SYMMETRY
BROKEN SYMMETRY QUANTUM

CASCADE
CATASTROPHIC
CATION ORDERING
CELL-MULTIPLYING
CHARGE DENSITY WAVE
CHARGE-ORDERING
CHARGE TRANSFER CONTROLLED
CHIRAL
CIVILIAN
CLASSICAL
COLLAPSE
COMMENSURATE
COMMENSURATE-INCOMMENSURATE
COMMENSURATE-INCOMMENSURATE DEPINNING
CONCENTRATIONAL
CONCENTRATIONAL METAMAGNETIC
CONDENSATIONAL
CONTINUOUS
CONTINUOUS STRUCTURAL

COOPERATIVE
COOPERATIVE JAHN-TELLER
CRITICAL POINT
CRYSTAL-FIELD
CUBIC-TETRAGONAL (and others defined by change in symmetry)
CURIE POINT
CURRENT-INDUCED SUPERCONDUCTING-RESISTIVE

DECONFINEMENT
DECORATION
DECORATION-ITERATION
DECOUPLING
DEFECT-MEDIATED
DEFORMATION STRUCTURAL
DENSITY-DRIVEN
DENSITY-DRIVEN MOTT
DENSITY-DRIVEN QUANTUM
DIFFUSE
DIFFUSION(AL)
DIFFUSIONAL-DISPLACIVE
DIFFUSIONLESS
DIFFUSIONLESS DISPLACIVE
DILATATIONAL
DILATATION DOMINANT
DIMERIZED-INCOMMENSURATE
DIPOLE
DISCOMMENSURATION
DISCONTINUOUS
DISCONTINUOUS DYNAMIC
DISORDER-DRIVEN METAL-INSULATOR
DISORDER-INCOMMENSURATE
DISPLACIVE
DISPLACIVE FERROELECTRIC
DISPLACIVE-REPLACIVE
DISSIPATIVE-QUANTUM
DISTORTIVE (DISTORTIONAL)
DRIVEN BY CONDUCTION ELECTRONS
DRIVEN BY INTERACTION OF LOCALIZED ORBITAL
        ELECTRONIC
STATES AND CRYSTAL LATTICE
DRIVEN BY SOFT-SHEAR ACOUSTIC MODE
DUAL
DUALITY-DECIMATION
DYNAMICAL LOCALIZATION-DELOCALOZATION
DYNAMICAL REORDERING
DYNAMICAL STRIPE ORDERING

ELECTRIC FIELD INDUCED STRUCTURAL
ELECTRONIC
ELECTRONICALLY DRIVEN
ELECTRONIC TOPOLOGICAL
ELECTROWEAK
ENTROPY-INDUCED
EQUITRANSLATIONAL
EXCHANGE INVERSION
EXTENDED DECORATION-ITERATION
EXTENDED STAR-TRIANGLE
EXTRINSIC
EXTRINSICALLY FERROELECTRIC

FERMION CONDENSATION QUANTUM
FERROELECTRIC TO SUPERIONICALLY CONDUCTING
FERRIELECTRIC
FERRIMAGNETIC
FERROIC
FERRO-ORDER
FERRODISTORTIVE
FERROELASTIC
FERROELECTRIC
FERROMAGNETIC
FERROMAGNETISM-INDUCED REENTRANT STRUCTURAL
FIELD-INDUCED
FIRST-ORDER
FIRST ORDER CHARGE DISPROPORTIONATION
FIRST-ORDER QUANTUM
FIRST-ORDER SUPER-RADIANT
FIRST-ORDER SYNCHRONIZATION
FLUCTUATION DRIVEN FIRST ORDER
FLUCTUATION INDUCED QUANTUM

GARDNER
GATE-INDUCED MOTT
GELATION
GENERIC
GEOMETRY-INDUCED
GRADUAL

HELICAL
HETEROGENEOUS
HEXAGONAT TO SQUARE
HIGHER-ORDER
HOMOGENEOUS
HYDROGEN-HOPPING

IMPROPER FERROELASTIC
IMPROPER FERROELECTRIC
IMPROPER LOCK-IN
INCOMMENSURATE-COMMENSURATE
INFINITE ORDER
INFINITELY-MANY ABSORBING-STATE NONEQUILIBRIUM
INTER-BAND
INTERFACE-INDUCED
INTERMARTENSITIC
INTRA-BAND
INTRINSIC
INTRINSIC FERROELASTIC
INTRINSICALLY ANTIDISTORTIVE
IONIC-TO-NEUTRAL
IRREVERSIBLE
ISING
ISOLATED POINT
ISOMORPHIC
ISOSTRUCTURAL
ISOTHERMAL

JAHN-TELLER
JUMP-LIKE

KASTELEYN
KINETICS DRIVEN COMMENSURATE-INCOMMENSURATE

LAMBDA
LATTICE DYNAMICAL
LIFSHITZ
LOCALIZED-COLLECTIVE ELECTRON
LOCK-IN
LOW SPIN TO HIGH SPIN
LOW SPIN TO INTERMEDIATE SPIN STATE

MAGNETIC
MAGNETIC-FIELD-INDUCED ORIENTATIONAL
MAGNETIC-NONMAGNETIC
MAGNETIC-NONMAGNETIC QUANTUM
MAGNETIC ORDERING
MAGNETIZATION-REVERSAL
MAGNETO-DIELECTRIC
MAGNETOELASTIC
MAGNETO-STRUCTURAL
MARTENSITIC
MARTENSITIC CIVILIAN
MARTENSITIC MILITARY

MASSIVE
METAL-INSULATOR
METAL-SEMICONDUCTOR
METAL-NONMETAL
METAMAGNETIC
MICROCANONICAL FIRST ORDER
MILITARY
MIXED TYPE
MIXED VALENCE
MORIN (SPIN-FLIP)
MOTT

NEAR-CRITICAL
NEARLY FIRST-ORDER
NEARLY SECOND-ORDER
NEEL
NEEL-DIMER
NEEL TO SPIN-PEIERLS
NEUTRAL REGULAR TO IONIC-DIMERIZED
NEUTRAL-TO-IONIC
NOISE-INDUCED ORDERING
NONADIABATIC
NONELECTRONIC
NON-EQUITRANSLATIONAL
NON-ISOTHERMAL
NON-LINEAR
NON-POLAR STRUCTURAL
NON-PROPORTIONAL
NON-WEAK
NORMAL
NORMAL-INCOMMENSURATE
NUCLEATION-GROWTH

OGUCHI
ORBITAL ORDERING
ORDER-DISORDER
ORDER-DISORDER FERROMAGNETIC
ORDER-DISORDER LAYERING
ORDER-DISORDER-ORDER
ORDER-ORDER FERROMAGNETIC
ORDINARY
ORIENTATIONAL ORDER-DISORDER
ORIENTATION-SWITCHING

PAIRING SYMMETRY
PARITY-CONSERVING KINETIC
PARTIAL LOCK-IN

PARTLY GRADUAL PARTLY ISOTHERMAL
PEIERLS
PERCOLATIVE
PHASON PINNING-UNPINNING
PHONON
PHOTOINDUCED MAGNETIC
PIEZODISTORTIVE
PLATEAU
POLARIZATION-FLOP
POLAR STRUCTURAL
POSITIONAL ORDER-DISORDER
POTTS
PREMARTENSITIC
PRESSURE-INDUCED
PRESSURE-INDUCED ELECTRONIC TOPOLOGICAL
PRESSURE-INDUCED LOCK-IN
PROPER FERROELASTIC
PROPER FERROELECTRIC
PROPER OPTIC FERROIC
PSEUDOSPIN-FLIP
PURE DISPLACIVE

QUADRUPOLAR
QUANTUM
QUANTUM HALL TO INSULATOR
QUASIMARTENSITIC

RADIATIVE
RECONSTRUCTIVE
REENTRANT
REENTRANT STRUCTURAL
REORIENTATION
REPLICA SYMMETRY BREAKING
REVERSIBLE
RIGIDITY
RINGS
ROTATIONAL

SALAM
SECOND-ORDER
SECOND-ORDER-LIKE INCOMMENSURATE-COMMENSURATE
SECOND-ORDER QUANTUM
SEMICONDUCTOR-METAL
SEMIRECONSTRUCTIVE
SHOCK-INDUCED
SIMPLE
SINGLET-TRIPLET

SOFT-MODE
SOFT-MODE QUANTUM
SPIN-CROSSOVER
SPIN-DENSITY-WAVE
SPIN-FLOP
SPIN-FREEZING
SPIN-PEIERLS
SPIN-REORIENTATION
SPIN STATE
SPINOIDAL-TYPE
STANLEY-KAPLAN
STAR-TRIANGLE
STRAIN-MEDIATED
STRESS-INDUCED
STRONGLY FIRST ORDER
STRUCTURAL
SUBSTITUTIONAL
SUBSTITUTIONAL ORDER-DISORDER
SUPERCONDUCTING
SUPERCONDUCTOR-METAL-INSULATOR
SUPERFLUID-INSULATOR
SUPERIONIC
SUPERPROTONIC
SYMMETRIC-ASYMMETRIC
SYMMETRY-BREAKING
SYNCHRONIZATION

THERMODYNAMIC
THERMOTROPIC
THIRD-ORDER
TOPOLOGICAL
TOPOLOGICAL KOSTERLITZ-THOULESS
TOPOTACTIC
TRIANGULAR TO SQUARE FLUX LATTICE
TRICRITICAL
TRIGGERED
TWO-DIMENSIONAL
TWO SPIN-STATE

UNUSUAL MAGNETIC FIELD-INDUCED
USUAL

VALENCE
VALENCE INSTABILITY
VERWAY
VOLUME-CHANGE
VOLUME-COLLAPSE

VORTEX DEPINNING
VORTEX-GLASS
VORTEX LATTICE MELTING
VORTEX LIQUID DECOUPLING
VORTEX-MATTER

WEAK-ORDER
WEAKLY FIRST ORDER
WILSON

YAFED-KITTEL

ZONE BOUNDARY
ZONE CENTER
ZONE CENTER LOCK-IN

APPENDIX 2: SHORTCOMINGS OF ADIABATIC CALORIMETRY AND AN UNNOTICED ADVANTAGE OF DIFFERENTIAL SCANNING CALORIMETRY*

A2.1 Introduction

One of the goals of contemporary calorimetry is to determine with the greatest possible precision the shape of specific heat anomalies, such as the "lambda-peaks" in the temperature ranges of phase transitions. In order to achieve this goal, calorimetric techniques, the adiabatic method being the most common one, have been refined to the point of providing very high precision and temperature resolution. Consequently, the scatter in the experimental points and the variance in data published by different authors have become more affected by variations in the state of the sample rather than by instrumental error [1-4].

In spite of the high precision of these techniques, it is shown here that they nevertheless suffer from certain shortcomings as far as measurements in the range of the phase transitions are concerned. These effects were first noted in a 1979 article by the present author [5], where it was shown that these effects may give rise to apparent specific heat peaks. A misleading role of such apparent peaks in the current theory of phase transitions is thus implied.

In the present paper this problem is discussed in detail and its simple resolution based on the application of differential scanning calorimetry (DSC) is described. In conclusion, the described method is illustrated by a re-examination of the NH_4Cl case, which is of special interest, since it has been used to exemplify specific heat λ-anomalies [6]; clear evidence is presented that no anomaly really exists in this case.

A2.2 Application of conventional techniques to the investigation of specific heat anomalies.

Modern calorimeters utilized for the measurements of specific heat anomalies are usually "one way" instruments in the sense that the measurements can be carried out only as a function of increasing temperature. It is vital, however, that the precise measurements be performed both as a function of increasing temperature (such an

---

* References are given at the end of this article.

experiment will be referred to as a temperature-ascending run) as well as decreasing temperature (it will be referred to as a temperature-descending run). Only in this manner it is possible to test whether thermodynamic equilibrium was attained in these measurements (and to inevitably get a negative answer).

Routinely, the experimental approach used to cope with the problem of the thermodynamic equilibrium is to utilize the intermittent measurement mode [7]. A "sufficient" amount of time is allowed to pass after each input of energy into the calorimeter, and it is assumed that if the time-intervals are long enough, the establishment of thermodynamic equilibrium is ensured. However, this mode of measurement is inadequate by reason of the non-equilibrium nature of the process of the phase transitions. In those rare calorimetric works, in which both temperature-ascending and descending runs were performed, *hysteresis* in the location of the λ-peak was observed. An example is shown in Fig. 1 [8]. The λ-peak in the NH_4Cl phase transition around -30° C was always found to be shifted in the direction of temperature change. The phenomenon could not be eliminated by increasing the observation time. What was left unnoticed in Ref. [8] and in later works is that such a hysteresis is incompatible with a heat capacity origin of the λ-peaks - if it is true that heat capacity is an unequivocal function of temperature and pressure.

A solution to the problem of λ-peaks which exhibit hysteresis has originally been suggested in Ref. [5]. It was shown that apparent "λ-peaks" can arise in conventional heat capacity measurements of *first-order* phase transitions in solids. These peaks represent a latent heat of phase transitions and are due to the inherent non-equilibrium nature of the phase transitions.

As is well known, all of the numerous λ-peaks and similar heat capacity anomalies reported in literature are always *positive* in graphs of specific heat C_p *vs.* temperature T (upward is positive, downward is negative), even when both temperature-ascending and descending runs were performed (Fig. 1). We now consider the sign, or direction, of an apparent λ-peak that is in fact caused by the latent heat of a phase transition.

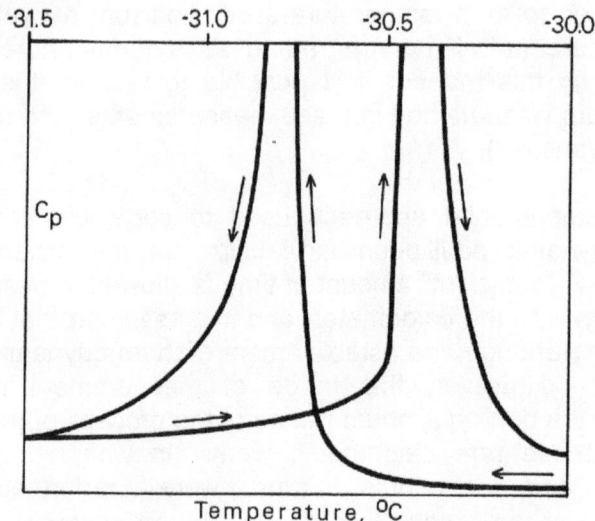

Fig. 1 (From Ref. 8) The specific heat λ-peak of the NH₄Cl phase transition in temperature-ascending and descending runs. A temperature hysteresis is observed, the peak being positive in both runs.

It should be recalled that phase transitions are *endothermic* when the temperature is increased and *exothermic* in the reverse direction. In differential thermal analysis (DTA) the corresponding recorded peaks are of opposite sign. This is not the case, however, in conventional calorimetric techniques that are used in specific heat measurements. Indeed, when larger amounts of heat Δq must be introduced into the sample in order to elevate its temperature by ΔT, one attributes this to the increase in the specific heat $C_p = (\Delta q)/(\Delta T)$; however, the same effect could also be caused by uptake of latent heat corresponding to an endothermic phase transition. Now, in the reverse run, larger amounts of Δq removed from the sample in order to lower its temperature by ΔT can mean that the $C_p = (-\Delta q)/(-\Delta T)$ increases, but could be also caused by release of the latent heat corresponding to an exothermic phase transition. Thus "latent heat λ-peaks" will always face upward in $C_p(T)$ plots, that is, be *positive regardless of the direction of the run.*

A2.3 *Application of DSC*

Our data and considerations presented here, as well as earlier [5], suggest that all of the previously reported specific heat peaks should be re-examined. This is hardly practical in terms of conventional techniques, even if one assumes that an irreversible calorimeter could be converted into a reversible one. There are three reasons: (1) Inconvenience, related to the modification of the instrument; (2) Hysteresis of the peak under examination may be the only indication of its latent-heat origin, and this might seem not sufficiently convincing to some workers in the field; (3) Point by point measurements (still on an intermittent basis) of many hundreds of the anomalies reported in the literature would be too expensive and time consuming.

DSC allows for a resolution of this predicament. The DSC technique is fast and simple, and recent improvements have made it rather precise. Carrying out temperature descending runs with DSC is as easy as ascending runs. Most important, however, is that DSC has inherited from DTA its feature of displaying endothermic and exothermic peaks with *opposite* signs in the chart recordings.

This latter feature results from the manner in which the signal is measured in DSC [9]; the differential heat flow required to maintain a sample and an inert reference at the same temperature is fed into a recorder, while the control circuit is programmed to increase or decrease the average temperature at a predetermined rate. In the absence of any thermal reaction, the recorder will trace the difference between the heat capacities of the sample and the reference. (With no reference in the reference holder the record will be a heat capacity plot of the sample.) This plot serving as the base line, a latent heat signal proportional to the additional arising difference between the heat input to the sample and that to the reference will appear in the recorder output as a peak. The *sign* of the signal, and of the corresponding peak, will evidently depend on the fact whether the thermal reaction in the sample is endothermic or exothermic.

To take advantage of this feature of DSC for the re-examination of heat capacity peaks, both temperature-ascending and descending runs should be performed. In a strip-chart recording with the temperature X-axis along the chart, the two plots must be symmetrical with respect to one another about the Y-axis, provided that no latent heat was involved (Fig. 2a). In contrast, a latent heat contribution will appear as an endothermic peak in a temperature-ascending run and as an exothermic one in a descending run, *i.e.* in positions which can loosely be called "antisymmetrical" (Fig. 2b). If the peak does change sign in the reverse run, the *latent heat* of the phase transition can be determined from the area under the peak. In such a case the latent heat peak must be eliminated from the specific heat plots.

DTA can also be used in the same way to verify the origin of specific heat peaks, but the interpretation of the results is less straightforward. Besides, DTA does not provide quantitative data which would permit a conclusion to be drawn on whether the area under the latent heat peak is equal to that of the reported "specific heat λ-anomaly," in other words, whether any residual specific heat anomaly still remains after the elimination of the latent heat contribution. On the other hand, a modern DSC instrument possesses the necessary characteristics; with such an instrument not only a rapid determination of specific heats can be done, but also latent heat contributions can be easily identified and thus subtracted.

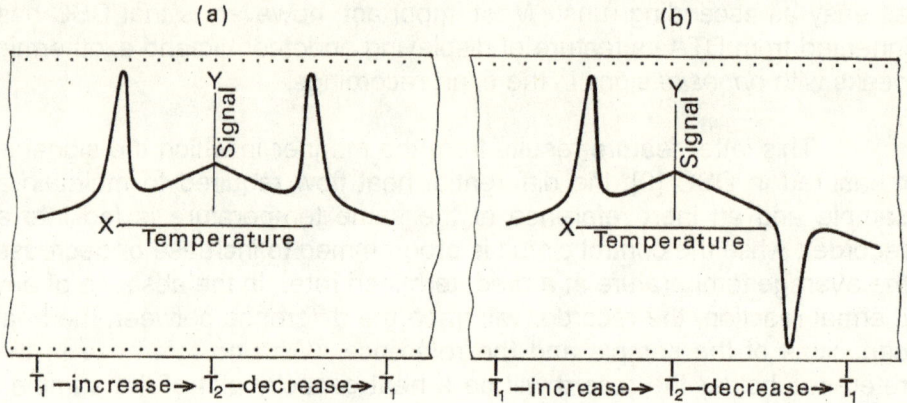

Fig. 2   DSC strip-chart recording of λ-peak in a phase transition cycle involving heating $T_1 \rightarrow T_2$ and cooling $T_2 \rightarrow T_1$ (schematic).
(a) The appearance of the peak if it would be a true specific heat peak. Plots of the two runs, including the peaks, are symmetrical to each other about the Y axis.
(b) The appearance of the peak caused by latent heat of phase transition. Only true heat capacity contributions in the plots are symmetrical about the Y axis, while the peaks are of opposite Y sign; also they occur in different temperature ranges due to hysteresis.

A2.4 *An example*

The phase transition in NH_4Cl around -30 °C has become an illustration of specific heat λ-anomalies. It was the subject of numerous calorimetric investigations [1,2,8,10-14], all of which confirmed the presence of the λ-peak.   This case was selected by us for re-examination by DSC.

Our strip-chart recordings made with a Perkin-Elmer DSC-1B instrument immediately revealed that the peak acquires an opposite sign in the reverse run (Fig. 3), the temperature-ascending peak being endothermic. There is therefore no doubt about the latent heat origin of the peak. All other experimental facts described in detail elsewhere [15] fall in line with this conclusion. Particularly, the value of the latent heat of the phase transition found by us corresponds exactly to the area of the reported [13] λ-peak.

The results of the above-described test with DSC seem to be quite representative. With conventional calorimetry the same peak was found to be positive in both temperature-ascending and descending runs; no doubt has ever been expressed about its heat capacity origin. This fact clearly shows that the commonly used calorimetry is incapable of detecting and eliminating latent heat effects. This suggests that numerous other specific heat anomalies should be re-examined, preferably with high quality modern DSC instruments.

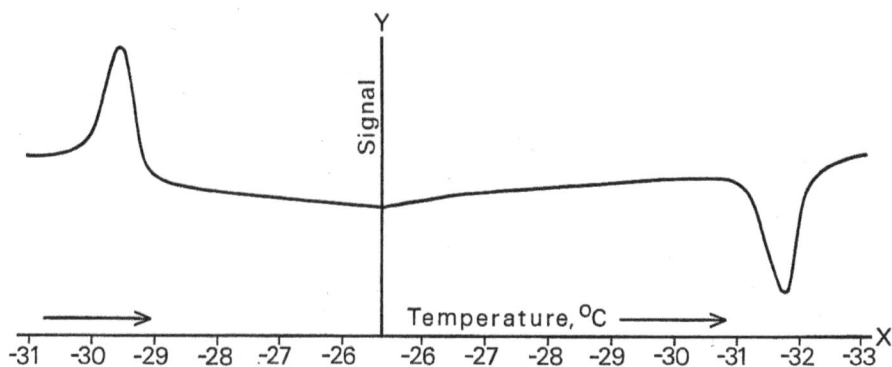

Fig. 3   An actual DSC recording of NH4Cl phase transition cycle, matching the Fig. 2b, rather than the Fig. 2a case, thus proving a latent heat origin of the peak.

REFERENCES

1. P. Schwartz, *Phys.Rev.B* **4**, 920 (1971).
2. H. Chihara and M. Nakamura, *Bul.Chem.Soc.Jap.* **45**, 133 (1972).
3. D.L. Connelly, J.S. Loomis, and D.E. Mapother, *Phys.Rev.B* **3**, 924 (1971).
4. W.E. Maher and W.D. McCormick, *Phys.Rev.* **183**, 573 (1969).
5. Y. Mnyukh, *Mol.Crist.Liq.Crist.* **52**, 163 (1979).
6. D.P. Shoemaker, C.W. Garland, and J.I.Steinfeld, Experiments in Physical Chemistry, 4th Ed. (McGraw-Hill, 1981).
7. E.F. Westrum and J.P. McCullough, *in* Physics and Chemistry of the Organic Solid State, v.1 (Wiley, New York 1965).
8. R. Extermann and J. Weigle, *Helv.Phys.Acta* **15**, 455 (1942).
9. J.L. McNaughton and C.T.Mortimer, Differential Scanning Calorimetry (Perkin-Elmer, 1975).
10. F. Simon, *Ann.Phys.* **68**, 241 (1922).
11. F. Simon, C.V.Simson, and M. Ruhemann, *Z.Phys.Chem.* **A129**, 339 (1927).
12. W.T. Ziegler and C.E. Messer, *J.Am.Chem.Soc.* **63**, 2694 (1941).
13. V. Voronel and S.R. Garber, *Sov.Phys.JETP* **25**, 970 (1967).
14. J.E. Callanan, R.D. Weir, and L.A.K. Staveley, *Proc.R.Soc.Lond.A* **372**, 489 (1980); **372**, 497 (1980); **375**, 351 (1981).
15. Y. Mnyukh, *Sp.Sci.Tech.* **6**, 275 (1983).

APPENDIX 3:   POLEMIC WITH AN ANONYMOUS OPPONENT

A3.1   *Introduction*

The concept of this book (put initially forward in a number of scientific articles) has not been openly challenged in the scientific press. The purpose of the present appendix is to make the reader acquainted with the type of the objections which the workers involved in the recording or treatment of the λ-peaks are able to produce. The present opportunity resulted from the fact that the content of Appendix 2 was initially submitted to the *Review of Scientific Instruments* journal as an article entitled "A Previously Unnoticed Feature of Differential Scanning Calorimetry." Given below is the adverse report on this manuscript by an anonymous reviewer, as well as the present author's reply sent to the editor. The scientifically illiterate reviewer report prevailed: the article was rejected.

A3.2   *The reviewer's report*

The paper entitled "A Previously Unnoticed Feature of Differential Scanning Calorimetry" is not appropriate for publication in the Review of Scientific Instruments because it represents neither a new idea in instrumentation nor a significant improvement in a familiar device and its application. It purports to suggest a new application for an instrument; however, the mode of operation of the instrument the author suggests is well-known and the application suggested is technically unsound. Three elements of the paper considered to be technically unsound will be discussed here; it does not necessary to be exhaustive.

The first concerns the nature of transitions. Very few transitions are considered to be wholly discontinuous (first-order) or wholly continuous (gradual, second-order). A great number of lambda transitions (gradual) end in a manner associated with a first-order transition. Conversely, first-order transitions that begin abruptly, without some degree of gradual transition leading up to the isothermal transition, are rare and considered to be remarkable. Hysteresis has historically been associated with first-order transitions; indeed it is the experimental observation of hysteresis which has often led to the identification of the partial first-order nature of a lambda transition. As most transitions are recognized to be of mixed order, they will frequently display some degree of thermal hysteresis.

| The author distinguishes between a lambda transition, which he
| refers to as a specific heat anomaly or apparent specific heat peak,
1 | and a phase transition or latent heat. Both of these effects are in
| fact transitions and manifest themselves in the calorimetric
| measurements. Though he says clear evidence is presented that no
anomaly exists in ammonium chloride, no such evidence is presented,
nor should one expect it to be.

| The purpose of the careful delineation of a transition is not to
2 | determine it's exact shape, as indicated by the author, but to
| determine the enthalpy of the transition as accurately as possible in
| order to obtain more reliable thermodynamic data. The author also
criticizes the procedures followed to ensure the attainment of equilibrium
after an energy input in adiabatic calorimetry. These procedures are not
related to elapsed time; the concern is the return of the system to the
same conditions of drift that existed before the energy input. For
substances which exhibit exceedingly sluggish transitions, where there
is no certainty that a return to an equilibrium state has been attained,
the error caused by the non-attainment is estimated to be less than 0.3
percent.

The second serious technical shortcoming concerns the nature
and use of adiabatic calorimeters and differential scanning calorimeters
(DSC). An adiabatic calorimeter is designed and constructed to be used
for precise work only in a heating mode; it is generally used in a cooling
mode in a qualitative sense only. A scanning calorimeter can be used in
both modes and because of its construction shows endothermic and
exothermic effects in opposite directions. It is unreasonable to expect
an adiabatic calorimeter, which is constructed differently, to behave
similarly. An adiabatic calorimeter used in the intermittent mode has an
accuracy and precision at least an order of magnitude or more greater
than that attainable with a DSC under the best of conditions. An
inherently less accurate instrument, which has none of the possibilities
for control that an adiabatic calorimeter does, can hardly be expected to
give more reliable results. There is no advantage to determining the
enthalpy of transition in a cooling mode rather than a heating mode.

| In the last paragraph the author proposed using the DSC to
| eliminate the isothermal part of the transition. He does not indicate
3 | why it is desirable to eliminate it.   If one is trying to assess the
| thermodynamic properties of the system, the elimination of any
| contribution to these properties is senseless and incorrect.
Furthermore, the peak area obtained with the cooling mode of the DSC

should be exactly equal in magnitude, though opposite in sign, to that
| obtained in the heating mode. I fail to see how obtaining the
4 | enthalpy in a cooling run does eliminate the lambda peak from
| consideration.

A third serious error lies in the treatment of the order-disorder transitions. It is probable that this is the most studied of all lambda transitions. It has long been recognized that this transition has an isothermal component as indicated in references cited by the author, among others. The statement by the author in the abstract that there is no "specific heat anomaly" associated with the order-disorder transition of ammonium chloride is not supported by the extensive literature dealing with this transition or by the evidence presented in this paper.

Rather than document each of the criticisms specified in this review, I should like to note in particular the papers by Schwartz, Frederick, and co-workers at Illinois, those by Garland and his colleagues at MIT, and two recent books. In "Disorder in Crystals by N.G. Parsonage and L. A. K. Staveley, there are detailed and well-documented discussions of the nature and classification of transitions, including a thorough discussion of magnetic and thermal hysteresis; in another section the extensive discussion of transitions on the ammonium halides existing in the literature is summarized. "The Molten State of Matter" by A. R. Ubbelohde has a chapter on solid-solid transitions that treats the theory of transitions and hysteresis.

A3.3   The author's reply

It is shown in my manuscript submitted to the Review of Scientific Instruments that many "specific heat λ-anomalies," if not all, are most probably pseudo-anomalies, and that DSC techniques can serve as a "truth test detector." Neither the former nor the latter are known to those numerous workers, worldwide, whose activities are related to producing and/or treatment of the λ-peaks. After a careful analysis of the adverse referee report, I found that none of the arguments given by the reviewer represent a serious challenge or criticism to the correctness of my work. This is demonstrated below.

The main difficulty with the reviewer's report is, on my opinion, that it is written in a particular loose style in which definite scientific terms are replaced by more ambiguous ones. Consequently distinctions between different scientific notions become difficult to discern. Upon substitution of the ambiguous terms by the exact ones, it becomes apparent, as shown below, that the reviewer is in error.

- 286 -

In the text which follows, the following abbreviations have been used: C_p = heat capacity or specific heat, Q = latent heat of phase transition, T = temperature. The numbers in brackets represent the quotations in the report, which I have marked for convenience.

(1) *"The author distinguishes between a lambda transition which he refers to as a specific heat anomaly or apparent specific heat peak, and a phase transition or latent heat. Both of these effects are in fact transitions and manifest themselves in the calorimetric measurements."* This excerpt from the reviewer report I consider first.

To begin with, I did not suggest such a classification of phase transitions; also, it is not appropriate to identify a *transition*, which is a *phenomenon*, with one of many its specific *parameters*, such as C_p or Q. I do distinguish between *specific heat* C_p and *latent heat* Q, in accordance with thermodynamics. There is absolutely no reason for blending these two concepts and presenting them under the amalgamated name "transitions." In thermodynamics such a value or function is unknown. All these λ-peaks are present not in the plots of "transitions vs. T," but they are presented as *specific heat* peaks in the plots of $C_p(T)$. Since Q is not C_p, only C_p must appear in plots of C_p! All authors without a single exception, measured and published their "specific heat λ-peaks" sincerely believing that a *true heat capacity and only a heat capacity* manifested itself in their calorimetric measurements. The theory of critical phenomena always considers the λ-peaks as strictly C_p-peaks.

(2) Another characteristic example: *"The purpose of the careful delineation of a transition is not to determine its exact shape, as indicated by the author, but to determine the enthalpy of the transition as accurately as possible in order to obtain more reliable thermodynamic data."*

This statement may seem to be a worthy criticism, but it will be shown below that it is incorrect in every respect:
(a) the first part is meaningless;
(b) it questions, without foundation, the text which is correct;
(c) it is mistaken on the concept of enthalpy of the transition;
(d) the common purpose for finding exact shape of the λ-peak is erroneously indicated.

The only meaning of the obscure words "delineation of transition" can be "determination of the exact shape of $C_p(T)$" since only the latter is under discussion. Upon substitution into the reviewer's text, we obtain "The purpose of the careful determination of the exact shape of $C_p(T)$ curve is not to determine its exact shape." This is a tautology of two mutually opposite statements.

Yet, we compare this with the statements made in original papers (see "References," in the manuscript):

Ref. 2. Title of the article: "Heat Capacity of Ammonium Chloride Between 8 and 300 K." Purpose of the work: "The earlier measurements...were not sufficiently closely spaced to reveal the shape of the heat capacity anomaly in the vicinity of T_c. We have undertaken the precision measurements of the heat capacity..."

Ref. 3. Title: "Specific Heat of Nickel Near the Curie Temperature." Purpose: "The results reported here are believed to give a more detailed picture of the form of the singularity in $C_p(T)$ for Ni than any of the previously reported investigations."

Thus, there was no reason for challenging my statement "One of the goals of contemporary calorimetry is to determine with the greatest possible precision the shape of specific heat anomalies, such as the "lambda-peaks" in the temperature ranges of phase transitions. It is also evident that it is a *specific heat* (heat capacity) which the measurements in the above cited works were concerned with.

The application of the term "enthalpy of the transition" to λ-transitions is a serious mistake. In thermodynamics enthalpy of phase transitions is known only as *latent* heat of first-order phase transitions; it is assumed to be isothermally absorbed or released heat which is not related to the local shape of $C_p(T)$. λ-Transitions have always been associated with second- or "higher-" order phase transitions of which a major property is Q = 0. Nonetheless the reviewer used the term "the enthalpy of the transition" without indication of its meaning. In order to clarify it, we do not need to go deeply into theoretical physics. Indeed, even if we would accepted the reviewer's point that a λ-transition is combined of first- and second-order transitions, we have to conclude the following: the contribution of the first-order transition is a *latent heat* Q which cannot be present in the $C_p(T)$ plots, while the contribution of the second-order transition is zero. Thus, the "careful delineation of a transition" has nothing to do with enthalpy of transitions known in thermodynamics. I can only speculate that the reviewer may have meant the area under a C_p - λ-peak, which some (only few) authors have calculated from their data, but in fact could not utilize, because it was thermodynamically unsound. It will acquire however a clear meaning after realization that the whole peak is a pseudo heat capacity resulting from the *latent heat* Q.

As is well known, the detailed determinations of the shape of the $C_p(T)$-λ-peaks were aimed at comparison with the theory of critical phenomena.

> (3) *"...The author proposed using the DSC to eliminate the isothermal part of the transition. He does not indicate why it is desirable to eliminate this. If one is trying to assess the thermodynamic properties of the system, the elimination of any contribution to these properties is senseless and incorrect."*

This assertion, being undoubtedly the central reviewer's point, is especially indicative of both his ambiguous style and erroneous approach to the problem. As everywhere in the report, the standard exact notions are abandoned in favor of vague ones, such as "isothermal part of the transition" (should be "latent heat"), "thermodynamic properties" (should be "specific heat"), and "any contribution" (should be "the contribution of the latent heat"). After substitution by the above definite terms the reviewer's text becomes self-explanatory and reads as follows:

"...The author proposed using the DSC to eliminate the latent heat Q from the plots of the specific heat $C_p(T)$. He does not indicate why it is desirable to eliminate this. If one is trying to measure $C_p(T)$, the elimination of the latent heat Q is senseless and incorrect"

From this an astonishing fact becomes evident that the reviewer is actually *suggesting* that the latent heat Q should be included in $C_p(T)$ plots. In fact, since previously both Q and C_p were *inadvertently* measured together, much confusion has been created in the field. It is time for these two quantities be measured separately.

> (4) The final reviewer's statement which is discussed here reads as follows: *"I fail to see how obtaining the enthalpy in a cooling run does eliminate the lambda peak from consideration."*

I must admit, in turn, that I fail to see a meaning in this sentence even after I replaced the incorrect "enthalpy" by the "enthalpy of phase transition," and the obscure "lambda- peak" by the "specific heat λ-peak." Why in the temperature-ascending run the peak in question is called *(specific heat) "λ-peak,"* while in the reverse run the same peak changes its nature to become *"enthalpy" (of phase transition)*? I can only conclude that the reviewer has failed to understand the main idea of my manuscript. Now I describe this idea in a shortened and simplified form.

A fundamental property of a specific heat C_p is that C_p is a unique function of T. It follows that any *true* $C_p(T)$ curve recorded with any calorimeter in a temperature-ascending run will be reproduced in the reverse run within the experimental error of the instrument. The shape of $C_p(T)$ curves cannot depend on the direction in which the temperature is varied: a *true specific heat peak* in the $C_p(T)$ curve should be, evidently, reproduced. The temperature descending run is a test: if a DSC instrument does not trace the same positive λ-peak and, instead, drastically changes its path to show the alleged "heat capacity" peak facing downward (and even protruding far into the area of negative (!) heat capacities), this is undeniable evidence that the peaks do not represent specific heats. Consequently, these peaks must be removed from the $C_p(T)$ plots.

In summary, after analyzing the comments of the reviewer, I find that changes in my manuscript are not necessary. After the foregoing examples of amazing scientific illiteracy of the reviewer, there is hardly any need to discuss his other arguments.

APPENDIX 4: ON STATUS OF THEORY OF FERROMAGNETISM AND FERROELECTRICITY AS STATED BY EXPERTS IN THE FIELD

S.V. Vonsovskii
["Magnetism," vol 1 & 2, Wiley, 1974]

[On magnetism:] "At first glance it might be assumed that everything has already been resolved in this field of physical science, and that the physical picture of the phenomena has been completed in every detail... However, magnetism has so far remained an intensively developing field of physics... in which new trends constantly appearing on the scene" (p.3).

B.I. Bleaney & B. Bleaney
["Electricity and Magnetism," Oxford, Clarendon Press, 1963]

"There is no doubt that ferromagnetism is due to the exchange forces first discovered by Heisenberg, but the quantitative theory of ferromagnetism contains many difficulties..." (p.601).

"...We have a broad understanding of the outlines of ferromagnetic theory, but not of the details. The exchange interaction between two electrons cannot be calculated *a priori* ... We cannot even be certain of its sign, and in this connection we may mention a suggestion of Zener... Zener's theory has not been generally accepted, and as yet there is little experimental evidence by which it can be tested" (p.606).

[On Curie-Weiss law:] "...For many substances no single equation represents the susceptibility variation adequately over a wide temperature range" (p.190).

K.P. Belov
["Magnetic Transitions," Boston Tech. Publ., 1965]

"...Certain extremely important problems associated with [ferromagnetic] phenomena, discovered long ago, have as yet, in spite of their importance, been studied very little" (p.X).

"...Many important questions connected with the behavior of materials in the region [of ferromagnetic transition] remain unsettled or in dispute to the present time. These include ...the actual temperature behavior of the spontaneous magnetization near the Curie point, the causes of the 'smearing out' of the magnetic transition... the existence of 'residual' spontaneous magnetization above the Curie temperature, and the nature of the temperature dependence of elastic, electric, thermal, and other properties near the Curie point. It even remains unsettled what we should take to be the Curie temperature, and how to determine it."

"The theory of Weiss and Heisenberg cannot be applied to the quantitative description of phenomena in the neighborhood of the Curie point... Even for such a 'simple' ferromagnetic substance as nickel it is not possible to 'squeeze' the experimental results into the Weiss-Heisenberg theory" (p.XII).

The New Encyclopedia Britannica
[(30th Ed.), v.6, 1979, p.560]

"In general, ferroelectricity is not well understood, and explanations developed for one particular type of crystal (e.g. $BaTiO_3$) fail when applied to another (e.g. KH_2PO_4)."

T. Mitsui, I. Tatsuzaki and E. Nakamura
["An Introduction to the Physics of Ferroelectrics," Gordon & Breach, 1976]

"...It seems to be practically impossible to write down a truly systematic theory of ferroelectrics today."

"It is not easy to answer the question why domain structure occurs in ferroelectrics... We have no reason for a domain structure to occur in ferroelectrics."

I. S. Zheludev
["Principles of Ferroelectricity," Atomizdat, Moscow (1973)]

"Unfortunately, thus far the nature of ferroelectric properties cannot be regarded as quite elucidated and this makes classification of ferroelectrics in terms of their nature impossible."

A. Hubert
["Theorie der Domanenwande in Geordneten Medien," Springer-Verlag, 1974]

[On the origin of domain structures in antiferromagnetics:] "While experimental observation of antiferromagnetic [domain] structures is common, the reason for their occurrence is still rather elusive. As opposed to the case of ferromagnetic domains, they cannot lower the energy of the leakage field, because the resultant magnetic moment of an antiferromagnetic in the ground state must be zero on definition."

"Still, all the above-mentioned phenomena does not allow us to understand the origin of domain structures observed in the absence of magnetic field in multiaxial antiferromagnetics."

C. Kittel
["Introduction to Solid State Physics," 4th ed., Wiley, 1971]

[On critical temperature of antiferromagnetic phase transitions] "The Neel temperatures T_N often vary considerably between samples, and in some cases there is large thermal hysteresis."

C. Kittel and J.K. Galt
["Ferromagnetic Domain Theory," *in* "Solid State Physics," Academic Press, 1956]

[On the coercive force and hysteresis:] "Several theories have been advanced (Becker, 1932; Kondorsky, 1937, 1949; Kersten, 1938, 1943; Neel, 1946; Stoner and Wohlfarth, 1947, 1948; Goodenough, 1954) and a certain amount of progress has been made, although the problem is beset with the usual difficulty in explaining any material property which is highly structure sensitive..."

"...Much more to be done to make possible a theory of coercive force grounded in experimental fact."

R. M. Bozorth
["Ferromagnetism," D. Van Nostrand Co., New York, 1951]

[On predictions of the Weiss theory:] "The data for iron and for nickel [at low temperatures] show that the Weiss theory in either its original or modified form is quite inadequate."

"The Curie point is not always defined in accordance with the Weiss theory but in other more empirical ways..."

J. Crangle
["The Magnetic Properties of Solids," Edward Arnold, London 1977]

"It seems difficult to be convinced that direct exchange between localized electrons can be the main origin of the ferromagnetism in metals of the iron group."

R. P. Feynman
[R. P. Feynman, R. B. Leighton, M. Sands: The Feynman Lectures on physics,
v.2, Addison-Wesley 1964]

"When it was clear that quantum mechanics could supply a tremendous spin-orienting force - even if, apparently, of the wrong sign - it was suggested that ferromagnetism might have its origin in this same force, that due to the complexities of iron and the large number of electrons involved, the sign of the interaction energy would come out the other way around."

"The most recent calculations of the energy between the two electron spins in iron... still give the *wrong sign*."

"Even the quantum theory deviates from the observed behavior at both high and low temperatures."

"But the exact behavior near the Curie point has never been thoroughly figured out. That's an interesting problem to work out some day if you want a problem that has never been solved."

"One of the challenges of theoretical physics today is to find an exact theoretical description of the character of the specific heat near the Curie transition - an intriguing problem which has not yet been solved. Naturally, this problem is very closely related to the shape of the magnetization curve in the same region."

"...To the theoretical physicists, ferromagnetism presents a number of very interesting, unsolved, and beautiful challenges. One challenge is to understand why it exists at all."

"...The theory of the sudden transition at the Curie point still needs to be completed."

APPENDIX 5: REVIEW ON "LIGHT SCATTERING NEAR PHASE TRANSITIONS"

(Ed. H.Z. Cummins and A.P. Levanyuk, North-Holland, 1983 - Volume 5 of "Modern Problems in Condensed Matter Sciences")

Note: This unpublished review was written in 1988. It was now subjected to a minor editing; also references on particular sections of the present book were added. It is still very much relevant to the issues of the day. Besides, it is instructive, illustrating what quite a few theoretical physicists do when the found solution of a problem does not fit their theoretical constructions.

As demonstrated in Sec. 3.6, the cause of the light scattering "central peaks" in the region of crystal phase transitions is trivial: the light is scattered by the nuclei and interfaces during this temporary two-phase state. This scattering depends on the morphology of nucleation and growth in the given transition. For example, the scattering in a one-centered and multi-centered transition, or in an epitaxial and non-epitaxial transition will not be the same. One frequent kind of the morphology was described in Sec. 2.9.6 and shown in Fig. 2.1. A cloudy two-phase band moves through the 'i' single crystal upon uniform heating. If light scattering was measured from a small crystal area, a "central peak" would undoubtedly be recorded when the band passes through it.

In terms of these considerations, the voluminous collective work *Light Scattering Near Phase Transitions* of 666 pages was not justified. Yet, it helps us to demonstrate the impasse that the *critical phenomena* concept has reached. On one hand, this volume is filled with analyses of the critical fluctuations that allegedly take place near phase transitions and have to manifest themselves in an increased light scattering. On the other hand, this volume is an expression of a collective frustration over the fact that no evidence of such effects has been found.

We start from a remark that the title itself is in error. No change in light scattering takes place *near* phase transition. The first change occurs with the appearance of the first nuclei of the new phase after the equilibrium temperature T_0 has been passed.

About 80% of the book is devoted to solid-state phase transitions, the subject of our discussion. A large fragment (125 pages) of it is Chapter 1 "General Theory of Light Scattering Near Phase

Transitions in Ideal Crystals" by Ginzburg, Sobyanin and Levanyuk (GSL). Considering the surprising admissions made by GSL in the beginning, a serious doubt arises whether the theory in question had to be developed. GSL apparently felt (and tried to circumvent) the flaws of the premises that their theory rested on. Let us turn to these premises.

(1) GSL state: "Thermal fluctuations of certain physical quantities and, first of all, those of the order parameter increase near second-order phase transitions." On the other hand, they admit that "second-order phase transitions in solids could be expected to occur, strictly speaking, only in some exceptional cases." This contradiction is reconciled in the usual way: the Landau theory of second-order transitions is applicable to "weak" first-order transitions, such as in NH_4Cl. But besides few "weak first-order" transitions there are plenty of "strong" ones to which their theory, thus, must be inapplicable. Like many other authors, GSL simply *postulate* that the "weak first-order" transitions can be treated as second-order. They argue that such transitions have very small "jumps in the thermodynamic quantities" and "temperature hysteresis effects." The fundamental distinction between first- and second-order phase transitions, which these authors were unaware of, is whether the process is cooperative in the bulk or a local rearrangement at interfaces between the coexisting phases. This distinction is valid toward *any* first-order transition, however "weak" it is deemed. GSL view first-order phase transitions differently. They believe that "inner deformation" is responsible for "those first-order phase transitions which are nearly second-order," choosing this "inner deformation" as the definition of the "order parameter" - a key parameter of their theory.

Thus, the theory by GSL is unconditionally inapplicable to most crystal phase transitions because they are "strong" first-order. It is inapplicable to a few remaining phase transitions thought to be "weak first-order" because they are not different from "strong" ones. It is inapplicable to "pure" second-order transitions because they are nonexistent. No phase transitions are left to which their theory can be applied.

(2) As is well known, "critical fluctuations" were initially introduced in order to account for the apparently singular behavior of crystals in a region near the critical point T_c. Now GSL acknowledge that "no critical region has been yet observed with confidence" in solid-state phase transitions with the exception of "critical points in magnetics." Considering that the exception made for magnetics is not valid (Sec. 4.2.2), the admission is of special value because the authors

would prefer the opposite to be the case. In an attempt to circumvent this truly "critical" predicament, "the main attention is paid to the temperature region where the Landau phase transition theory is applicable (it is often referred to as the classical region), although some specific features of light scattering in critical (scaling) region are discussed too." In spite of "the absence of evidence at the present time for the existence of a critical region for structural phase transitions," GSL maintain that "the critical region is often very narrow." Such a definitive style in regard to a phenomenon which no one ever detected is typical for the whole literature on the topic.

GSL do not assert any more that the increase in light scattering due to critical fluctuations must be as great as 10^4 times. Now they maintain that the effect is extremely weak and has to be extracted from the "integral intensity." As to the wider "classical region," there is no evidence as well that there are any thermal fluctuations specifically related to a phase transition. GSL lay the blame on poor experimental data: "The existing experimental data on the integral intensity of the thermal (molecular) light scattering near phase transition points in solids are neither reliable nor full enough to be compared with the above theoretical predictions." It turns out that the theory under consideration was not intended to account for the observed phenomena. Rather, it *postulates* one unobserved phenomenon (specific thermal fluctuations) to *predict* another unobserved phenomenon (singular light scattering). This does not seem rational, considering that the really observed "central peaks" remained unexplained by GSL. The origin of these peaks is almost self-evident (see Sec. 3.6); rather, the issue is to comprehend how it was possible to skirt around the experimentally proven light scattering by nuclei and interfaces in the quartz, NH_4Cl, KDP and other phase transitions.

Chapter 2 "Light Scattering Anomalies due to Defects" (fourty pages) by Levanyuk, Sigov and Sobyanin (LSS) is apparently aimed at further development of the theory expounded in Chapter 1. It is asserted in the introduction that the facts regarding anomalous light scattering near phase transitions can be accounted for after crystal defects are taken into account. Which facts? "...That the 'critical opalescence'...near the transition in quartz is mainly no more than elastic scattering, *i.e.*, it is not connected with thermal fluctuations of the order parameter." Which defects? The following are considered:

- Interstitial ion;
- Interstitial impurity;
- Substitutional impurity ion;
- Concentration fluctuations of the above defects;
- Space-inhomogeneous stresses;
- Static local lattice distortions;
- "Hopping defects."

Where are, however, the nuclei, interfaces and a heterophase structure in a range of transition, all of which have been experimentally identified as the actual cause of the central peaks? The possibility of such a source of light scattering was mentioned, but not included into the analysis. Why was the correct solution put forward by Bartis (see Sec. 3.6.4) ten years earlier not given a due consideration? No reason was given; his work was not alluded to.

The light scattering by heterophase matter has been abandoned in favor of thermal fluctuations and crystal defects. Since neither fluctuations nor defects alone give rise to the central peaks, the theory assumes that their coupling does. This idea, however improbable, allowed the fluctuation concept to be retained. It should be remarked that the defects in question permanently reside in the crystal and scatter light (if they are not too small to do) at all temperatures, and not the defects induced by the interface motion. There is no need to consider the theory itself - only its premises. First-order phase transitions were assumed to be second-order; a continuous-medium approximation was applied in spite of a heterophase structure of the scattering matter; the concept of critical point T_c was used in spite of the existence of a heterophase temperature range. This list of wrong premises is sufficient for the theory to not be taken seriously.

The comparison with the experimental data has turned out to be disappointing: "The results obtained do not correspond..."; "At present it is hard to speak about an agreement..."; "A disagreement between theory and experiment still exists"; "We are not able to list direct experimental corroborations of the 'dynamic' peak existence." Since the theory was unable to account for the actually observed light scattering, in particular in NH_4Cl and quartz, these cases were commented on in a quite speculative manner: "In many real crystals the defects are inhomogeneously distributed throughout a sample owing to irregular nonisotropic growth of a crystal. It is possible in particular that peculiarities of light scattering in NH_4Cl and quartz are caused just by macroscopic imperfections in their structure." How can the established role of a heterophase structure in the light scattering in these phase transitions [161,182] be so openly ignored? A scientific article can be correct or incorrect, but it must not leave the reader misinformed.

Although no correlation was found between the theory and experiment, LSS contend, but without supporting arguments, that there are plenty of different mechanisms contributing to the integral intensity of the central peaks and that the main contribution is due to elastic scattering by fluctuations in the defect concentration. In the end they recommend that joint efforts of both experimentalists and theoreticians are to be undertaken to study the anomalies of light scattering. The degree of their rejection of the already experimentally proven cause of the central peaks is beyond reasonable limits.

Chapter 7 (55 pages) "Central Peaks near Structural Phase Transitions" was written by Fleury and Lyons (FL). They write about the central peaks in an emphatic manner: "the most striking feature of the scattered spectrum," "striking increase in the light scattered near transition," "puzzle," "question," "problem." They refer to the failure of the soft mode theory to account for the central peaks and infer from this that the initial expectation of a single explanation of all central peaks was inappropriate. This argument is flawed; the explanation does not exist on the basis of the fluctuation concept, but it already existed: the light is scattered by nuclei and interfaces.

Let us now recall that according to the authoritative admission by Ginzburg, Levanyuk and others no effect of "critical" or "thermal" fluctuations in the light scattering near phase transition has been found. On the other hand, a heterophase structure was the source that has been positively identified by several authors. Never mind: FL write affirmatively: "There are a number of mechanisms responsible for central peak components in various frequency domains, and often two or more of these are active in a single material at the same time." The following mechanisms are listed:

- Entropy fluctuations;
- Phonon density fluctuations;
- Dielectric relaxations;
- Molecular reorientations;
- Phasons;
- Over-damped soft-modes;
- Solitons;
- Degenerate electronic transitions;

- Dynamic and static clusters and domains;
- A host of phenomena related to:
  - impurities
  - vacancies,
  - strains,
  - dislocations, etc.

The real cause and its experimental demonstrations (see Sec. 3.6) were not mentioned. Such a consistent disregard of scientific facts by theoretical physicists may only be explained by the intention to defend their failed dynamic theories at any cost.

FL pose a question: "Whether the central peaks observed reflected the intrinsic critical dynamics of the phase transitions?." They do not answer this question outright, but, in fact, do their best to answer it positively in the process of consideration of the following seven "prototypical systems":

- Cooperative Jahn-Teller transition;
- Order-disorder ferroelectric;
- Displacive transition;
- Plastic crystal;
- Superionic conductor;
- Incommensurate phase transition;
- Defected crystal.

There is, however, a system "prototypical" to all the above: nucleation-growth transition over a temperature range of two-phase coexistence. This system accounts for all central peaks, which the authors failed to consider. As a result, they were unable to explain central peaks, as seen from the following unfounded conjecture: "Apparently static components, presumably related to frozen defects, have been observed in lead germanate, KDP, and α - β transition in SiO_2." Whether it was their unawareness about experimental facts, or unwillingness to accept them, the conclusion by FL that "any vestige of the notion that central peaks can be attributed to a single phenomenon has been clearly laid to rest" was wrong.

The last chapter "Light Scattering in Quartz and Ammonium Chloride..."(30 pages) was written by Yakovlev and Shustin (YS), the first experimentalists who observed central peaks, respectively, in quartz and NH_4Cl. Initially YS had attributed the phenomenon to "critical opalescence." Now they refuse to recognize that it was caused simply by a heterophase state of the crystals. They describe their own experiments in great detail, some subsequent experiments in less detail, and the crucial evidence (such as the light scattering due to the coexistence of two phases in quartz) in the least detail. YS are certainly unaware that a heterophase structure appears in every first-order phase transition. In a direct contradiction with the available evidence they conclude: "Thus, the state of the [quartz] crystal in the vicinity of the phase transition cannot be considered as a heterophase state of coexistence of and phases." They believe that the increase of light scattering in the first-order phase transitions in crystals takes place because they are "close to second order." Quite the opposite is true, however. The central peaks arise exactly owing to a "first-order" nature of phase transitions, namely, such their features as nuclei, interfaces and two-phase transition range. The fact that two-phase temperature range in these phase transitions is in practice rather narrow does not make them "less first order": they proceed at interfaces, and not homogeneously. But YS maintain that "the effect is a more complicated process than was expected" and suggest that "new experiments and also the theoretical developments are clearly needed for a complete explanation of the phenomenon."

Such is the instructive story of the "critical opalescence" in solid-state phase transitions. Even after the error has been discovered, the experimentally established source of the "central peaks" is not generally recognized owing to the resistance of the authors of the insolvent theories. The search for the solution to an already solved problem is continuing...

ADDENDUM A: THE THEORY OF PHASE TRANSITIONS: HISTORY OF DISTORTIONS AND IGNORANCE

Physicists in the beginning of the twentieth century knew that phase transitions in solid state are not "continuous" in their nature. L. Landau, the original creator of the theory of second-order (continuous) phase transitions, also recognized that fact, but suggested that second-order phase transitions "*may also exist.*" Landau defined second-order phase transitions as the antithesis of first-order, describing the latter as a "jump-like rearrangement of the crystal lattice," at which latent heat is absorbed or released, the symmetry of the new phase is not related to the original phase, and overheating or overcooling is possible. As for second-order phase transitions, they occur homogeneously, without any overheating or overcooling, at "critical points" where only the crystal symmetry changes, but structural change is infinitesimal. Landau left no doubt that his theory is that of *second-order* phase transitions only.

Leaving alone the theory itself, there were several shortcomings in the Landau presentation:
- He had not answered the arguments of some contemporaries, Max von Laue among them, that second-order phase transitions cannot exist;
- The only examples he used to illustrate second-order phase transitions, NH_4Cl and $BaTiO_3$, both turned out to be first order;
- Description of first-order phase transitions left a false impression that the "jump-like" changes occur simultaneously over the bulk;
- It was not specified that the only way first-order phase transitions can occur is *nucleation and growth* (Sec. 1.2);
- He remained silent when other theorists began to further "develop" his theory by violating and distorting the conditions separating the two antipodal types of phase transitions.

In order to properly evaluate the ensuing chain of events, we need to expand on Landau's characterization of a first-order phase transition (Chapter 2): it is an intrinsically *local* process which has nothing to do with bulk "critical fluctuations" bringing about an instant overall change at a fixed "critical point"; it starts from nucleation in a crystal defect; nucleation temperature is not a "critical point" and not exactly reproducible; nucleation lags are inevitable (hysteresis); it proceeds by "molecule-by-molecule" rearrangements at interfaces over a temperature range where the two phases coexist. The underlying conclusions are impeccable:

(1) *There are no first-order phase transitions that are "weak," "almost," or "close to" second-order, for they occur by rearrangement at interfaces, and not homogeneously in the bulk*

(2) *In no way can first-order phase transitions be described or approximated as "continuous" and "critical phenomenon" in order to become subject to statistical mechanics.*

The Landau theory initiated an avalanche of theoretical papers and books, that were presented not as a "theory of *second-order* phase transitions," but as a "theory of phase transitions"; first-order phase transitions were either incorporated as a "critical phenomenon" or simply disregarded. No attention was paid to Landau's actual position. The available new data that *any* solid-state phase transition, when carefully investigated, turned out to be *first order* were ignored. Every ferroelectric phase transition turned out of first order (Sec. 4.2.1). Every ferromagnetic phase transition, including in Fe, Ni and Co, turned out of first order (Sec. 4.2.2, 4.2.3, 4.4, 2.6.8). Every "order-disorder" phase transition turned out of first order (Sec. 2.7). Even magnetization proceeds as a first-order phase transition (Sec. 4.13). While there are still claims that one or another phase transition is of second order, not a single well-proven case exists.

The next theoretical step was the "scaling renormalization group" theory of the 1970's, completely devoid of first-order phase transitions. The understanding that existed earlier that phase transitions are "usually" first order, while second order only "may exist," has vanished. It is not to say that the theory in question was necessarily not good: it may have other applications, but it has nothing to do with phase transitions - at least in solid state. Nucleation-and-growth does not need to be described by a "scaling" theory.

The theoretical work did not stop there. As one author recently stated, "the scaling theory of critical phenomena has been successfully extended for classical first order transitions..." [M. A. Continentino, cond-mat/0403274]. To any one who understands the real physical nature of first-order phase transitions, and not only as those exhibiting "jumps," it should be evident that such an "extension" cannot be justified (Sec. 1.3).

The next section of the theoretical chain was "quantum phase transitions," put forward in the last decade of the previous century. The theory rests entirely on the wrong premise that solid-state phase transitions are critical phenomenon, rather than occurring by nucleation and growth.

The incorporation of first-order phase transitions into the theory of "quantum" phase transitions followed: once again, nucleation and crystal growth became a homogeneous process and a "critical phenomenon."

As of today, the last link in the chain is the aforementioned article by Continentino, who applied "scaling ideas to quantum first order transitions," thus creating a conglomeration of utmost scientific falsehood.

To make the above picture more complete, two more theoretical branches are to be mentioned, both of which disregard the fact that first-order phase transitions are a process of nucleation and crystal growth. One, "soft-mode" concept (Sec. 1.6, 1.7), is compatible only with instant (cooperative) phase transitions over the bulk (see Addendum G). The other is the theory of "topological" phase transitions which erroneously assumes that polymorphs are somehow related (Sec. 1.4.3, 2.2.2, 2.2.4) and transform into one another by a kind of displacement / deformation (see Addendum C).

Is there another scientific field in modern science where the theory is so totally wrong?

ADDENDUM B: QUANTUM PHASE TRANSITIONS: WRONG CONCEPT

The invention of "quantum phase transitions" was not legitimate, insofar as it was based on a misrepresentation of well established facts. To demonstrate this, the review article "Quantum Phase Transitions" by M. Vojta [cond-mat/0309604] is used here. It is helpful for two reasons. 1) It is very authoritative, for S. Sachdev, who had published the canonical book on quantum phase transitions, "contributed enormously to the writing of this article," and many other authorities also had "illuminating conversations and collaborations." 2) Its author tried to be logical in justifying the new class of phase transitions – and thus made it easier to find where he failed. Our analysis is structured mostly as comments on the Vojta's article.

Excerpt: "The [non-quantum] phase transitions ... occur at finite temperature; here macroscopic order ... is destroyed by thermal fluctuations."
Comment: Those phase transitions are nucleation and growth of a new phase in the solid medium of the original phase, rather than "destroying" the original phase by thermal fluctuations as assumed by "second-order" (critical phenomena) theory. This fact is a major point of the present book.

Excerpt: "[Quantum phase transitions take] place at zero temperature. A non-thermal control parameter such as pressure, magnetic field, or chemical composition, is varied to access the transition point. There, order is destroyed solely by quantum fluctuations."
Comment: Since the previous assumption is wrong, the deduction from it hangs in the air. Vibration energy – including quantum – is only a part of crystal free energy. If changing one of the above control parameters gives rise to a lower free energy, there is no reason for the phase transition not to proceed in the same way at low temperatures as well. In other words, quantum effects may contribute to the free energy of competing phases, but have nothing to do with the mechanism of phase transition: it is still nucleation-growth.

To the number of classifications of solid-state phase transitions, discussed in Chapter 1, one more was added by Vojta: "*classical* -

quantum." How do "quantum" phase transitions differ from "classical?" First, the latter need to be defined.

Excerpt: "[Classical] phase transitions are traditionally classified into first-order and continuous transitions. At first-order transitions the two phases co-exist at the transition temperature – examples are ice and water at 0 °C, or water and steam at 100 °C."

Comment: It is not accidental that the chosen examples of first-order phase transitions are not solid-to-solid, even though "quantum" phase transitions are. The reason becomes evident since all "classical" solid-state transitions are declared "continuous" and a "critical phenomenon," conveniently omitting the fact that almost all of them have already been recognized as first order in experimental literature. Moreover, it is in direct contradiction with L. Landau, the "father" of "continuous" phase transitions. In Statistical Physics by Landau and Lifshitz we read: *"Transition between different crystal modifications occurs usually by phase transition at which jump-like rearrangement of crystal lattice takes place and state of the matter changes abruptly. Along with such jump-like transitions, however, another type of transitions may also exist related to change in symmetry."* Thus, according to Landau, phase transitions between crystal modifications are first order, but "continuous" phase transitions *may* also exist (and, consequently, do not necessarily exist). In reality, they don't exist; the remaining few would be re-classified to first order upon careful reexamination. Landau used only two particular examples of "second-order" phase transitions - in NH_4Cl and $BaTiO_3$ - and both turned out to be first order.

The introduction of "quantum phase transitions" by Vojta can be briefly summarized as follows. All solid-solid "classical" phase transitions occur at critical points in which the previously existing order is destroyed by thermal fluctuations. Not far from 0 °K, the "classical" critical point becomes "quantum" and so does the phase transition.

Comment: The introduction of "quantum phase transitions" as the antithesis to first-order phase transitions and as an extension of "critical" behavior of all other solid-state phase transitions (called "classical") was based on an erroneous premise and, therefore, is not valid. Still, it is useful to extend our dissection somewhat further.

Excerpt: "In contrast, at continuous transitions the two phases do not co-exist. An important example is the ferromagnetic transition of iron at 770 °C, above which the magnetic moment vanishes. This phase transition occurs at a point where thermal fluctuations destroy the regular ordering of magnetic moments – this happens continuously in the sense that the magnetization vanishes continuously when approaching the transition from below. The transition point of a continuous phase transition is also called critical point."

Comment: Ferromagnetic phase transitions have become the last resort for conventional theory to exemplify continuous phase transitions and critical phenomena. Vojta's confusing explanation (magnetization changes continuously at critical point) illustrates that the theory has reached an impasse. Currently, even this last resort has been eliminated by innumerous experimental examples of first order ferromagnetic phase transitions, including those in Fe and Ni. But the introduction of "quantum" phase transitions required ignoring this fact and describing ferromagnetic phase transition in Fe in a quite erroneous way. It ignores evidence [Preston, Refs.192, 193] that the transition is first order (therefore, nucleation-and-growth). Contrary to the Vojta's description, the magnetization curve is not sharp when measured in either direction, leaving the "critical phenomenon" without its "critical point." The "continuous" change of magnetization is due to multiple nucleation over a temperature range of two-phase coexistence. See Chapter 4 for details.

To complete the picture, there were numerous publications, both theoretical and experimental, where certain "quantum" phase transitions were stated to be first order. Who is wrong: those arguing that "quantum" phase transitions are an antithesis to first-order and a "critical phenomenon," or those embracing "first-order quantum phase transitions?" The answer is: all of them are. The experimentalists, who concluded the "quantum" phase transitions they observed were first order, are less erroneous: the transitions are first order indeed. (In some such works the conclusions were based on the observed hysteresis). But they still believe in specific "quantum" phase transitions rather than accepting that they are nucleation and growth.

Conclusion: the concept of "quantum" phase transitions is wrong, as is the description of "classical" solid-state phase transitions by theory of critical phenomena.

ADDENDUM C: NO DISPLACEMENTS IN "DISPLACIVE" AND "TOPOLOGICAL" PHASE TRANSITIONS

"Displacive" phase transitions (PTs) have been misinterpreted: there are no displacements. The idea of "displacive" PTs was put forward by Buerger in the 1950's as a cooperative deformation/distortion of the original structure by *displacements* of atoms/molecules *in* the crystal lattice without breaking their chemical bonding. When such structural modification could not be imagined, the alternative was postulated to be "reconstructive" PTs, but how they can occur remained unknown. The classification was not based on an experimental investigation of the process. Rather, it was assumed from comparisons of the initial and final structures. The comparisons frequently resulted in "hybrid" cases with some bonds being broken; these cases were deemed "displacive" anyway. The underlying specific structural orientation relationship (OR) was not verified.

Presently, the notion "displacive" PTs is loosely used by scientific literature in those cases where the structures of polymorphs seem to be "sufficiently similar." Theorists sometimes claim "displacive" PTs to be second order (simultaneous participation of all particles in the bulk). Experiments usually find them to be first order (a rearrangement at interfaces). Some consider them to occur by a "soft-mode" mechanism (which requires instant cooperative movement of atoms/molecules to their final positions - in contradiction with actual observations). Some combine all these contradictory notions by identifying "martensitic" PTs (first order) not only as "displacive," but even as resulting from soft-modes (see Addendum E). In many instances, especially when dealing with ferroelectrics, "displacive" PTs are placed against "order/disorder," rather than "reconstructive" PTs. This classification is plagued by observations of "crossovers" of the two; at that, "order/disorder" PTs are believed to be a cooperative process, which it is not.

There are PTs, however, plenty of them, that even the most inventive theorists were unable to "squeeze" into the "displacive" category. The molecular mechanism of these "reconstructive" PTs cried for explanation. If not by displacement, then how? The answer was: still by displacement/deformation/distortion of the original structure. In the abstract world of theoretical thinking, not attentive to the available contradicting facts, the "natural" idea of atomic/molecular displacements in the original structure was advanced further by "topological" theories in which "reconstructive" PTs proceed through several intermediate, more or less "displacive," imaginary stages. Even though this must lead to a certain fixed OR, the idea to verify it rarely comes about. What is more, it is becoming difficult to ignore the fact that both "displacive" and "reconstructive" PTs involve *nucleation*; hence the need to reconcile flame (nucleation) and water ("continuum" modeling).

This cumbersome state of affairs has a radical solution, which was initially presented in my experimental articles in scientific journals, and then in this book. It turns out, a transition to another polymorph is like building a new house from the bricks obtained by disassembling the nearby abandoned brick house; the new house may, or may not, resemble the old one: the only feature they share is the construction material. All PTs in crystals (believed to be "displacive," "martensitic," "reconstructive," "order/disorder," or otherwise) proceed exclusively by *nucleation* and (temperature-dependent) *crystal growth*, constituting two inseparable stages of the general molecular mechanism of the PTs. However attractive the idea of a *modification* of the original structure can be, PTs are simply a *crystal growth* from a *solid* medium, very much like that from liquids. The new crystal phase is built molecule-by-molecule which are taken from the original phase and fill molecular "kinks" at *contact* interfaces. The function of the old phase is to supply "bricks" (atoms, molecules); its actual crystal structure, however similar the polymorphs may sometimes be or seem, is no more relevant than the structure of a liquid phase upon crystal growth from a solution or melt. The original structure is responsible only for nucleation in its crystal defects - *microcavities* - where different nucleation temperatures and orientations of the new crystals are "encoded." In certain cases, *e.g.* in layered structures, the nucleation is *epitaxial*, leading to a certain OR, in other cases it is not (for example, OR in *p*-dichlorobenzene is different in every sample). The event of a phase transition is determined by two factors: the free energies of the polymorphs and the nucleation lags (manifesting themselves as hysteresis).

The idea of crystal phase transitions by displacements / distortion / deformation may seem self-evident, but it simply does not materialize. The phase transitions are always crystal growth.

ADDENDUM D: ON PHYSICS OF MAGNETIZATION*

Experimental facts reported in three recent articles in *Nature*[1,2,3] indicate the need for an important correction to the current interpretation of magnetization by a magnetic field. It is believed to be a "rotation" ("switching," "reversal") of spins *in* the intact crystal lattice. The new interpretation, put forward earlier[4], states that the spin vector is an orientation characteristic of its carrier (atom, molecule), therefore magnetization involves turning the carriers. The latter proceeds by the universal mechanism of crystal growth in liquids and solids: by nucleation and then filling "kinks" (steps) at the interfaces, molecule-by-molecule, to complete layers, and building successive layers in this manner.

Lavrov *et al.*[1] note that while the magnetic field affects the orientation of spins, it, it seems, should have little impact on the crystal structure; nevertheless, changes of crystal orientation were recently reported. But their own observation of structural rearrangement in an *antiferromagnet* turned out to be even more unexpected. According to the current views, spins in antiferromagnets strongly interact with each other, making a magnetically neutral system and should not be affected by a magnetic field. Yet, the authors observed the generation and motion of crystallographic twin boundaries and kinks moving along the crystal surfaces, resulting in a reorientation of the crystal. While their findings are inconsistent with magnetization by "switching," they are in accord with the magnetization mechanism presented in Ref.4. Actually, Lavrov *et al.* dealt with the *antiferromagnet → ferromagnet* phase transition when every second *spin carrier* was turned during the relocation at twin domain interfaces to make all spins parallel to the field. Evidently, spins were strongly bound to their carriers rather than to each other.

Novoselov *et al.*[2] recorded magnetization with a high resolution never before attained. They found that the ferromagnetic domain interface propagated by distinct jumps matching the lattice periodicity, the smallest being only a single lattice period. Some results also

---

* The article was submitted to journal *Nature* in July 2004. It was rejected by editor Rosalind Cotter while stating that "we are not questioning the validity of your innovative theory, or its interest to others in the field." The same editor recently promoted publication in *Nature* of an article on the quantity of ways to tie up shoe laces.

suggested that "kinks" were running along the interface. The authors interpreted the interface movements as following the Peierls potential of crystal lattice and stated that further theoretical and experimental work is needed to understand the unexpected dynamics of domain walls. The phenomenon, however, had been described, predicted to be traced to the molecular level, and illustrated with a molecular model in Ref.4 as a *crystal rearrangement* by the above-mentioned mechanism. In fact, the same mode of interface movement (running kinks and layer-by-layer) was observed by Lavrov *et al.*[1], only on a more macroscopic scale, and there crystal rearrangement was firmly established.

Tudosa *et al.*[3] estimated experimentally the ultimate speed of "magnetization switching" in tiny single-domain particles - an important issue in developing magnetic memory devices. The speed turned out three orders of magnitude lower than was predicted and, besides, was not the same in the different particles. The error of that prediction is hidden in the term "switching," in other words, in the assumption of spin rotation *in* the crystal structure. The lower speed had to be expected, considering that magnetization is not a "switching," but occurs by nucleation and growth in every individual domain. Nucleation is heterogeneous, requires specific crystal defects and is not simultaneous in different particles. It is nucleation that controls re-magnetization of small single-domain particles [4].

Recognition that magnetization results solely from structural rearrangement will be an essential advancement in understanding magnetism.

1. A. N. Lavrov, Seiki Komiya & Yoichi Ando, *Nature* 418, 385 (2002).
2. K. S. Novoselov, A. K. Geim, S. V. Dubonos, E. W. Hill & I. V. Grigorieva, *Nature* 426, 812 (2003).
3. I. Tudosa *et al.*, *Nature* 428, 831 (2004).
4. Y. Mnyukh, *Fundamentals of Solid-State Phase Transitions, Ferromagnetism and Ferroelectricity*, Authorhouse, 2001.

ADDENDUM E: BLENDING THE MECHANISMS INTO A PULP

According to the literature, there are numerous kinds of solid-state phase transitions (PTs); their number keeps growing with time and thus far has exceeded 300. Usually they are no more than labels reflecting a change of a physical property, even though their authors may believe that they deal with the PT mechanism. But some types were originally introduced theoretically as a particular physical process. They can be distinguished by their purported characteristics. Their failure (see Chapter 1) is not the point of the present context. The following is a brief reminder on how they have been conceived.

First-order PTs are characterized by unrestricted symmetry and properties changes ("jumps"), temperature dependence, phase coexistence, latent heat and hysteresis.

Second-order PTs should occur at a strictly fixed critical point (*i.e.* no hysteresis), without latent heat and jumps other than of C_p and symmetry.

Martensitic PTs are first order, exhibit hysteresis, start from nucleation and proceed by structural rearrangement at specific "habit" planes moving with the speed of sound under the action of strains arising at the planes.

Displacive PTs are assumed to occur by simultaneous displacements of atoms and molecules to new positions without breaking bonds in the original structure, and in that way reforming it.

Topological PTs are a combination of several displacive steps in the case when a one-step displacive PT cannot be imagined.

Order-disorder PTs are a cooperative (simultaneous) loss of orientation order, otherwise preserving long-range crystal order.

Soft-mode and incommensurate PTs are modulation (distortion) of original crystals by temperature-dependent vibration modes according to their frequencies at critical points.

Ferromagnetic PTs are second-order PTs resulted from spin rearrangement in the original crystal structure giving rise to a change of magnetization at critical points.

Quantum PTs occur at specific "quantum" critical points near 0 °K where the crystal order is destroyed exclusively by quantum fluctuations. All other features are inherited from "classical" second-order PTs.

To these "basic" mechanisms suggested in the literature, *nucleation and growth*, as described in the book, should be added.

The above reminder is necessary due to the very casual use of these "mechanisms" in literature, paying no regard to the original ideas and purported characteristics. This alone is harmful to the science of PTs. If a "displacive" PT requires nucleation, it is not displacive; if a "martensitic" PT is temperature-dependent, it is not martensitic; if a "quantum" PT reveals hysteresis, it is not quantum, and so on. What is worse, two or more notions are frequently used together. Blending them in contradictory combinations has become common. Examples:

- Blending the first and second order PTs produced the "first close to (or almost) second order" - even though the first order PTs are nucleation and rearrangement at interfaces, while "cooperative" second order PTs must proceed homogeneously.
- "Crossovers" of displacive" and "order / disorder" PTs are reported.
- Many ferromagnetic PTs, assumed to be "critical phenomenon," are classified as first order.
- Blending "reconstructive" and "displacive" PTs produced a "semireconstructive" hermaphrodite.
- Some "displacive" PTs are claimed to be first order, thus blending "displacive" mechanism with nucleation-and-growth.
- Nucleation-and-growth PTs are called "diffusional" even though they have little to do with diffusion.
- Some "reconstructive" PTs are claimed to be second order, blending two mutually exclusive notions.
- "Reconstructive" PTs, the antithesis to "displacive" PTs, are treated as a sequence of "topological" atomic/molecular displacements even when experimental data are available indicating nucleation and growth.
- "Topological" PTs are somehow made compatible with nucleation and subsequent growth.
- "Martensitic" PTs are called "displacive" and "soft mode"-triggered.
- "The martensitic transformation ... is a classical cooperative phenomenon similar to ferromagnetism" [MSR Bulletin, 2002].
- "Incommensurate", "magnetic", "soft-mode", "quantum", and "superconducting" were invoked in description of one particular case.
- "First-order magnetic phase transition from incommensurate phase to antiferromagnetic state" [ACNS, 2006].
- Some "incommensurate" PTs are claimed to be "displacive."
- Some "displacive" PTs are claimed to be "diffusional."
- Many "quantum" PTs are classified as first order.
- And so on...

ADDENDUM F: GUIDE TO THE SHAPE MEMORY EFFECT

Introduction

It is generally accepted that understanding the physics of shape memory effect in alloys is vital to widen its practical applications. To date, valuable empirical information has been accumulated, but the basic phenomenon leading to the effect, called "martensitic transformation," was misinterpreted. Even though the shape memory effect, as such, was not included in the 1st edition of this book, it contained all the basic data and new important specifics for its correct interpretation. This Addendum is a guide to the shape memory effect.

The current views

There is no specific "martensitic" mechanism of solid-solid phase transitions (Sec. 1.5.1). Not a single property was proven to make it unique. Except for a general consensus that it is "nucleation and growth" and "first order," all initial assumptions of the concept, such as a "habit plane" moving with speed of sound, or specific "martensitic" temperatures, fail in reality. The concept was fading until its more recent renaissance related to the phenomenon of shape memory alloys; its failings are now disregarded or forgotten, or both.

Here are two examples of how "martensitic transformations" are presented nowadays. Ahlers [1]: "A cooperative movement in which the atom arrangements can change their volume and shape, but neighbors remain neighbors"; "kinetics and morphology is dominated by the strain energy"; "diffusionless"; "distortion" of the original "matrix." Otsuka and Kakeshita [2]: "The martensitic (also called displacive or diffusionless) transformation is a classical cooperative phenomenon in solids similar to ferromagnetism" [probably, "to ferromagnetic phase transition" - Yu. M.].

Neither of the above descriptions is correct. They would be appropriate for a homogeneous process, but they do not fit the rearrangements at interfaces, even when presented as propagation of the "habit plane" as a whole. Solid-state phase transitions (PTs) - "martensitic," "displacive," or otherwise - are not a "cooperative," much less "classical cooperative," phenomenon. The term "cooperative" is defined as "all together" and was used by theorists in its correct capacity to describe non-existent homogeneous "second-order" PTs. The only alternative to that homogeneous process is "molecule-by-molecule" rearrangement (Sec. 1.2); propagation of the "habit plane" as a whole is outside of the allowed possibilities. PTs are not a "distortion" of the original "matrix." "Martensitic" PTs are not "diffusionless," for "diffusional" PTs do not exist*). Their kinetics depends on temperature and the presence of crystal defects. Strain energy is not their moving force and affects kinetics only as a secondary effect. The comparison with ferromagnetic PTs is in sharp contrast to the theory presenting the latter as a second-order, but is partially correct in a sense that, contrary to the theory, ferromagnetic PTs occur by nucleation and growth (Chapter 4).

Finally, they are not "displacive." Introduction of "martensitic" and "displacive" mechanisms of PTs in the first half of 20th century had nothing to do with each other. The former was described as first-order, nucleation-and-growth, driven with speed of sound by strains that was caused by a mismatch at the interfaces. The latter, on the other hand, was believed to be second-order, a cooperative, homogeneous distortion/deformation of the original structure by displacement of its atoms/molecules to the new positions without bond-breaking. There is no scientific reason to blend them (see Addenda C and E).

What are "martensitic transformations"?

As indicated in [2], they are divided in two categories: *thermoelastic* and *non-thermoelastic*. Only the former can produce a shape memory effect. This is well correlated with the statement of this book that all solid-solid PTs are nucleation and growth, but divided in two categories: *epitaxial*, when nucleation of the new phase is oriented by the original structure, *and non-epitaxial*, when the nuclei result in different unspecified orientations. *The "thermoelastic martensitic transformations" are, in fact, epitaxial nucleation-and-growth phase transitions.* They are characteristic of layered structures or those with a pronounced cleavage. Their major features, studied on a number of organic crystals, were published in 1975 (Ref. 7, p.12) and summarized again in this book (Sec. 2.8). They are:

- strict structural orientation relationship prior to and after PTs,
- small temperature hysteresis due to low nucleation barriers,
- formation of the new structure with different stacking of almost (but not exactly) the same molecular layers,
- molecule-by-molecule rebuilding of every layer rather than the displacement of whole layers to the new type of layer stacking.

---

* We leave notion "phase transition" to a process with chemically identical phases.

The source of shape

A growing single crystal tends to acquire an external shape reflecting its internal structure. This is true even when crystals of the new phase grow within a single-crystal medium of the original phase during PTs (see picture on the book cover and Figs. 2.2 to 2.10). As opposed to the "displacive," "soft mode," "martensitic," "topological," "second order," *etc.*, theories, these pictures reflect the general nature of solid-solid PTs. Their molecular mechanism, illustrated in Fig 2.17, is similar to that of melt crystallization: changing their phase affiliation at a *contact* interface, molecule-by-molecule, to fill "kinks" on the crystal face of the new growing crystal. There are no cooperative deformations, distortions or displacements, no propagation as a whole of "habit" planes. Strains always emerge, but they are a secondary effect frequently obscuring the underlying molecular mechanism.

As mentioned above, PTs are a crystal growth of two kinds. In the case of *non-epitaxial* PTs, crystal growth of the new phase is limited by the external frame of the original crystal; it stops after the new structure fills out the old frame (see photo, Fig. 2.6g). This case corresponds to the alloys with PTs not producing a shape memory effect.

The outcome in *epitaxial* PTs is different. In this case the crystal layers grow quickly in lateral directions through the whole original single crystal and come to its surface. After many successive layers that surface becomes the crystal face of the new structure. In other words, the previous shape is replaced by the new one. The new structure can be similar to the original one, differing basically by the manner of layer stacking; it resulted from molecule-by-molecule crystal growth rather than mutual displacements of the whole layers. The details are in Sec. 2.8. A shape change resulted from *epitaxial* PT is shown in Figs. 2.37c and 2.44, even though the original purpose of these figures was not related to shape memory effect.

The source of memory

The memory is hidden in the location, structure, and the number of the crystal defects serving as the heterogeneous nucleation sites of the new phase. The results of experimental study of nucleation in solid-solid PTs were published in 1976 (Ref. 8, p.12) and summarized again in this book (Sec. 2.5). Basic features of nucleation are:

- It is heterogeneous: without at least one proper crystal defect serving as a nucleation site the PT cannot occur at any temperature.
- Nucleation lags are solely responsible for PT hysteresis which is inevitable and larger in more perfect and smaller crystals.
- Temperature of a PT is "pre-coded" in the individual structure of the particular defect; it is not the same in different nucleation sites (Fig. 2.24).
- In the case of *epitaxial* nucleation, the hysteresis level is drastically lower.
- Orientation of a nucleus (and the orientation and shape of the new crystal) is "pre-coded" in the structure of the defect/nucleation site (Sec. 2.52).
- *Epitaxial* nucleation in a layered structure gives rise to a strict structural orientation relationship with the same direction of layers before and after the PT (Fig. 2.37).
- Nucleation sites exhibit different degrees of stability upon cyclic PTs: some survive only a single PT, others - several cycles, and some can practically become permanent nucleation sites (Sec. 2.52).
- Only one type of defect serves as the nucleation sites: *microcavities of a certain optimal size* (Sec. 2.5.4). In case of layered structures, they are interlayer wedge-like microcracks predominantly located at the faces composed of the layers' ends (Sec. 3.52 and Fig. 3.10a).

Shape + Memory

Let us consider a PT in a single crystal of rectangular lattice and pronounced cleavage (Fig. 2.45a). Usually there is an abundance of wedge-like microcracks on its side faces. This is why the PT will probably turn out to be multi-nucleus, resulting in the formation of a stack of laminar domains (Fig. 2.45b). If the resultant phase has a lower symmetry, the stack of alternating domains will exhibit zigzag side surfaces (which one should expect to see if Fig. 2.45b is enlarged). The number of different domain orientations (called "variants" in literature) is determined by the number of equally probable symmetry-equivalent ways of epitaxial nucleation (three in Fig. 2.45c).

In principle, a desirable shape memory effect, apart from a number of practical parameters, should consist of two elements: (a) a significant shape change upon PT and (b) a single stable nucleation site serving as such in both directions upon cyclic PTs. One can imagine a 'single crystal→single crystal' PT resulted from a single nucleus (crystal defect) that, luckily, turned out to be stable in both directions of the PT, or also a lucky situation of two stable nucleation sites, one acting upon heating, the other upon cooling. In more realistic cases of multiple nucleation, the number of nucleation sites can be reduced by eliminating less stable sites by cyclic back-and-forth PTs (Sec. 2.52). Ultimately, elimination of all "unwanted" sites can sometimes be achieved, making possible a "two way" shape memory effect (reiteration of the same shapes in cyclic PTs).

Application of stress

The last point to mention here is the application of stress to force a stack of alternating domains to acquire the same orientation. This effect, not touched on in the book, is easy to understand. Stress makes domains of a certain crystallographic orientation energetically preferable, causing their growth at the expense of others. Macroscopically, the preferable domains become thicker by means of movement of the twin boundaries separating them from the "unfavorable" neighbors. Microscopically, rearrangement of every successive layer at the twin boundary occurs by filling "kinks," molecule by molecule. (Running little steps along the twin boundary in a similar process called "mechanical twinning" have been observed). For the same reason, application of sufficient stress can produce a change of shape by causing PT in the original crystal at the temperature where the latter was stable. This will also occur by nucleation and growth, but will give rise to a single crystal orientation independent of the number of activated nuclei.

REFERENCES

1. M. Ahlers, The martensitic transformation, www.materials-sam.org.ar, (2005).
2. Kazuhiro Otsuka and Tomoyuki Kakeshita, Science and technology of shape-memory alloys: new developments. *MRS Bulletin*, February 2002.

ADDENDUM G: INCOMMENSURATE APPROACH TO PHASE TRANSITIONS

The International Union of Crystallography has a Commission on Aperiodic Crystals with a major task to investigate "incommensurately modulated crystals." Emergence of this peculiar solid matter in literature was described in Sec. 1.7. It allegedly results from modulation ("distortion") of the higher-temperature "prototype" ("mother") phase by a vibrational mode which is irrational toward the crystal parameters. The modulating wave becomes "frozen-in" in the resultant "incommensurate" phase. It was explained that "the new phase does not at all possess any periodicity...The fundamental feature of the crystal state is lost" [R. Pynn, *Nature* **281**, 433 (1979)]. *There are compelling reasons for such solid matter not to exist.*

- The essence of a periodic nature of crystal state is disposition of its constituent particles at particular equilibrium distances to provide minimum free energy to the system. The modulation in question violates these distances by shrinking or stretching *all* of them. In this respect, the "incommensurately modulated crystals" differ from any kind of imperfect crystals, amorphous material and liquids. The modulation is a phase transition to the system of a *higher* free energy - in violation of thermodynamics. Besides, nothing would prevent this higher-energy "frozen-in" system from transferring itself to the lower-energy normal crystal by usual nucleation-and-growth mechanism.

- Suppose that the "incommensurate" phase was produced by growing it from a solution. There is no reason for it to be a distorted version of another ("prototype") crystal existing under different *p, T* and be "modulated" by an optical mode of that unrelated crystal phase.

- Theoretically, phase transition to "incommensurately modulated" phase must be a critical phenomenon (second-order phase transition), since the modulation occurs simultaneously over the bulk. But structural changes in second-order phase transitions at their critical points are expected to be *infinitesimal* (Sec. 1.2), while the modulation would produce a *macroscopic* structural jump.

- In many experimental cases the transition to "incommensurately modulated" phase has been classified as first order, actually meaning its *nucleation-and-growth* mechanism rather than "modulation."

Still, the crystallographic studies of "commensurately" and "incommensurately" modulated crystals cannot be ignored. The terminology is objectionable, however. First, it is frequently used with no regard to the definitions of "commensurate" and "incommensurate" phases as the products of "soft-mode" actions. Second, intentionally or not, it renders support to the erroneous theories of "soft-mode" and "incommensurate" phase transitions, while all solid state phase transitions occur by nucleation and growth. Third, in most instances the structures are not modulated at all, as in the cases when magnetic structure covers two or more crystal-structure sub-sells. A key to the interpretation of the latter cases is that *spin distribution and orientation is imposed by the crystal packing.* The magnetic-structure unit cell therefore represents the true size of the structural unit cell. Setting those cases aside, there are still structures that seemed to be modulated indeed. The case in point is one or another kind of imperfect crystals with "long" periods and imprecise sub-cells. "Self-modulated crystals" would be a better term for them, for these imperfect crystals *result from crystal growth* - whichever the initial phase was: vapor, liquid or solid.

How can crystals become "self-modulated?" There are crystals with roughly periodical long-range imperfections that occurred during their growth. Thus, a phenomenon comes to mind of a "rhythmical" crystal growth from a liquid phase, caused by accumulation of latent heat, demonstrated by late A. V. Shubnikov. An analogous effect was observed upon phase transition in thin elongated plates of hexachloroethane: strips across the plate, developed due to periodical release of internal strains during interface propagation. Another example is "long periods" produced by the folding of long-chain molecules.

The following type of "self-modulated" crystals can also be envisioned. In rare cases the shape of molecules may prevent them from packing to a strictly periodical lattice, because formation of somewhat distorted structure provides a lower free energy. Let us consider a hypothetical example of crystallization of nearly ball-shaped asymmetric molecules. In a strictly periodical cubic lattice only few intermolecular contacts will be optimal; the structure is not sufficiently closely packed. The crystallization may take another route, producing a strained cubic structure with the majority of intermolecular contacts close to optimal at the expense of some being compressed; the structure is more densely packed and can be energetically preferable. As crystallization proceeds, the accumulating strains distort the unit cell until an interruption of the continuity occurs; the cycle then repeats itself.

Crystallography would reveal an approximate "long" periodicity and a roughly reproducible short-range unit cell. *Then the case will be reported as "incommensurate phase."* With good probability, another phase of this substance - strictly cubic - would exist at higher temperatures. *The case will be reported as "commensurate-incommensurate phase transition"* with the word "commensurate" meaning "non-incommensurate."

ADDENDUM H: LESSON ON BARKHAUSEN EFFECT FOR DURIN AND ZAPPERI

Dear Dr. G. Durin and Dr. S. Zapperi,

This is a response to your cond-mat/0404512, 51-page, review article "The Barkhausen effect." The article supplies a reader with lots of experimental observations, statistical treatments, guessing (like the wrong parallel with second-order phase transitions), and vast list of references - except the only one where the phenomenon was explained. This was done in my book "*Fundamentals of Solid-State Phase Transitions, Ferromagnetism and Ferroelectricity*" published in 2001. You knew about it, and that the Barkhausen effect was considered there, from my e-mail sent to both of you on 10.15.2003.

Let me explain to you the physics of the phenomenon.

1.   First, you should recall how crystals grow from liquids and gases: molecule-by-molecule filling the "kinks" (steps) on a crystallographic plane to complete the layer and then building next layers by the same way - or a cascade of layers simultaneously. The process has been investigated in literature in detail: steps of different height, irregularities, avalanches... - the same as in Barkhausen effect.
2.   As I discovered in 1966 [Ref. 76] crystal-crystal phase transitions proceed by rearrangement at interfaces in the very similar way. At that, formation of "kinks" is controlled by *nucleation* in specific crystal defects. Again, small running steps, larger steps, layer-by-layer growth, irregularities, avalanches, etc., were observed.
3.   According to the fundamentals of ferromagnetism put forward in the book, magnetization is never simply "rotation" ("switching," "reversal") of spins *in* the crystal, but always resulted from *crystal rearrangement* at the domain interfaces. Sec. 4.10 in the book is entitled "Barkhausen effect as manifestation of crystal growth."

Isn't it simple? You could challenge it, but not even mentioning it in a very comprehensive review is not appropriate if you care about scientific truth.

Sincerely, Yuri Mnyukh

ADDENDUM I: FERROMAGNETISM EXPLAINED

The cause of magnetic hysteresis (MH) has not been found by the current theory. Sixteen authors (cond-mat/0611542) start their article by asking "What causes MH?" and state that MH is fundamental to magnetic storage technologies and a cornerstone to the present information age. They found that all the "beautiful theories of MH based on random microscopic disorder" failed to explain their data. So, they produced new theories. Their answer to the above question was: "New advances in our fundamental understanding of MH are needed." Neither that article, nor other extensive work on the topic, such as 2160 pages of the 3-volume book "The Science of Hysteresis" (Academic Press, 2006) were able to find the cause of the phenomenon. Further attempts based on the current theory will be futile.

The search, however, continues – without any reference to the detailed explanation of the cause of MH put forward in this book.. This explanation became possible in terms of the new fundamentals of ferromagnetism that emerged as a result of the investigation of solid-state phase transitions. The latter were found to be nucleation in crystal defects and growth by rearrangement at interfaces in all instances. The concept of the new fundamentals of ferromagnetism is simple. When crystal structure changes, so do its physical properties, magnetization included. Change in magnetization **resulted from** the change of crystal structure, and not the other way round. **Orientation of spins in a crystal lattice is set by the orientation of particles carrying them.** In the case of structural rearrangements at domain interfaces (e. g., under the action of external magnetic field), the new state of magnetization resulted from the new orientation of the crystal lattice and its constituent particles. This new concept instantly eliminated major problems ferromagnetism faced:

● **FERROMAGNETIC STATE.** "*One challenge is to understand why it exists at all*" (R. P. Feynman). In order to explain why a system of parallel spins is stable the (very strong) Weiss/Heisenberg's molecular/exchange field was introduced. But it has failed on many levels, including producing the wrong sign of the exchange integral. *SOLUTION*: The molecular field does not exist. The contribution of magnetic interaction to the total crystal free energy is small compared to that of chemical bonding and other components. A ferromagnetic crystal is stable due to lower **total** energy **in spite** of a small destabilizing effect of the magnetic interaction.

- **DISPARITY WITH FERROELECTRICITY.** Ferromagnetism and ferroelectricity are very similar phenomena with an analogous set of manifestations. The standard theory was unable to find a unified approach to them since the Weiss/Heisenberg molecular field was applied only to ferromagnetism. No analog to it was found (or even needed) for ferroelectricity. *SOLUTION:* This profound inconsistency disappeared after the Weiss/Heisenberg molecular field was eliminated from consideration. Now the two phenomena can have parallel explanations.

- **FIRST-ORDER FERROMAGNETIC PHASE TRANSITIONS**. The current theory treats ferromagnetic phase transitions as second-order (critical phenomenon), but sufficiently accurate experimental studies find them to be first order (consequently, not a critical phenomenon). The attempts to incorporate first-order phase transitions into the conventional theory defy logic. *SOLUTION:* A magnetic phase transition is simply a reflection of the structural phase transition. All phase transitions in solid state, ferromagnetic or not, are nucleation and crystal growth. As to the few remaining "second-order" phase transitions, they are destined to be reclassified.

- **ORIGIN OF MH**. The interpretation of magnetization as a rotation of spins **in** the crystal structure makes a reasonable explanation of MH impossible. *SOLUTION:* MH is that of structural rearrangements at interfaces, both in ferromagnetic phase transitions and in magnetization of domain systems. Structural phase transitions require 3-D nucleation to begin; both processes require 2-D nucleation to proceed. The nucleation is heterogeneous, localized in specific crystal defects – microcavities – where nucleation lags are encoded. **These nucleation lags are the cause of MH.** Hysteresis in ferroelectrics is not different.

- **MAGNETIZATION MECHANISM.** The common terms "*switching*" and "*reversal*" imply instant change of spin orientation in the crystal lattice. They are inconsistent even with the conventional concept of "Bloch wall" where the changing of spin orientation **in** crystal structure is localized. As for the "Bloch wall" concept, it itself is wrong (Sec. 4.8). It could not explain why experimentally estimated ultimate speed of "magnetization switching" in single-domain particles turned out orders of magnitude lower than theoretically predicted. *SOLUTION*: Spin orientation is a fixed property of every particular crystal structure. "Magnetization switching" and "magnetization reversal" *result from changing the crystal structure itself* into that of a different spin orientation. The structural rearrangement can be activated by a change of temperature, pressure, or an external magnetic field. In any case the process involves nucleation and propagation of interfaces, as presented in this book. Data in Addendum D further validate this solution.

ADDENDUM J: MAGNETOCALORIC EFFECT IN TERMS OF CRYSTAL GROWTH

The magnetocaloric effect (MCE) in a ferromagnetic material F is the heat emanated when a magnetic field H is applied. From a thermodynamic point of view, the application of H to F makes it unstable, creating the condition for its rearrangement that brings spin directions S closer to H. The only exception is when H is applied along S. If the rearrangement actually occurs, the corresponding energy gain is exothermic MCE. In certain cases the subsequent reduction in the H strength leads to endothermic MCE - the phenomenon allowing the creation of magnetic refrigeration technique.

The actual physics of MCE was only superficially understood due to misinterpretation of solid-state phase transitions in general, ferromagnetic ones in particular. Thus, endothermic MCE was explained in terms of "critical phenomena" notions as a result of "randomization of domains similar to the randomization at Curie temperature". At the same time, hysteresis (nucleation lags in a nucleation-and-growth process) was cited as one of the problems in the magnetic refrigeration technique.

The basics of MCE are interpreted here in terms of the crystal growth concept. Let us analyze the application of H to a single ferromagnetic domain or a polydomain "single crystal". The material becomes simply an unstable crystal phase. Its rearrangement in magnetic field H is considered in Sec. 4.7 to 4.13. We deal with a structural rearrangement by nucleation-and-growth rather than a rotation of S toward H. The resultant crystal has the same crystal structure, but a new spatial orientation in which S directions are closer to H. The difference between the two free energies manifests itself as exothermic MCE. Subsequent H decrease to zero does not produce MCE.

No meaningful endothermic MCE is possible until phase transition *ferromagnetic-paramagnetic* (F-P) comes into play. Even though it is called a "magnetic" phase transition, it proceeds by nucleation-and-growth. Paramagnetic phase is an orientation-disordered crystal phase (ODC) where arbitrary orientation of spins results from thermal rotation of their carriers. Application of H at temperatures not too far from T_0 shifts T_0 upward in favor of F with S preferably directed along H, triggering $P \rightarrow F$ nucleation-and-growth phase transition. Its exothermic latent heat is MCE, Turning H off makes the paramagnetic phase preferable again, giving rise to the reverse phase transition F → P; its latent heat manifests itself as the endothermic MCE.

Conclusion: Every ferromagnetic can exhibit magnetocaloric effect. If a large effect is observed, it most probably involves the latent heat of the *ferromagnetic-paramagnetic* phase transition.

REFERENCES

1. Phase Transitions and Critical Phenomena, ed. C. Domb *et al.*, v.1 (1972), v.2 (1973), v.3 (1974), v.5a (1976), v.5b (1976), v.6 (1976), v.7 (1983), v.8 (1983), v.9 (1983), v.10 (1987), v.11 (1987), v.12 (1988), v.13 (1990), v.14 (1991), v.15 (1992), v.16 (1994), v.17 (1995), Acad. Press.
2. P.A. Fleury, Science **211,** No.4478, 125 (1981).
3. L. Onsager, Phys.Rev. **65,** 117 (1944).
4. R. Brout, Phase Transitions, New York, (1965).
5. M. Fisher, The Nature of Critical Points, Univ. Colorado (1965).
6. Critical Phenomena (proceedings of a conference, Washington, D.C., 1965), ed. M.S. Green and J.V. Sengers, Natl. Bur. Std., Washington, D.C. (1966).
7. K.G. Wilson, Phys.Rev. **B4,** 3174 (1971); ibid., 3184 (1971).
8. H.E. Stanley, Introduction to Phase Transitions and Critical Phenomena, Clarendon Press (1987).
9. Critical Phenomena (proceedings, Intern. School of Phys. Enrico Fermi), Acad. Press (1972).
10. M. Fisher, Rev.Mod.Phys. **46,** 59 (1974).
11. Fluctuations, Instabilities and Phase Transitions, ed. T. Riste, Plenum Press (1975).
12. Chang-Keng Ma, Modern Theory of Critical Phenomena, Benjamin Co. (1976).
13. Local Properties at Phase Transitions, ed. K.A. Muller and A. Regomonty (proceedings, Scuola intern. di fisica, Varenna, Italy), North-Holland (1976).
14. Critical Phenomena (proceedings, Sitges Intern. School of Statistical Mechanics, 1976), Springer-Verlag (1976).
15. P.C. Hohenberg and B.I. Halperin, Rev.Mod.Phys. **49,** 435 (1977).
16. P. Pfeuty, and G. Toulose, Introduction to the Renormalization Group and to Critical Phenomena, Wiley (1977).
17. Ordering in Strongly Fluctuating Condensed Matter Systems, ed. T. Riste (proceedings, NATO Advanced Study Inst. on Strongly Fluctuating Condensed Matter Systems, Geilo, Norway, 1979), Plenum Press (1980).
18. Structural Phase Transitions, ed. K. Muller and H. Thomas, Springer-Verlag (1981).
19. A.D. Bruce and R.A. Cowley, Structural Phase Transitions, Taylor and Francis (1981).
20. Multicritical Phenomena, ed. R. Pynn and A Skjeltorp, Plenum Press (1983).
21. Magnetic Phase Transitions, ed. M. Ausloos and R.J. Elliot, Springer-Verlag (1983).
22. J.C. Toledano and P. Toledano, The Landau Theory of Phase Transitions, World Sci. (1986).
23. M.F. Collins, Magnetic Critical Scattering, Oxford (1989).

24. G.A. Baker, Quantitative Theory of Critical Phenomena, Acad. Press (1990).
25. H. Dosch, Critical Phenomena at Surfaces and Interfaces, Springer-Verlag (1992).
26. K. Okada, Catastrophe Theory and Phase transitions: Topological Aspects of Phase Transitions and Critical Phenomena, Trans. Tech. Publ. (1994).
27. O.G. Mouritsen, Computer Studies of Phase Transitions and Critical Phenomena, Springer-Verlag (1984).
28. D.I. Uzunov, Introduction to the Theory of Critical Phenomena: Mean Fields, Fluctuations, and Renormalization, World Sci. (1993).
29. Critical Phenomena (proceedings of summer school in S. Africa, 1982), ed. F. Hahne, Springer-Verlag (1983).
30. P.W. Anderson, Science **218**, 763 (1982).
31. M. Azbel, *preface to* Phase Transitions, R. Brout (Russian ed.), Mir, Moscow (1967).
32. P. Ehrenfest, Leiden Comm.Suppl., No.75b (1933).
33. E. Justi and M. von Laue, Physik Z. **35**, 945; Z.Tech.Physik **15**, 521 (1934).
34. L. Landau, *in* Collected Papers of L.D. Landau, p.193, Gordon & Breach (1967); [Phys.Z.Sowjet **11**, 26 (1937); **11**, 545 (1937)].
35. L. Landau and E. Lifshitz, *in* Collected Papers of L.D. Landau, p.101. Gordon & Breach (1967); [Phys.Z.Sowiet **8**, 113 (1935)].
36. L.D. Landau and E.M. Lifshitz, Statistical Physics, Addison-Wesley (1969).
37. M.J. Buerger, Kristallografiya, **16**, 1048 (1971) [Soviet Physics - Crystallography **16**, 959 (1971)].
38. D.R. Moore, V.J. Tekippe, A.K. Ramdas, and J.C. Toledano, Phys.Rev.B **27**, 7676 (1983).
39. H.D. Megaw, Crystal Structures: A Working Approach, Saunders Co. (1973).
40. Yu.M. Gufan, Structural Phase Transitions, Nauka, Moscow (1982, Rus.).
41. Structural Phase Transitions and Soft Modes, Ed. E.J. Samuelson and J. Feder, Universitetsfurlaget, Norway, (1971).
42. C.N.R. Rao and K.J. Rao, Phase Transitions in Solids, McGraw-Hill (1978).
43. M.J. Buerger, *in* Phase Transformations in Solids, Wiley (1951).
44. W.C. McCrone, *in* Physics and Chemistry of the Organic Solid State, v.2, Wiley (1965).
45. J.P. McCullough, Pure and Appl. Chem. **2**, 221 (1961).
46. E.F. Westrum and J.P. McCullough, *in* Physics and Chemistry of the Organic Solid State, v.1, Wiley (1965).
47. A.B. Pippard, The Elements of Classical Thermodynamics, Cambridge Univ. Press (1964).
48. W. Cochran, Adv.Phys. **9**, 387 (1960).
49. In Ref.42, pp.204-223; in Ref.57, pp.98-101.

50. P.W. Anderson, *in* Fizika Dielectrikov, Ed. G.I. Skanavi, Akad. Nauk SSSR, Moscow (1959).
51. R. Blinc and B. Ziks, Soft Modes in Ferroelectrics and Antiferroelectrics, North-Holland (1974).
52. J.F. Scott, Rev.Mod.Phys. **46**, 83 (1974).
53. G. Shirane, Rev.Mod.Phys. **46**, 437 (1974).
54. N.G. Parsonage and L.A.K. Staveley, Disorder in Crystals, Clarendon Press (1978).
55. M.E. Lines and A.M. Glass. Principles and Applications of Ferroelectrics and Related Materials, Clarendon Press (1977).
56. E.R. Bernstein and B.B. Lal, Mol.Cryst.Liq.Cryst. **58,** 95 (1980).
57. Incommensurate Phases in Dielectrics, v.2, North-Holland (1985).
58. Light Scattering Near Phase Transitions, Ed. H.Z.Cummins and A.P. Levanyuk (Ch. 3 & 7), North-Holland (1983).
59. R. Pynn, Nature **281**, 433 (1979).
60. H. Cailleau, *in* Incommensurate Phases in Dielectrics, North Holland, v.2 (1985).
61. F. Denoyer and R. Currat, *in* Incommensurate Phases in Dielectrics, v.2, North-Holland (1985).
62. A.I. Kitaigorodskii, Organic Chemical Crystallography, Consultants Bureau (1960).
63. A.I. Kitaigorodskii, Molecular Crystals and Molecules, Acad. Press (1973).
64. K.P. Belov, Magnetic Transitions, Boston Tech. Publ. (1965).
65. R.M. Bozorth, Ferromagnetism, Van Nostrand (1951).
66. K.C. Russell, Nucleation in Solids, *in* Nucleation III, ed. A.C. Zettlemoyer, M. Dekker (1977).
67. J. Housty and J. Clastre, Acta Cryst. **10**, 695 (1957).
68. E. Frasson, Acta Cryst. **12**, 126 (1959).
69. J.A. Goedkoop and C.H. MacGillavry, Acta Cryst. **10**, 125 (1957).
70. J.D. Morrison and J.M. Robertson, J.Chem.Soc., 1001 (1949).
71. L. Brockway and J.M. Robertson, J.Chem.Soc., 1324 (1939).
72. T. Watanabe, Y. Saito and H. Chihara, Sci. Papers Osaka Univ., No.2, 9 (1950).
73. A. McL. Mathieson, Acta Cryst. **6,** 399 (1953).
74. J.W. Christian, *in* Physical Metallurgy, ed. R.W. Cann, North-Holland (1965).
75. W. Bollmann, Crystal Defects and Crystalline Interfaces, Springer-Verlag (1970).
76. Y. Mnyukh, N. Petropavlov and A. Kitaigorodskii, Docl.Acad.Nauk SSSR. **166,** 80 (1966); [Sov.Phys.-Doklady **11**, 4 (1966)].
77. W. Kossel, Nachr. Ges. Wiss. Goetingen, 135 (1927).
78. I.N. Stranski, Z.Phys.Chem. **136,** 259 (1928).
79. N.H. Hartshorne and M.N. Roberts, J.Chem.Soc., 1097 (1951).
80. N.H. Hartshorne, G.S. Walters and W. Williams, J.Chem.Soc., 1860 (1935).
81. N.H. Hartshorne, Disc.Farad.Soc. **5**, 149 (1949).
82. W.E. Garner, Disc.Farad.Soc. **5**, 194 (1949).

83. R.S. Bradley, N.H. Hartshorne and M. Thackray, Naturte **173,** 400 (1954).
84. R.S. Bradley, J.Phys.Chem. **60,** 1347 (1956).
85. C. Briske and N.H. Hartshorne, Disc.Farad.Soc. **23,** 196 (1957).
86. N.H. Hartshorne, Disc. Farad. Soc. **23,** 224 (1957).
87. N.H. Hartshorne and M. Thackray, J.Chem.Soc., 2122 (1957).
88. N.H. Hartshorne, *in* Recent Work on the Inorganic Chemistry of Sulfur, Symp. Chem. Soc., Bristol (1958).
89. C. Briske, N.H. Hartshorne and D.R. Stransk, J. Chem. Soc., 1200 (1960).
90. M. Thackray, Nature **201**, 674 (1964).
91. C. Briske and N.H. Hartshorne, Trans.Farad.Soc. **63**, 1546 (1967).
92. Yu. Mnyukh, Dokl.Akad.Nauk SSSR **201**, 573 (1971); [Sov.Phys.Doklady **16**, 977 (1972)].
93. E.G. Cox, Proc.R.Soc.Lond. **A247**, 1 (1958).
94. A.I. Kitaigorodskii and K.V. Mirskaya, Crystallogr. (USSR) **6**, 507 (1961); **9**, 174 (1964).
95. R.W. Cahn, *in* Physical Metallurgy, Ed. R.W. Cann, North-Holland (1965).
96. C. Domb and J. Lebowitz, *preface to* Phase Transitions and Critical Phenomena, ed. C. Domb and J. Lebowitz, **v.8** (1983).
97. A.R. Verma and P. Krishna, Polymorphism and Polytypism in Crystals, Wiley (1966).
98. Phase Transitions in Molecular Crystals, Far.Div.Chem.Soc., London (1980).
99. Light Scattering Near Phase Transitions, Ed. H.Z.Cummins and A.P. Levanyuk, North-Holland (1983).
100. R.W. Munn, Chem.Brit. **14,** 231 (1978).
101. P. Toledano and V. Dmitriev, Condensed Matter News **2**, 9 (1993).
102. R. Becker, Ann. der Physik **32**, 128 (1938).
103. P. Schwartz, Phys.Rev.B **4**, 920 (1971).
104. D.L. Connelly, J.S. Loomis and D.E. Mapother, Phys.Rev.B **3**, 924 (1971).
105. Y. Mnyukh, Mol.Cryst.Liq.Cryst. **52**, 163 (1979).
106. L.D. Landau, A.I. Akhiezer and E.M. Lifshitz, General Physics, Pergamon Press (1967).
107. A.R. Ubbelohde, Melting and Crystal Structure, Clarendon Press (1965).
108. A.R. Ubbelohde, *in* Dinamica reazioni chim., p.85, Roma (1967).
109. D.G. Thomas and L.A.K. Staveley, J.Chem.Soc. 2572 (1951).
110. P. Dinichert, Helv.Phys.Acta **15**, 462 (1942).
111. L.J. Soltzberg, S.J. Bobrowski, P.A. Parziale, E.C. Armstrong and V.E. Cohn, J.Chem.Phys. **71**, 1652 (1979).
112. P. Weiss and R. Forrer, Compt.rend. 178, 1670.
113. 0. Bloch, Arch.sci.phys.nat. **33**, 293.
114. T. Petrovic and M. Napijalo, Fizika (Zagreb) **10** (Suppl.2), 186 (1978).
115. V.P. Vygovskii and Yu.V. Ergin, Fiz.Metal.Metalloved. **34**, 491 (1972).
116. C. Kittel, Introduction to Solid State Physics, 4th ed., Wiley (1971).
117. G.M. Drabkin, E.I. Zabidarov, and A.V.Kovalev, Sov.Phys-JETP **42**, 916 (1976); [Zh.Eksp.Teor.Fiz. **69**, 1804 (1975)].

118. F.J. Bartis, Phys.Lett. **61a**, 48 (1977).
119. W.E. Maher and W.D. McCormick, Phys.Rev. **183**, 573 (1969).
120. L.V. Kirenskii, A.I. Drokin, and D.A. Laptey, Temperature Magnetic Hysteresis of Ferromagnets and Ferrits, Novosibirsk (1965, Rus.).
121. S.V. Vonsovskii, Magnetism, vol. 1 & 2, Wiley (1974).
122. J.G. Aston, *in* Physics and Chemistry of the Organic Solid State, v.1, Wiley (1965).
123. Ya.I. Frenkel, Kinetic Theory of Liquids, Acad.publ.(1945,Rus.)
124. J.G. Marshall, L.A.K. Staveley and K.R. Hart, Trans.Farad.Soc. **52**, 19 (1956).
125. J.H. Colwell, E.K. Gill and J.A. Morrison, J.Chem.Phys. **39**, 635 (1965).
126. V.L. Broude, Zh.Eksp.Teor.Fiz. **22,** 600 (1952).
127. S. Seki and H. Chihara, Sci.Papers Osaka Univ. No.1, 1 (1950).
128. A.McL. Mathieson, Acta Cryst. **5**, 332 (1952).
129. Yu. Mnyukh, J.Phys.Chem.Solids **24,** 631 (1963).
130. J. Housty and J. Clastre, Acta Cryst. **10**, 695 (1957).
131. E. Frasson, Acta Cryst. **12,** 126 (1959).
132. F.H. Herbstein, Acta Cryst. **18,** 997 (1965).
133. C. Dean and M. Pollak, Acta Cryst. **11**, 710 (1958).
134. I. Taguchi, Bull.Chem.Soc.Jap. **34,** 392 (1961).
135. I. Nitta, T. Watanabe, and I. Taguchi, X-Rays **5,** 31 (1948).
136. I. Nitta, T. Watanabe, and I. Taguchi, Bull.Chem.Soc.Jap. **34,** 1405 (1961).
137. P. Coppens and T.M. Sabine, Molec.Cryst. **3,** 507 (1968).
138. L. Agmon and F. Herbstein, Proc.R.Soc.Lond. **A387,** 311 (1983).
139. Y. Mnyukh, N. Panfilova, N. Petropavlov, and N. Uchvatova, J.Phys.Chem.Solids **36,** 127 (1975).
140. R.H. Doremus, Rates of Phase Transformations, Acad.Press (1985).
141. M. Avrami, J.Chem.Phys. **7**, 1103 (1939); **8**, 212 (1940).
142. A.W. Lawson, Phys.Rev. **57,** 417 (1940).
143. G.E. Fredericks, Phys.Rev.B **4,** 911 (1971).
144. A.A. Boiko, Sov.Phys.-Cryst. **14,** 539 (1970).
145. F. Simon, Ann.Physik, **68,** 241 (1922).
146. Collected Papers of L.D. Landau, Gordon & Breach (1967).
147. A. Euchen, Physik Z. **35**, 954; Z.tech.Physik **15,** 530 (1934).
148. R. Extermann and J. Weigle, Helv.Phys.Acta **15,** 455 (1942).
149. A.R. Ubbelohde, Quart.Rev.Soc.(London) **11,** 246 (1957).
150. In Ref.107, Chapter 4.
151. V. Voronel and S.R. Garber, Zh.Eksperim.Teor.Fiz. **52,** 1464 (1967); [Sov.Phys.- JETP **25,** 970 (1967)]
152. H. Chihara and M. Nakamura, Bull.Chem.Soc.Jap. **45,** 133 (1972).
153. J.E. Callanan, R.D. Weir, and L.A.K. Staveley, Proc.Roy.Soc. Lond. **A 372,** 489 (1980); **A 372,** 497 (1980); **A 375,** 351 (1981).
154. A.B. Pippard, Phil. Mag. **1**, 473 (1956); also F.J. Bartis, J.Phys.C **6**, L90 (1973), as well as Ref. [54], p.326.
155. R. Guillien, Compt.Rend. **208,** 1561 (1939).
156. K. Kamiyoshi, Sci.Rep.Res.Inst.Tohoku Univ. **A8,** 252 (1956).

157. I.S. Zheludev, The Principles of Ferroelectricity, Atomizdat, Moscow (1973, Rus.).
158. G. Busch, Helv.Phys.Acta **11**, 269 (1938).
159. S. Hoshino, K. Vedam, Y. Okaya, and R. Pepinsky, Phys.Rev. **112,** 405 (1958).
160. K. Kamiyoshi and T. Miyamoto, J.Chem.Phys. 22, 756 (1954).
161. J.P. Pique, G. Dolino, and M. Vallade, J. de Physique **38,** 1527 (1977).
162. R. Guillien, Compt.Rend. **208,** 980 (1939).
163. R. Guillien, Ann.Phys. **17,** 334 (1942).
164. S.T. Bayley, Trans.Faraday Soc. **47,** 518 (1951).
165. M. Freymann, Compt.Rend. **233,** 1449 (1951).
166. L. Couture, S. Le Montagner, J. Le Bot, and A. Le Traon, Compt.Rend. **242,** 1804 (1956).
167. K. Kamiyoshi, J.Chem.Phys. **26,** 218 (1957).
168. H.G. Unruh, Phys.Lett. **17,** 8 (1965).
169. H. Ohshima and E. Nakamura, J.Phys.Chem.Solids **27,** 481 (1966).
170. T. Ikeda, K. Fujibayashi, T. Nagai, and J. Kobayashi, Phys.stat.sol.(a) **16**, 279 (1973).
171. K.B. Lyons and P.A. Fleury, *in* Light Scattering in Solids, ed. J. Birman and H.Cummins, Plenum Press (1979) p.357.
172. P.A. Fleury and K. Lyons, *in* Structural Phase Transitions, ed. K. Muller and H. Thomas, Springer-Verlag (1981).
173. I.A. Iakovlev, T.S. Velichkina, and L.F. Mikheeva, Sov.Phys. -Dokl. **1**, 215 (1956).
174. S.M. Shapiro and H.Z. Cummins, Phys.Rev.Lett. **21**, 1578 (1968).
175. N. Lagakos and H.Z. Cummins, Phys.Rev.Lett. **34**, 883 (1975).
176. L.N. Durvasula and R.W. Gammon, Phys.Rev.Lett. **38**, 1081 (1977).
177. O.A. Shustin, JETP Letters **3**, 320 (1966).
178. F.J. Bartis, Phys.Lett. **43a**, 61 (1973).
179. V.L. Ginzburg, A.P. Levanyuk, A.A. Sobianin, and A.S. Sigov, in Light Scattering in Solids, ed. J.L. Birman and H.Z. Cummins, p.331, Plenum Press (1979).
180. L.N. Durvasula and R.W. Gammon, *in* Light Scattering in Solids, ed. M. Balkanski, R.C.C. Leite and S.P.S. Porto, p.775, Flammarion Press (1975).
181. F.J. Bartis, J.Phys.C. L295 (1973).
182. G. Dolino, J.Phys.Chem.Solids **40**, 121 (1979).
183. T. Riste, E.J. Samuelson, K. Otnes and J. Feder, Solid State Commun. **9**, 1455 (1971).
184. J.D. Axe and G. Shirane, Phys.Rev.B **8**, 1965 (1973).
185. J. Crangle, The Magnetic Properties of Solids, Edward Arnold, London (1977).
186. The New Encyclopedia Britannica (30th Ed.), v.6 (1979), p.560.
187. T. Mitsui, I. Tatsuzaki and E. Nakamura, An Introduction to the Physics of Ferroelectrics, Gordon & Breach (1976).
188. J. Grindlay, An Introduction to the Phenomenological Theory of Ferroelectricity, Pergamon Press (1970).

189. E. Fatuzzo and W.J. Merz, Ferroelectricity, North-Holland & John Wiley (1967).
190. F. Jona and G. Shirane, Ferroelectric Crystals, Pergamon Press (1962).
191. G.G. Leonidova, Docl.Acad.Nauk SSSR **196**, 335 (1971).
192. R.S. Preston, S.S. Hanna and J. Heberle, Phys.Rev. **128**, 2207 (1962).
193. R.S. Preston, Phys.Rev.Let., **19**, 75 (1967).
194. A.A. Hirsch, J.Magn.Magn.Mater. **24**, 132 (1981).
195. J.C. Hamilton and T. Jach, Phys.Rev.Let., **46**, 745 (1981).
196. R.P. Feynman, R.B. Leighton and M. Sands, The Feynman Lectures on Physics, v.2, Addison-Wesley (1964).
197. C. Kittel, Introduction to Solid State Physics, 4th ed., Wiley (1971).
198. A. Hubert, Theorie der Domanenwande in Geordneten Medien, Springer-Verlag (1974).
199. T.G. Gray, The Magnetic Properties of Solids, *in* The Defect Solid State, Interscience (1957).
200. J.H. Van Vleck, Phys.Rev. **52**, 1178 (1937).
201. R.W. De Blois, Final Report AFCRL-67-0107, Bedford, Mass., March 1967.
202. Ya.S. Shur, *in* Physical Encyclopedic Dictionary, v.1, p.156, Soviet Encyclopedia (1960, Rus.).
203. K. Honda and S. Kaya, Sci.Rep.Tohoku Imp.Univ. **15**, 721 (1926).
204. S. Kaya, Sci.Rep.Tohoku Imp.Univ. **17**, 639 (1928).
205. S. Kaya, Sci.Rep.Tohoku Imp.Univ. **17**, 1157 (1928).
206. M. Sachs, Solid State Theory, McGraw-Hill (1963).
207. E. Sawaguchi and K. Kittaka, Phys.Soc.Japan **7**, 336 (1952).

www.ingramcontent.com/pod-product-compliance
Lightning Source LLC
Chambersburg PA
CBHW021426180326
41458CB00001B/146